Percolation Theory in Reservoir Engineering

Other Related Titles from World Scientific

The Imperial College Lectures in Petroleum Engineering
Volume 1: An Introduction to Petroleum Geoscience
by Michael Ala
ISBN: 978-1-78634-206-5

The Imperial College Lectures in Petroleum Engineering
Volume 2: Reservoir Engineering
by Martin J Blunt
ISBN: 978-1-78634-209-6

The Imperial College Lectures in Petroleum Engineering
Volume 3: Topics in Reservoir Management
by Deryck Bond, Samuel Krevor, Ann Muggeridge, David Waldren and Robert W Zimmerman
ISBN: 978-1-78634-284-3

The Imperial College Lectures in Petroleum Engineering
Volume 4: Drilling and Reservoir Appraisal
by Olivier Allain, Michael Dyson, Xudong Jing, Christopher Pentland, Marcel Polikar and Vural Sander Suicmez
ISBN: 978-1-78634-395-6

The Imperial College Lectures in Petroleum Engineering
Volume 5: Fluid Flow in Porous Media
by Robert W Zimmerman
ISBN: 978-1-78634-499-1

Percolation Theory in
Reservoir Engineering

Peter R King
Imperial College London, UK
Mohsen Masihi
Sharif University of Technology, Iran

World Scientific

NEW JERSEY • LONDON • SINGAPORE • BEIJING • SHANGHAI • HONG KONG • TAIPEI • CHENNAI • TOKYO

Published by

World Scientific Publishing Europe Ltd.

57 Shelton Street, Covent Garden, London WC2H 9HE

Head office: 5 Toh Tuck Link, Singapore 596224

USA office: 27 Warren Street, Suite 401-402, Hackensack, NJ 07601

Library of Congress Cataloging-in-Publication Data
Names: King, P. R. (Peter R.), (Petroleum engineer), author. | Masihi, Mohsen, author.
Title: Percolation theory in reservoir engineering / by Peter R King
(Imperial College London, UK), Mohsen Masihi (Sharif University of Technology, Iran).
Description: [Hackensack] New Jersey : World Scientific, 2018. |
Includes bibliographical references.
Identifiers: LCCN 2017060702 | ISBN 9781786345233 (hc : alk. paper)
Subjects: LCSH: Oil reservoir engineering. | Soil percolation. | Percolation (Statistical physics)
Classification: LCC TN870.57 .K56 2018 | DDC 622/.3382--dc23
LC record available at https://lccn.loc.gov/2017060702

British Library Cataloguing-in-Publication Data
A catalogue record for this book is available from the British Library.

For any available supplementary material, please visit
http://www.worldscientific.com/worldscibooks/10.1142/Q0154#t=suppl

Desk Editors: Sundar Madhan/Jennifer Brough/Koe Shi Ying

Typeset by Stallion Press
Email: enquiries@stallionpress.com

Printed in Singapore

Preface

Porous media are ubiquitous in nature. Sedimentary rocks, such as sandstones and carbonates, are permeable to the flow of water and hydrocarbons. Most of the world's freshwater lies in underground aquifers in such rocks. The oil and gas that we rely on for energy and petrochemicals are to be found in these permeable rocks. They also form a potential trap for carbon dioxide that would otherwise be released to the atmosphere. Other natural systems, such as our lungs or termite nests, can be considered as porous media. They are also commonly found in industrial processes. Catalysts are normally constructed as permeable media. Fuel cells are another example of an artificial porous medium. Understanding fluid flow in these systems is imperative in, e.g., maximising hydrocarbon recovery, improving pollution control in aquifers or the design of efficient fuel cells. Although the focus of this book is on hydrocarbon extraction, the methods described are equally applicable to other cases where flow or transport in porous media are important.

Modelling flow in porous media is a very difficult task. At the pore scale, the geometry and topology or the pore space network is enormously complex. Multiphase flow is influenced by the balance of the three main forces: capillary or interfacial tension, the viscous drag on the rock and gravity. The interfacial forces are dependent on the surface chemistry (and hence mineralogy) as well as the ionic constitution of the fluids. Oil reservoirs vary enormously in extent but a "typical" North Sea field for example may be 10 km by 10 km by tens of metres and so contains around 10^{23} (similar to Avogadro's number) of such pores. It would be impractical to model the flow at this scale for the whole system, even if we had precise knowledge of all the relevant properties at the pore scale. This means that we have to make use of a "coarse-grained" or "upscaled" representation of the flow at the continuum level. This is based on the experimental work by the French hydrologist Henry Darcy in the mid-19th century with further extensions in the mid-20th century to account for multiphase flow. These basic laws are the mainstay of modern reservoir flow modelling. They may be inadequate to cover all eventualities, but they have proved reasonably robust for the last half century. Recent advances in pore-scale modelling have greatly improved our understanding of flow at the pore scale. However, this is beyond the scope of this book. Apart from a brief survey of results in the final chapter, we shall focus entirely on the macroscopic, continuum modelling.

This in itself is also a very challenging task. In oil reservoirs, geological heterogeneities exist on a very broad range of length scales from centimetres to tens of kilometres. If we take the core plug as the basic scale at which properties are measured. These are around 10 cm in linear dimension, so there are of the order of 10^{13} such plugs in a "typical" reservoir. Again this renders it impractical to model the flow at this scale. This is exacerbated by the fact that we have very sparse knowledge of the property distribution throughout the reservoir, so we need to generate a large number of plausible models, or stochastic realisations, to estimate the uncertainty in our predictions. It is essential to estimate this uncertainty so that development decisions can be properly risked. Failure to do so can lead, and has in the past led, to very expensive mistakes. There is, therefore, a very great need to develop efficient and accurate ways to estimate the large-scale reservoir forecasts without recourse to prohibitively expensive numerical simulations. That is the aim behind this book: to present some tools to enable the practising engineer to estimate flow, and its uncertainty, within practical timescales. Many of the results presented here can be used in a simple spreadsheet.

The fundamental assumption on which this book is based is that the flow is strongly controlled by connectivity or continuity. That is the continuity of flow paths or of flow barriers. In many cases, this is clearly and obviously reasonable. Examples would include low to intermediate net-to-gross reservoirs where the high permeability sands form a connected network in an otherwise low-permeability background. Another classic example would be fractured carbonates where the fractures are the main pathways for flow. The interconnectivity of the fracture network is key to understanding how easily oil may be recovered and at what rates. Both of these examples are characterised by a wide difference in permeability between the flow units and the rest of the material in the reservoir. These extreme cases form the framework for the studies in most of the book, although in later chapters we describe how the methods described can be extended to a much wider range of examples.

In statistical physics, there is a standard model of connectivity called percolation theory. It was introduced by Broadbent and Hammersley in 1957 and has since been applied to an enormous range of applications from the spread of epidemics, the spread of forest fires to flow in porous media. The literature on percolation theory is now very extensive and some is highly theoretical and mathematical. However, the basic concept is very simple. Its principle beauty lies in the fact that otherwise complex set of results can be reduced down to a few simple algebraic relationships (scaling laws). Furthermore, many of the parameters that appear in these laws are "universal", i.e., they don't depend on many of the small-scale details in the system. This is extremely powerful and enables us to make predictions about the large-scale behaviour based on a very simple characterisation of the geological disorder and the use of these simple algebraic relations. As stated before, these results can be encoded in a simple spreadsheet. It should be emphasised that to gain this level of simplification, a variety of approximations and assumptions need to be made. However, the results are good enough to be used for simple screening studies before having to exploit computationally expensive detailed calculations. For example, the percolation approach we describe can be used to get a "back of the envelope" estimate of the optimal well spacing (with associated uncertainty) in a matter of seconds before investing a large amount of time and effort in constructing the detailed reservoir models typically employed by oil companies worldwide today.

The primary aim of this book is to introduce practising engineers to the subject of percolation theory and how it can be used to make useful forecasts of reservoir performance, and its uncertainty, in practical timescales. To that end, the book is written as a textbook with example problems given at the end of each chapter. We have deliberately not given the solutions as many of the questions are open ended, and we hope to encourage the student to think more broadly around the problems. The other thing to emphasise is that this is not a closed problem. There are many issues we have not studied, and so the other aim of the book is to encourage further research in this area. We believe that percolation theory has the potential to solve many practical problems in reservoir modelling. If this book is able promote research in this area, it will have succeeded.

About the Authors

Peter King completed his PhD in theoretical statistical physics at Cambridge in 1982. This was under the supervision of Prof. Sir Sam Edwards to whom he owes a debt of gratitude for passing on not only his deep understanding of the subject but also his understanding of how to apply the methods of theoretical physics to problems in industry. He then started work for BP at their technology centre in Sunbury-on-Thames, UK. There he applied the methods of statistical physics to problems of how to model the heterogeneity in an oil reservoir and in particular the flow through it at a variety of length scales. He developed a real-space renormalisation approach to this "upscaling" as well as applying more abstract field theoretic approaches. He was made a Royal Academy of Engineering visiting professor in the Principles of Engineering Design in the Department of Engineering, University of Cambridge in 1997 and a visiting scholar in the Department of Physics, Boston University in 1996. He took the chair of Petroleum Engineering at Imperial College in 2000, not only continuing research into reservoir modelling but also studying the physics of granular systems and the application of statistical physics to optimising decision making in the oil industry. He is a fellow of the Institute of Physics and the Institute of Mathematics and its Applications (having served on its governing council from 1991 to 1994).

Mohsen Masihi is a professor of Petroleum Engineering at the Sharif University of Technology, Tehran, Iran. His main research interests are reservoir characterisation and modelling with focus on fractured reservoirs, application of new techniques in improving oil recovery and extending new approaches such as percolation approach in reservoir studies.

Acknowledgements

The genesis of the ideas for this book came from a chance conversation when one of the authors worked for an oil company. In discussions with a geologist about a small gas field, it became clear that the connectivity of the good sands was key. From this, it became clear that some simple ideas from percolation theory were useful in estimating what fraction of the sands were connected and what an optimal well spacing could be. This developed into an expanding programme of work which evolved into an extremely fruitful collaboration with Prof. Gene Stanley's group in the Department of Physics at Boston University. This collaboration lasted many years and involved a large number of other academics, post-doctoral researchers and PhD students. The list is long but includes Shlomo Havlin, Sergei Buldyrev, Nikolay Dokholyan, Youngki Lee, Jose Andrade, Gerry Paul, Eduardo Lopez and many others. This collaboration laid the groundwork for much of the material in this book and is looked back on with great appreciation for the many convivial and stimulating meetings held in Boston.

On taking up the position at Imperial College, Prof. King continued research on the application of percolation theory, in particular, with two highly productive students, Peyman Nurafza and Mohsen Masihi. On graduation, Mohsen (now Prof. Masihi) returned to Sharif University of Technology where he pursued research in this area continuing the collaboration with Prof. King. Many students (Sara Masoum, Behnam Tavagh-Mohammadi, Maryam Ghorbani,

Hossein Barati, Omid Pourali, Saeid Sadeghnejad and Sara Shokrollahzadeh Behbahani) developed the ideas that form the material of this book. During this time, Prof. Masihi also gave a course "A percolation approach to petroleum reservoir modelling" to masters and PhD students. It is the course notes that formed the basis for this book. The percolation ideas were further extended to a broader range of systems by post-doctoral workers Haibao Wen and Paula Gago at Imperial College. The funding for all this research came from a variety of sources including BP, BG, Statoil, Petrobras and the UK Department of Trade and Industry. The authors also acknowledge their colleagues Martin Blunt, Branko Bijeljic and Gary Hampson for permission to use their original figures. In particular, we would like to thank Gary Hampson for permission to use his fine photograph of the Blackhawk formation Utah for the cover of this book.

So, as always in science, the work reported here has come from collaborations with a very large number of people (and we apologise to those whose names are omitted here). The errors and defects in the book may be ours but the inspiration and hard work is theirs.

Contents

Chapter 1

Introduction

The nature of fluid flow in hydrocarbon reservoirs is very complex because of the complicated geology formed as a result of the intricate sedimentary processes, which deposited them over time. The flow depends strongly on the spatial distribution of the heterogeneities, which appear on all length scales from microns to tens of kilometres. These have to be modelled to make reliable predictions of reservoir performance. The spatial distribution of reservoir facies, and more precisely the facies associations corresponding to major flow units (such as rock types), is probably the single most important heterogeneity factor for reservoir performance estimation (Haldorsen and Damsleth, 1990). For example, a typical low to intermediate net-to-gross reservoir mainly consists of a mixture of good sandstones with high permeability and poorer siltstones, mudstones and shales with low permeability. The good sandstones (with a higher permeability and porosity) are the main bodies containing oil within their pore and can be considered to be the primary flow units. Obviously the volume of oil recoverable between a pair of wells depends on the connectivity of these good quality sands. Percolation is a mathematical theory of connectivity which can aid the investigation of the impact of the geological disorder on reservoir performance. It is a very fast method which can assist management decisions and risk analysis. The aim of this book is to describe the methods of percolation theory and how they can be used in reservoir modelling.

1.1 Geological uncertainty

An important issue for reservoir characterisation is to capture the geological uncertainty because of its impact on the uncertainty in production forecasts. The uncertainty arises from our lack of knowledge of the precise structure of the subsurface. This is especially true during the early stages of reservoir development, where the available data are limited. Figure 1.1 shows a schematic of the interwell volume. A reservoir model is constructed by interpolating rock properties between the known values at the two wells. There are insufficient data to constrain this interpolation precisely which leads to uncertainty as shown by the multiple possible locations of the white layer shown on the left well in Figure 1.1. Assuming that this is a shale layer, this uncertainty may influence significantly the vertical permeability and flow. This is important in some process such as water coning or gravity drainage. We use probabilities to quantify this uncertainty.

Successful reservoir modelling requires the integration of all kinds of data which are obtained from various scales. We have very few direct measurements of the flow properties. Core plugs (samples of the rock taken from the reservoir or outcrop) directly measure the permeability but they represent a very small volume of the entire

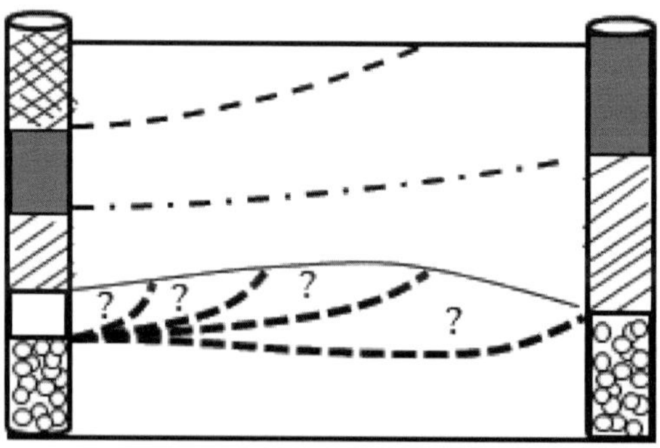

Figure 1.1. Schematic for the uncertainty in interwell properties.

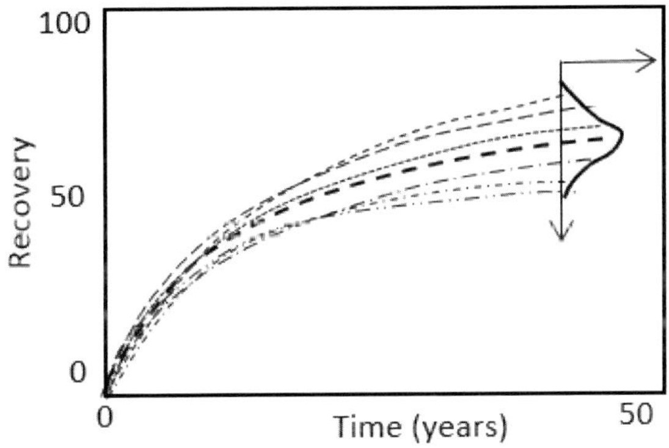

Figure 1.2. Illustration of the effect of the geological uncertainty on the cumulative recovery by showing different realisations of the reservoir model.

reservoir. Typically, a core plug has a volume of $10^{-3}\,m^3$, whereas a reservoir could have a volume of around $10^{10}\,m^3$. So a core plug represents roughly 10^{-13} of the volume of a typical reservoir.

Other types of data, such as well logs and well tests measure larger volumes (10^{-4} and 10^{-7} of the reservoir volume respectively) but the measurements have to be interpreted to infer the permeability. The consequence is that there is a great deal of uncertainty about the spatial distribution of the heterogeneities in a reservoir.

These geological uncertainties feed through to uncertainty in reservoir performance (Lasseter *et al.*, 1986). Figure 1.2 shows the effect that the geological uncertainty has on the forecast of cumulative recovery. Each curve represents the production from different possible realisations of what the reservoir could look like.

1.2 Reservoir modelling and heterogeneities

The conventional approach to estimating the impact of geological uncertainties on the prediction of future reservoir performance is to build a detailed reservoir model using geophysical and geological data. Typical, conventional reservoir models have around 10^7–10^8 grid blocks, although models with up to 10^9 grid blocks have been constructed. To give some idea of scale, consider a reservoir of size $10\,km \times 10\,km \times 100\,m$ (i.e., a volume of $10^{10} m^3$) modelled with 10^7 grid blocks. Then each block has a volume of $1{,}000\,m^3$

(corresponding to $30\,\text{m} \times 30\,\text{m} \times 1\,\text{m}$) which is still very coarse compared with the scale of heterogeneities ($\sim$10 cm), so they fall very short of the actual level of heterogeneity that we know to be present.

Given the uncertainty in rock properties discussed previously it is important to construct many models consistent with the known data. Each of these models needs to be assigned a probability. This is called stochastic modelling and the mathematical tools are collectively known as geostatistics. We will not deal with these explicitly in this book as there is a large literature already (e.g., Journel (1989), Yarus and Chambers (1994), Olea (1999), and Deutsch (2002)) but just give a brief summary.

Geostatistical techniques are used to incorporate the spatial relationships of the reservoir properties. The most commonly used methods employ the covariance (or variogram) which are based on the relationship between data at pairs of points separated by a distance vector, known as the lag. More recently higher order statistics have been used to recreate more complex and geologically realistic models (Journel, 1989; Deutsch, 2002; Mariethoz and Caers, 2014). Geostatistical modelling also provides a framework to integrate various data types for subsequent estimation and simulation of reservoir properties (Yarus and Chambers, 1994; Olea, 1999; Kelkar and Perez, 2002; Caers, 2005).

An alternative to variogram based geostatistics is to use so called "object-based models". In this approach the geological entities (like fracture, shale barriers, fluvial channels) are represented as geometrical objects. These are parameterised by shape, size and location in space and placed in the model according to statistical rules. Again we do not go into detail but suggest the following references: Deutsch (2002) and Mallet (2002).

These geological models are then upscaled for fluid flow models which have around 10^4–10^5 grid blocks. Upscaling, or coarse graining, is the process of determining the correct representation of the physics of the flow at successively larger length scales. Typically, one calculates the unique flow property required in a large sample of the reservoir that correctly reproduces the same flow rate for the same pressure boundary conditions that would have been found for the same volume but with all the small-scale details included. As stated previously, in order to have a reliable estimation of uncertainty in reservoir performance, it is necessary to construct a number of possible reservoir models and then run flow simulations for each one.

We now explore the implications of running many realisations. A model with 10^7 grid blocks has some 10^8 parameters (such as the permeability components in each direction, porosity, relative permeability end points). To sample from these ranges of possible values for each of these would require in excess of 10^9 different realisations, which shows the typical computational scale of effort. Clearly brute-force Monte Carlo analysis to examine the full range of uncertainty in these reservoir models is completely impractical.

The problem with this approach is that it is computationally very expensive although there has been some progress to reduce the calculation time. For example, streamline-based simulation is able to decrease the run time by more than one order of magnitude (di Donato and Blunt, 2004; Huang *et al.*, 2004) compared to conventional grid-based simulation. There are also efficient Monte Carlo sampling methods but this approach is still intractable for realistic sized problems. Therefore, there is a great incentive to produce much simpler approaches to predict the uncertainty in reservoir performance. This should be based on the dominant physics that controls the displacement process.

1.3 Connectivity as the predominant feature of heterogeneous media

It has long been understood that many systems in nature have disordered structures with a complex topology and geometry (Sahimi, 1994). The topology or interconnectivity of individual constituent elements of the system has a significant role which needs to be characterised. It is known that flow in heterogeneous porous media, as in other disordered systems, is largely controlled by the continuity of permeability contrasts. Examples are high permeability streaks, flow barriers (e.g., shales), conductive fractures or sealing faults. Although there are other influences, in some cases these are the predominant features affecting the flow behaviour. This simplification enables us to model such reservoirs by using percolation theory, the principle mathematical model of connectivity. Examples of such reservoir structures include sands and channels distributed in fluvial (King, 1990) and turbidite (Mayall *et al.*, 2006) environments which will be discussed later.

An important property in reservoir engineering which affects the reservoir performance is the interwell connectivity. This can be defined as the probability that two wells will be in the same sand cluster at a given well spacing. This could be an important criterion for well-pattern design. There are methods to make estimates based on production and injection data analysis (Yousef *et al.*, 2006). However, they are usually very time consuming compared with the percolation approach. This gives the oil in place in the connected volumes, but the connectivity also controls the dynamic flow behaviour and hence the recoverable fraction (Hunt and Ewing, 2005). With this in mind, we are looking for the ways to model flow at the reservoir scale based on the connectivity of permeability contrasts. To show this explicitly, we consider a simple overlapping sandbody model.

1.4 Overlapping sandbodies models

Geologists frequently use the idea that a reservoir consists of geometrically complex connected and disconnected objects of different shapes. This leads to the use of the term "object-based modelling" or more simply "object modelling" as this approach to modelling subsurface heterogeneity (Lantuéjoul, 1997; Holden *et al.*, 1998). The geometrical and size statistics of these objects are obtained from outcrops or well data. These models are used by engineers to estimate hydrocarbons in place and the expected recoverable oil (Haldorsen *et al.*, 1988). They can also be used to predict interwell connectivity for infill drilling programs (Heffer *et al.*, 1997). In many cases, the connectivity of these heterogeneities is the key control for recovery. This is a function of sand/shale ratios (net to gross), sandbody geometry (Mijnssen, 1997; Ainsworth, 2005).

In those cases where the connectivity of flow units is of prime importance, percolation theory can be a very efficient way to predict performance. Examples where this has been done include fluvial (King, 1990; Kuuskraa *et al.*, 1997) and turbidite depositional environments (de Ruig and Hubbard, 2006; Mayall *et al.*, 2006). Other examples are fluvial point bar deposits which can be described by flow units (Berg, 1986) with nearly square sedimentary body shapes (Figure 1.3).

We consider a simple example where the percolation approach seems most appropriate. First, meandering rivers which deposit sand over the meander belt as the meanders move downstream across the meander belt (Figure 1.4). Over time, ancient channels may be filled with permeable sands to give permeable sandbodies. Over geological time, the river may change its course due to some event upstream (this is avulsion) and deposit new sandbodies that may overlap with those previously deposited (Smith *et al.*, 1989; Heller and Paola, 1996; Slingerland and Smith, 2004). This depositional process is not entirely random as it depends on external factors like

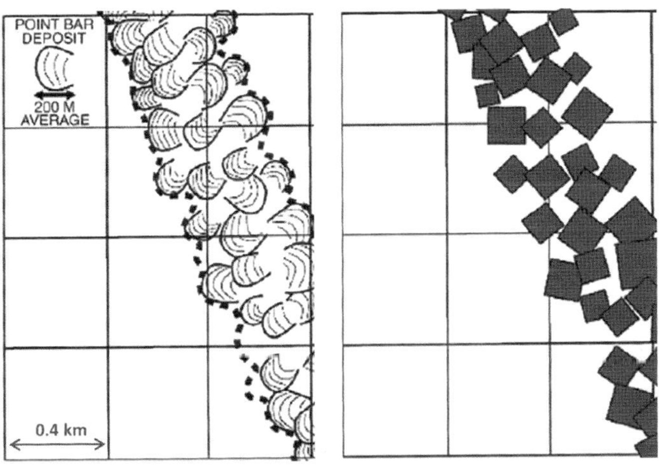

(a) (b)

Figure 1.3. Illustration of the use of simple geometric shapes for modelling geological flow units: (a) fluvial point bar deposits and (b) approximate equivalent squares (Li *et al.*, 2009).

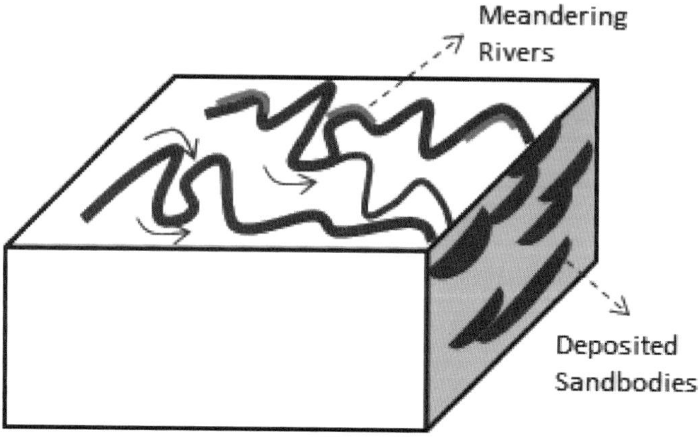

Figure 1.4. Schematic of a meandering river depositing sandbodies over time.

sea level (and its rate of change), compaction rates and so on. The process continues to form a system of sandbodies embedded in an impermeable background. This is a simple schematic representation of the complicated process of sedimentation over millions of years (Sun *et al.*, 1996).

An example of such structures is shown in Figure 1.5. The top part is an outcrop with channelised fluvial sandbodies (red in interpreted image in Figure 1.5(b)) within fine-grained floodplain deposits (green in the interpreted image). The outcrop is the Cretaceous Blackhawk Formation, Wasatch Plateau, central Utah, USA. The channelised sandbodies are interpreted to be nearly linear in 3D, with their axes oriented nearly perpendicular to the cliff face. These sandbodies overlap and amalgamate with each other in places; this spatial arrangement is attributed to the intrinsic switching in position of the river channels that deposited the succession. Finally a representative set of oblong objects, as would be used in an object-based model, are shown in Figure 1.5(c) indicating their interconnectivity, which is key to the percolation approach.

1.5 The percolation theory approach

The idea behind the percolation theory approach is to use percolation theory to make rapid estimates of global physical properties of the

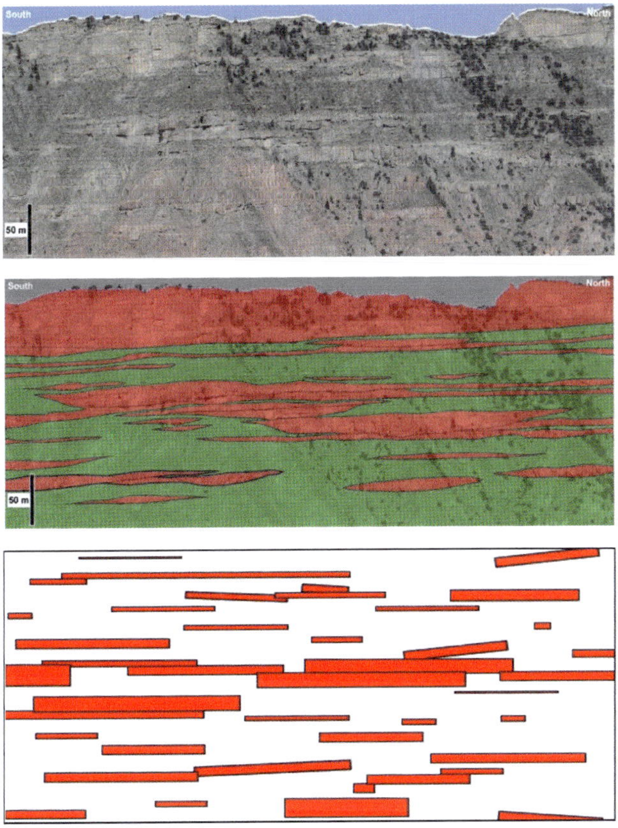

Figure 1.5. Schematic of the process of generating an interconnected object-based model from outcrop data. (a) An outcrop with channelised fluvial sandbodies (Cretaceous Blackhawk Formation, Wasatch Plateau, central Utah, USA), (b) interpreted image of the outcrop with channelised fluvial sandbodies (red) within fine-grained floodplain deposits (green) and (c) a representative set of oblong objects, as would be used in an object-based model, indicating their interconnectivity. (Modified from Hampson *et al.* (2012). Copyright © 2012 The Authors.)

reservoir model. This approach, in the first instance, is based on the assumption that the rock is either permeable or impermeable. The permeability of the permeable zones is assumed to be constant. With these simplifications, the geometrical and flow properties (that is, the connectivity and conductivity) can be estimated directly by using percolation theory. The advantage is that it reduces many results to some standard curves from which all possible outcomes can be predicted by simple algebraic transformations. In addition, this approach is able to estimate the uncertainty in these properties which is not possible with a single realisation model. These transformations mean that the effects of the complex geometry can easily be estimated in a fraction of a second on a spreadsheet. The disadvantage is that some of the flow physics and subtleties of the heterogeneity distribution are missed. Percolation theory is the basic mathematical model of connectivity in geometrically complex systems. The mathematical background to the theory is presented by Stauffer and Aharony (1992). The essential elements of the theory required for the application to reservoir modelling are reviewed in Chapter 2. The strategy adopted when using this approach includes the following steps:

- Build an appropriate object-based model for the reservoir type.
- Develop the appropriate mathematical formulation to account for the effect of finite boundaries of the system.
- Produce the prototype curves for the average value of each property of interest as well as its associated uncertainty.
- Encode this formulation in a spreadsheet.
- Validate predictions from the spreadsheet against commercial flow simulations.
- This spreadsheet can then be used to estimate reservoir performance very quickly.

We expand on this framework in subsequent chapters. The quantities that can be predicted this way include the connected fraction (the connectivity), swept fraction, effective permeability and breakthrough times. In this book, we first assume an isotropic system with isotropic objects through an object-based modelling technique. Then we extend the model to consider the effects of anisotropy, size distribution as well as orientation disorder as schematically shown in Figure 1.6. We then extend these simple model cases to more complicated well trajectories and non-strictly percolating systems.

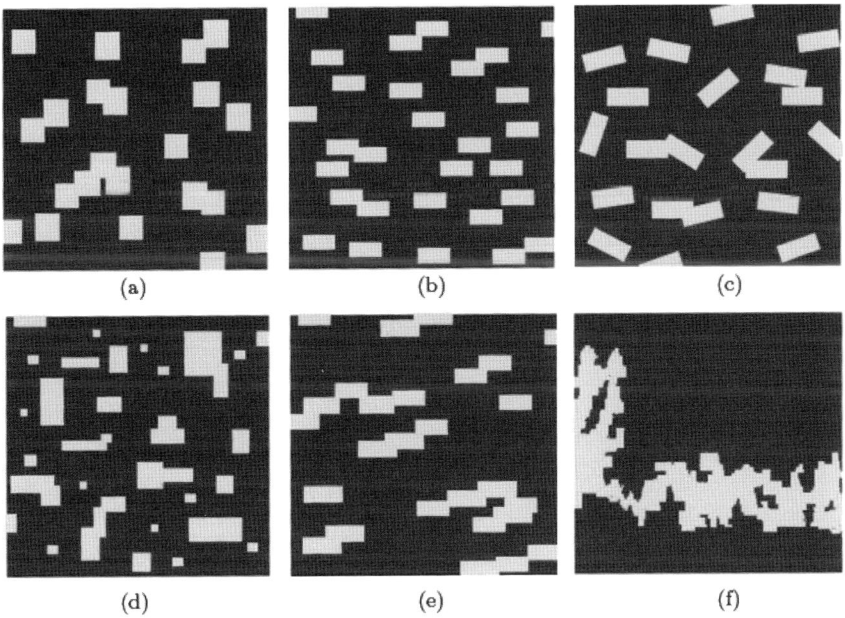

Figure 1.6. Illustration of complexity added to the simple sandbody model in 2D: (a) isotropic model with sands as squares, (b) anisotropic model with sands as rectangles, (c) sands with orientation distribution, (d) sands with size variation, (e) sands with spatial correlation and (f) a realistic channelised pattern.

1.6 Variants of percolation

The most simple percolation model is based on an infinite lattice of sites (called *site percolation*). On this lattice, sites are occupied randomly with a probability p. Clusters are formed when neighbouring sites are occupied (as shown in different colours in Figure 1.7(a)). Percolation theory describes mathematically how the number and the size of these clusters and consequently the percolation quantities vary as a function of the occupancy probability. The key feature of percolating systems is that as the occupancy probability, p, increases the clusters grow larger and merge until at a critical value one large cluster spans the whole system. This is known as the *percolation threshold* (p_c) and it plays a crucial role in the mathematical description of the behaviour. It is directly analogous to a second-order phase transition in thermodynamic systems (such

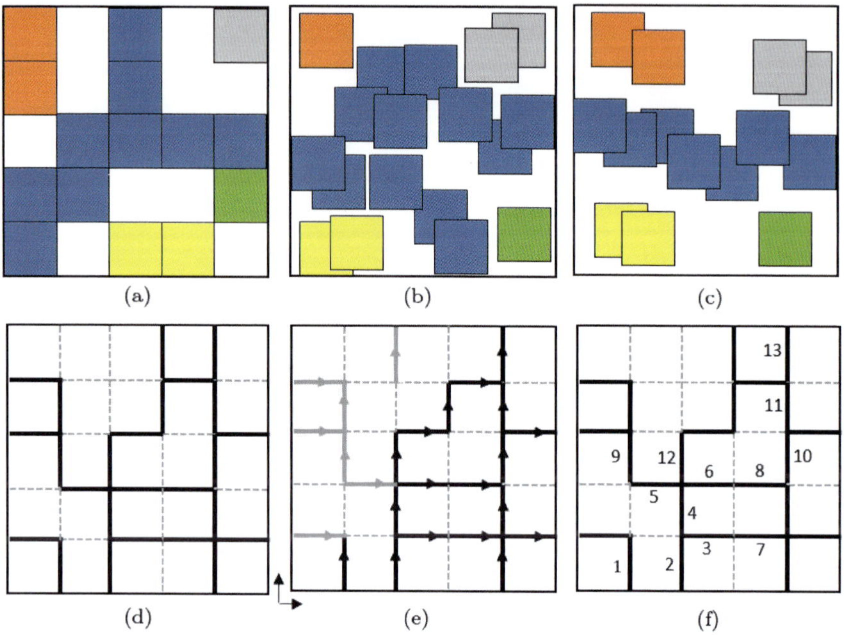

Figure 1.7. Schematic of various percolation models in 2D: (a) site percolation, (b) continuum percolation, (c) correlated percolation, (d) bond percolation, (e) directed percolation and (f) invasion percolation. Models (a)–(c) consider percolation cluster from left to right and models (d)–(f) from bottom to top.

as the Curie temperature in ferromagnets) and hence we can discuss systems close to the percolation threshold as in the critical regime. This will be discussed in more detail in Chapter 2. The lattices can be of different structure such as square, triangular, hexagonal or cubic. Similarly, it is possible to occupy the bonds instead of the sites (which is known as *bond percolation*) (Stauffer and Aharony, 1992). This model can provide some new applications, e.g., for modelling fracture networks (Figure 1.7(d)).

Percolation can also be applied to systems without an underlying grid (which is how we describe conductive fractures or sands represented with the geometrical objects). This is shown by squares (in Figure 1.6(b)) distributed randomly in space. This variant is called *continuum percolation* (Stauffer and Aharony, 1992; Stell, 1996).

The objects may also have some spatial correlation appropriate to represent stacking patterns for channels, e.g., of squares in Figure 1.7(c). This is an example of *correlated percolation* (Prakash *et al.*, 1992; Ioannidis *et al.*, 1993; Sahimi and Mukhopadhyay, 1996; Lee and Torquato, 1990; Guin and Ritzi, 2008; Ewing and Gupta, 1993; Manzocchi, 2002; Mourzenko *et al.*, 2005; Harter, 2005; Masihi and King, 2007a, 2007b).

Another variation is to include a direction to define connection between sites or bonds. In Figure 1.7(e), connection is only considered from bottom to top or left to right; this is called *directed percolation*. The configuration of the system could also change with time so that the probability that a bond or a site is open or closed depends on the time (Druger *et al.*, 1985; Sahimi, 1986; Granek and Nitzan, 1990). This is called *dynamic percolation*. Another dynamic percolation model is *invasion percolation* (Wilkinson and Willemsen, 1983; Diaz *et al.*, 1987; Maier and Laidlaw, 1993; Kueper and McWhorter, 1992; Ioannidis *et al.*, 1996; Glass *et al.*, 2000; Al-Futaisi and Patzek, 2003; Pesheva *et al.*, 2010; Blunt and King, 1990; Blunt *et al.*, 1992; Bo *et al.*, 2006; Mourhatch *et al.*, 2011) used to describe pore-scale displacement. In invasion percolation, the sites are filled in order according to a random number given to each site. In Figure 1.7(f), the labels show the filling order. The extensions, and their applications, will be described more fully in subsequent chapters.

1.7 Layout of the book

The outline of this book is organised as follows: The relevant fundamentals of percolation theory are presented in Chapter 2, which contains the mathematical background required for the subsequent chapters.

In Chapter 3, we describe continuum percolation and its application to hydrocarbon reservoir modelling. In particular, we present two idealised continuum percolation models, the overlapping sand-body model and the fracture network model. We shall use squares (and line segments) to represent sandbodies (or fractures) in 2D. Then the computational algorithm to produce realisations of these models is presented as are the techniques to collect statistics. These are used to generate the idealised curves for the average value of connectivity as well as its associated uncertainty.

In order to make the idealised overlapping sandbody models more realistic, the effects of anisotropy, sandbody size and orientation distribution, sand shapes as well as the spatial correlation and well pattern on percolation quantities need to be studied. This is discussed in Chapter 4. These models are also extended to 3D where different geological objects such as cubes, cuboids or ellipsoids are used to represent sandbodies. The rules described are then examined on several case studies.

Chapter 5 concentrates on the connectivity of realistic fracture networks. In particular, the effects of fracture length and orientation distribution as well as the spatial correlation and anisotropy of complex fracture networks on the fracture network connectivity are discussed. These models are extended to 3D and validated against field data.

Chapter 6 describes how the study of sub-networks such as the backbone and dangling ends can be used to estimate swept oil fractions. The calculation of effective permeability is also presented.

As the scaling laws presented are only valid for the critical region close to the percolation threshold and on binary permeability models, Chapter 7 addresses how to make predictions in real system without these limitations.

Chapter 8 focuses on the dynamic behaviour on percolating systems. In particular, the methods for prediction of the breakthrough and post-breakthrough behaviour between an injector–producer pair are described. These are validated against some field data.

Whilst this book focuses on the field-scale application of percolation theory, it should be emphasised that this is a general theory of connectivity which is widely applied in many disciplines (Sahimi, 1994; Hunt and Ewing, 2005). In particular, it has been used successfully to describe pore scale in porous media. Hence, a summary of small-scale applications of this theory is given in Chapter 9.

References

Ainsworth, R.B., Sequence stratigraphic-based analysis of reservoir connectivity: influence of depositional architecture. A case study from a marginal marine depositional setting. *Petroleum Geoscience*, 11: 257–276, doi: 10.1144/1354-079304-638, 2005.

Al-Futaisi, A. and Patzek, T.W., Impact of wettability alteration on two-phase flow characteristics of sandstones: a quasi-static description. *Water Resources Research*, 39(2), doi: 10.1029/2002WR001366, 2003.

Berg, R.R., *Reservoir Sandstones*. Prentice-Hall Inc., Englewood Cliffs, NJ, pp. 481, 1986.

Blunt, M.J. and King, P.R., Macroscopic parameters from simulations of pore scale flow. *Physical Review A*, 42(12): 4780–4789, 1990.

Blunt, M.J., King, M. and Scher, H., Simulation and theory of two phase flow in porous media. *Physical Review A*, 46(12): 7680–7699, 1992.

Bo, Z., Loggia, D., Xiaorong, L., Vassueur, G. and Ping, H., Numerical studies of gravity destabilized percolation in 2D porous media. *European Physics Journal B*, 50: 631, 2006.

Caers, J., *Petroleum Geostatistics*. Society of Petroleum Engineers, Richardson, TX, pp. 96, 2005.

de Ruig, M.J. and Hubbard, S.M., Seismic facies and reservoir characteristics of a deep-marine channel belt in the Molasse foreland basin, Puchkirchen Formation, Austria. *AAPG Bulletin*, 90(5): 735–752, 2006.

Deutsch, C.V., *Geostatistical Reservoir Modelling*. Oxford University Press, Oxford, UK, pp. 384, 2002.

di Donato, G. and Blunt, M.J., Streamline based dual porosity simulation of reactive transport and flow in fractured reservoirs. *Water Resources Research*, 40: W04203.1–W04203.12, 2004.

Diaz, C.E., Chatzis, I. and Dullien, F.A.L., Simulation of capillary pressure curves using bond correlated site percolation on a simple cubic network. *Transport in Porous Media*, 2: 215–240, 1987.

Druger, S.D., Ratner, M.A. and Nitzan, A., Generalized hopping model for frequency-dependent transport in a dynamically disordered medium, with applications to polymer solid electrolytes. *Physical Review B*, 31: 3939, 1985.

Ewing, R.P. and Gupta, S.C., Percolation and permeability in partially structured networks. *Water Resource Research*, 29(9): 3179–3188, 1993.

Glass, R.J., Conrad, S.H. and Peplinski, W., Gravity-destabilized nonwetting phase invasion in macro heterogeneous porous media: experimental observations of invasion dynamics and scale analysis. *Water Resources Research*, 36: 3121–3137, 2000.

Granek, R. and Nitzan, A., Dynamic bond percolation theory for diffusion of interacting particles: tracer diffusion in a binary mixture lattice gas. *The Journal of Chemical Physics*, 93: 5918, doi: 10.1063/1.459502, 1990.

Guin, A. and Ritzi, R., Studying the effect of correlation and finite-domain size on spatial continuity of permeable sediments. *Geophysical Research Letters*, 35(10): L10402, 2008.

Haldorsen, H.H. and Damsleth, E., Stochastic modelling. *Journal of Petroleum Technology*, 42(4): 404–412, 1990.

Haldorsen, H.H., Brand, P.J. and Macdonald, C.J., Review of the stochastic nature of reservoirs. In: Edwards, S. and King, P.R. (eds.), *Mathematics in Oil Production*. Clarendon Press, Oxford, pp. 109–209, 1988.

Hampson, G.J., Gani, M.R., Sahoo, H., Rittersbacher, A., Irfan, N., Ranson, A., Jewell, T.O., Gani, N.D.S., Howell, J.A., Buckley, S.J. and Bracken, B., Controls on large-scale patterns of fluvial sandbody distribution in alluvial-to-coastal plain strata: Upper Cretaceous Blackhawk Formation, Wasatch Plateau, central Utah, USA. *Sedimentology*, 59: 2226–2258, 2012.

Harter, T., Finite size scaling analysis of percolation in three dimensional correlated binary Markov chain random fields. *Physical Review E*, 72: 026120.1–026120.7, 2005.

Heffer, K.J., Fox, R.J., McGill, C.A. and Koutsabeloulis, N.C., Novel techniques show links between reservoir flow directionality, earth stress, fault structure and geomechanical changes in mature waterfloods. *SPE Journal*, 2(2): 91–98, 1997.

Heller, P.L. and Paola, C., Downstream changes in alluvial architecture: an exploration of controls on channel-stacking patterns. *Journal of Sediment Research*, 66(2): 297–306, 1996.

Holden, L., Hauge, R., Skare, A. and Skorstod, A., Modelling of fluvial reservoirs with object models. *Mathematical Geology*, 30(5): 473–496, 1998.

Huang, W., di Donato, G. and Blunt, M.J., Comparison of streamline based and grid based dual porosity simulation. *Journal of Petroleum Science and Engineering*, 43: 129–137, 2004.

Hunt, A.G. and Ewing, R., *Percolation Theory for Flow in Porous Media*. Lecture Notes in Physics, vol. 771. Springer, Berlin, 2005.

Ioannidis, M.A. and Chatzis, I., The effect of spatial correlations on the accessibility characteristics of three-dimensional cubic networks as related to drainage displacements in porous media. *Water Resources Research*, 27: 1777, 1993.

Ioannidis, M.A., Chatzis, I. and Dullien, F.A.L., Macroscopic percolation model of immiscible displacement: effects of buoyancy and spatial structure. *Water Resources Research*, 32(11): 3297–3310, 1996.

Journel, A.G., *Fundamentals of Geostatistics in Five Lessons*. American Geophysical Union, Washington, DC, USA, 1989.

Kelkar, M. and Perez, G., *Applied Geostatistics for Reservoir Characterization*. Society of Petroleum Engineers, Richardson, TX, pp. 264, 2002.

King, P.R., The conductivity and connectivity of overlapping sandbodies. In: Buller, A.T. *et al.* (eds.), *North Sea Oil and Gas Reservoirs — II*. Graham & Trotman, London, pp. 353–362, 1990.

Kueper, B.H. and McWhorter, D.B., The use of macroscopic percolation theory to construct large-scale capillary pressure curves. *Water Resources Research*, 28: 2425–2436, 1992.

Kuuskraa, V.A., Barrett, T., Mueller, R., Hansen, J., Reservoir characterization for development of Mesaverde Group sandstones of the Piceance Basin, Colorado. In: Coalson, E.B., Osmond, J.C. and Williams, E.T. (eds.), *Innovative Applications of Petroleum Technology in the Rocky Mountain Area*. Rocky Mountain Association of Geologists, Denver, CO, pp. 61–72, 1997.

Lantuéjoul, C., Conditional simulation of object-based models. In: Jeulin, D. (ed.), *Advances in Theory and Applications of Random Sets*. World Scientific Publishing Company, Singapore, pp. 289–318, 1997.

Lasseter, T.J., Wagoner, J.R. and Lake, L.W., *Reservoir Characterization, Reservoir Heterogeneities and Their Influence on Ultimate Recovery*. Academic Press, London, 1986.

Lee, S.B. and Torquato, S., Monte Carlo study of correlated continuum percolation: universality and percolation thresholds. *Physical Review A*, 41(10): 5338–5344, 1990.

Li, W., Jensen, J.L., Ayers, W.B., Hubbard, S.M. and Heidari, M.R., Comparison of interwell connectivity predictions using percolation, geometrical, and Monte Carlo models. *Journal of Petroleum Science and Engineering*, 68: 180–186, 2009.

Maier, R. and Laidlaw, W.G., Fluid topology for invasion percolation in 3-D lattices. *Transport in Porous Media*, 10(3): 221–234, 1993.

Mallet, J.L., *Geomodeling*. Oxford University Press, New York, pp. 596, 2002.

Manzocchi, T., The connectivity of two-dimensional networks of spatially correlated fractures. *Water Resource Research*, 38(9): 1162, 2002.

Mariethoz, G. and Caers, J., *Multi-point Geostatistics: Stochastic Modeling with Training Images*. Wiley-Blackwell, New York, pp. 376, ISBN: 978-1-118-66275-5, 2014.

Masihi, M. and King, P.R., A correlated fracture network: modelling and percolation properties. *Water Resources Research*, 43, doi: 10.1029/2006WR005331, 2007a.

Masihi, M. and King, P.R., Connectivity of spatially correlated fractures: simulation and field studies. Paper SPE 107132, Proceedings of the the SPE Europec, London, UK, 11–14 June, 2007b.

Mayall, M., Jones, E. and Casey, M., Turbidite channel reservoirs — key elements in facies prediction and effective development. *Marine and Petroleum Geology*, 23(8): 821–841, 2006.

Mijnssen, F.C.J., Modelling of sandbody connectivity in the Schooner Field. In: Ziegler, K., Turner, P. and Daines, S.R. (eds.), *Petroleum Geology of the Southern North Sea: Future Potential*. Geological Society, London, 1997.

Mourhatch, R., Tsotsis T.T., Sahimi, M., Determination of the true pore size distribution by flow permporometry experiments: an invasion percolation model. *Journal of Membrane Science*, 367(1–2): 55–62, 2011.

Mourzenko, V.V., Thovert, J.F. and Adler, P.M., Percolation of three dimensional fracture networks with power law size distribution. *Physical Review E*, 72: 036103.1–036103.14, 2005.

Olea, R.A., *Geostatistics for Engineers and Earth Scientists*. Kluwer Academic Publishers, Boston, pp. 303, 1999.

Pesheva, N., Stefanov, I. and Slavtchev, S., Application of the invasion percolation model to water-gas flows in artificial soils with plants. *Transport in Porous Media*, 83(2): 319–331, 2010.

Prakash, S., Havlin, S., Schwartz, M. and Stanley, H.E., Structural and dynamical properties of long-range correlated percolation. *Physical Review A*, 46: R1724, 1992.

Sahimi, M., *Applications of Percolation Theory*. Taylor and Francis, London, pp. 258, 1994.

Sahimi, M., Dynamic percolation and diffusion in disordered systems. *Journal of Physics C: Solid State Physics*, 19: 1311–1317, 1986.

Sahimi, M. and Mukhopadhyay, S., Scaling properties of a percolation model with long-range correlations. *Physical Review E*, 54: 3870, 1996.

Slingerland, R. and Smith, N.D., River avulsions and their deposits. *Annual Review of Earth and Planetary Sciences*, 32: 257–285, 2004.

Smith, N.D., Cross, T.A., Dufficy, J.P. and Clough, S.R., Anatomy of an avulsion. *Sedimentology*, 36: 1–23, 1989.

Stauffer, D. and Aharony, A., *Introduction to Percolation Theory*. Taylor and Francis, London, pp. 181, 1992.

Stell, G., Continuum theory of percolation. *Journal of Physics: Condensed Matter*, 8, A1–A17, doi: 10.1088/0953-8984/8/25A/002, 1996.

Sun, T., Meakin, M. and Jossan, T., A simulation model for meandering rivers. *Water Resources Research*, 32(9): 2937–2954, 1996.

Wilkinson, D. and Willemsen, J.F., Invasion percolation: a new form of percolation theory. *Journal of Physics A: Mathematical and General*, 16(16): 3365–3376, 1983.

Yarus, J.M. and Chambers, R.L., *Stochastic Modelling and Geostatistics: Principles, Methods, and Case Studies*. American Association of Petroleum Geologists, Tulsa, OK, pp. 379, 1994.

Yousef, A.A., Pablo, G., Jensen, J.L. and Lake, L.W., A capacitance model to infer interwell connectivity using production and injection data. *SPE Reservoir Evaluation and Engineering*, 9(6): 630–646, 2006.

Chapter 2

Percolation theory — the basics

The mathematical analysis of percolation theory was first developed by Broadbent and Hammersley in 1957, and since then the topic has been intensively studied, primarily in the physics literature (Stauffer and Aharony, 1992 and references therein). This theory is particularly well adapted to describe the global physical properties, such as the connectivity and conductivity, of geometrically complex systems. These properties are linked to the number density of objects placed randomly in space through some universal laws. By universality we mean that the large scale behaviour of the system is independent of the details (specifically the local geometries) of the system. The interested reader is referred to the book by Stauffer and Aharony (1992) for a detailed discussion and derivations of the theory. Percolation has been applied successfully to many subjects from the spread of forest fires and epidemics to composite materials and flow in porous and fractured media (Moore and Newman, 2000; Lagorio *et al.*, 2009; Christensen *et al.*, 1993; Coleman *et al.*, 1998; Zhang *et al.*, 2009; Glass *et al.*, 2001; Mourhatch *et al.*, 2011; Golden *et al.*, 2011; Adler and Thovert, 1999; Mourzenko *et al.*, 2005; Havlin and ben-Avraham, 1987; Sahimi, 1994; Belayneh *et al.*, 2006; Masihi *et al.*, 2007). It also provides a useful conceptual framework for understanding flow and transport in porous media (Selyakov and Kadet, 1996; Hunt and Ewing, 2005).

As a background for the subsequent discussions, we describe some of the principal characteristics of percolation theory related to the problem of the connectivity and conductivity. The theoretical framework was developed for infinite-size systems, so it is critical to understand the impact of finite boundaries on the mathematical behaviour. We put particular emphasis on this in the ensuing discussion.

2.1 Percolation on infinite lattices

The simplest model for introducing percolation theory is an infinite lattice of sites which are occupied with a probability p. This is called *site percolation.* By "infinite", we mean the asymptotic limit where the number of lattice sites tends to infinity, in which case the effects of boundaries can be ignored. It is assumed that the occupation of one site is independent of the occupation of any other site. Clusters are formed when neighbouring sites are occupied. Figure 2.1 illustrates the clusters formed on a small lattice and Figure 2.2 shows results for a much larger system.

Percolation theory describes mathematically how the number and the size of these clusters vary as a function of the occupancy probability p. As this increases, the clusters grow in size. The mechanisms are growth of individual clusters and merging of clusters by bridging. At one particular value (called the percolation threshold, p_c), one large cluster becomes infinite in size (the red cluster in Figure 2.2(b)).

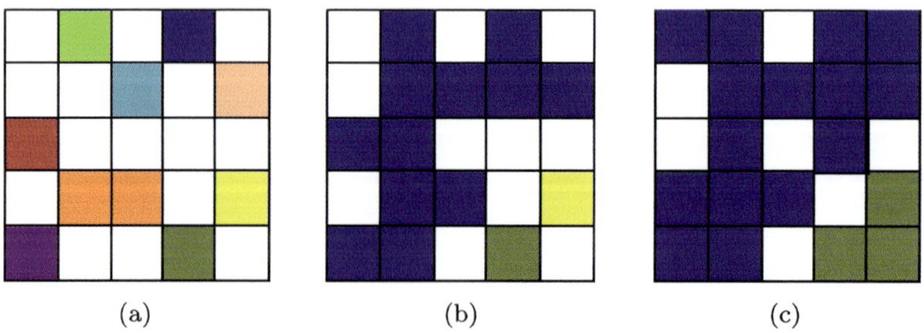

Figure 2.1. An illustrative example of *site* percolation on a lattice of size 5×5 at different occupancy: (a) below the threshold, $p = 0.4$, (b) at the threshold, $p = 0.56$ and (c) above the threshold, $p = 0.72$. Different colours represent different clusters. The spanning cluster is shown in blue.

(a)

(b)

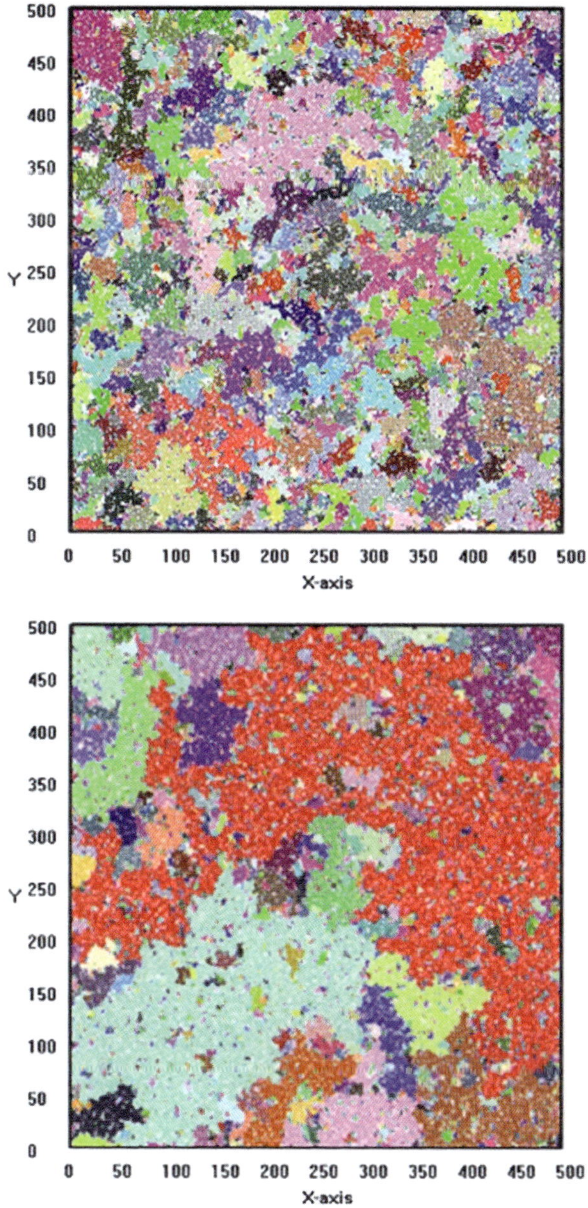

Figure 2.2. Examples of two realisations of *site* percolation at different occupancy p. Different colours represent different clusters: (a) below the threshold ($p = 0.56$) and (b) just above the threshold ($p = 0.593$). The spanning cluster which is shown in red contains ~33% of occupied sites.

Standard algorithms such as that of Hoshen and Kopelman (1976) are used to identify the various clusters. In many practical applications, discussed later in this book, we are interested in the largest cluster, which is called the *spanning*, *infinite* or *percolating* cluster. The existence of the threshold is a fundamental characteristic of percolation theory. It is analogous to second-order phase transition in thermodynamics. We define three regimes: below the threshold, close to the threshold (the critical region) and well above the threshold. Much of percolation theory concerns the critical region where behaviour is dominated by the spanning cluster and the threshold itself.

The numerical value of the threshold depends on both the dimensionality of the grid (2D or 3D) and the lattice topology (e.g., whether the lattice is made of tessellating triangles or squares). It also depends on the type of percolation considered (i.e., bond percolation, continuum percolation or directed percolation which were introduced in Section 1.6). The threshold values for some geometries in 2D and 3D lattices are given in Table 2.1 (for more complete tables, see Stauffer and Aharony (1992) and Berkowitz and Balberg (1993)). It can be seen that the threshold values in 3D are generally much lower than the ones obtained in 2D, implying that connected clusters form more easily in 3D compared with 2D.

Table 2.1. Thresholds of different lattice shapes for site and bond percolation (Stauffer and Aharony, 1992; Berkowitz and Balberg, 1993; Hunt and Ewing, 2005; Sahimi, 1994; Vogel *et al.*, 2010).

	Lattice shape	**Site percolation**	**Bond percolation**
1D	—	1	1
2D	Square	0.5927	1/2
	Triangular	1/2	$2\sin(\pi/18) \approx 0.347$
	Diamond	0.430	0.388
	Hexagonal	0.6962	$1 - 2\sin(\pi/18) \approx 0.6527$
3D	Simple cubic	0.3116	0.2488
	Body centred cubic	0.246	0.1803
	Face centred cubic	0.198	0.119

2.2 Percolation quantities and their properties

There are several percolation quantities that will be described in this section. These have many important applications for estimating quantities of interest in reservoir engineering.

2.2.1 *Percolation probability*

The first percolation property is the *percolation probability* $P(p)$. This is a measure of the *connectivity* (or *connected fraction*). It is defined as the probability that any occupied site belongs to the spanning cluster for a given value of p. An alternative, and equivalent, definition is the fraction of the occupied sites that are in the spanning cluster. Note that lowercase p denotes the occupancy probability, whereas uppercase P is the fraction of occupied sites belonging to the percolating cluster. This may be slightly confusing but is the standard nomenclature used throughout the literature.

Below the percolation threshold, there is no global connectivity, while above the threshold, the percolation probability $P(p)$ increases rapidly in the critical region (Figure 2.3). This is described by the following scaling law (Stauffer and Aharony, 1992) in the vicinity of

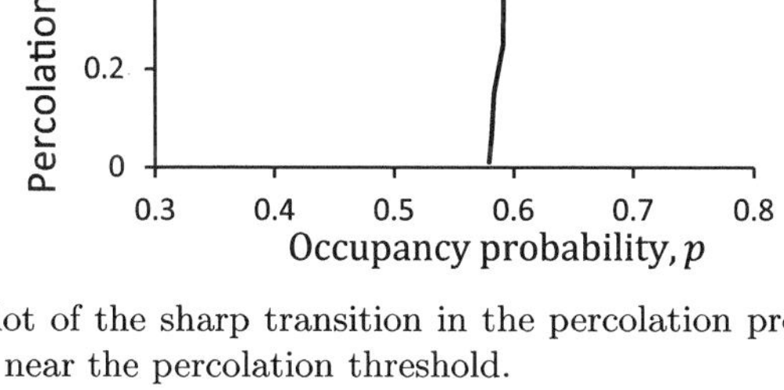

Figure 2.3. Plot of the sharp transition in the percolation probability $P(p)$ of infinite systems near the percolation threshold.

the threshold:

$$\begin{aligned} P(p) &= 0, & p &< p_c, \\ &\propto (p - p_c)^{\beta}, & p &> p_c. \end{aligned} \tag{2.1a}$$

The exponent β is the first of many used in percolation theory and is called the connectivity exponent.

It should be noted that the power law above the threshold is strictly only true very close to the threshold. As we move away, this is seen as the most "singular" behaviour; there may well be other analytic terms (as shown in Eq. (2.1b)) that vanish more rapidly as we approach the threshold.

$$P = a(p - p_c)^{\beta} + b(p - p_c) + c(p - p_c)^2 + \cdots. \tag{2.1b}$$

It is clear from the definition that as $p \to 1$, $P(p) \to 1$. In other words, as all the sites become occupied, they must all belong to the one spanning cluster. However, in practice, the singular behaviour represented by Eq. (2.1) is applicable over quite a broad range and is the key to understanding the behaviour for real, finite-size systems, which is described later. Extensions away from this critical regime are discussed in Chapter 7.

The simple power law (Eq. (2.1a)) is typical of the behaviour of many percolation properties and is known as *scaling*. We will see scaling laws again and again, and they are crucial to the application of percolation to reservoir modelling. It is a remarkable fact that the complex behaviour of these systems can be reduced to such simple scaling laws. Even more remarkable is that the exponents (such as β) only depend on the dimensionality of the system and not the details of the lattice structure, or indeed whether there is an underlying lattice, as in continuum percolation. This phenomenon is known as *universality* and again is a key to the application of these ideas, which we shall describe later. It should be remembered that whilst the exponents may be universal, the other parameters (like the percolation threshold and the pre-factors, coefficients a, b, c in Eq. (2.1b)) are not but must be determined for each system separately.

All of these scaling laws have to be understood in the same way as Eq. (2.1b) and there are corrections as you move outside the critical regime. Again this is something we shall return to later. Finally these scaling laws have been observed from large-scale numerical simulations, except in some very limited special cases; there are no theoretical derivations of them.

There is a similar quantity to the percolation probability (P) that is called spanning probability (Π). This is the probability that a connected cluster exists across the system. For an infinite system, it is a step function:

$$\begin{aligned} \Pi(p) &= 0, \quad p < p_c \\ &= 1, \quad p > p_c. \end{aligned} \tag{2.2}$$

2.2.2 *Correlation length*

The correlation length $\xi(p)$ represents the linear size of a typical finite cluster. It is another percolation property required later for reservoir engineering applications. The correlation length can be defined as the radius of gyration of the largest cluster. Assume that the positions of the n sites belonging to a given cluster are r_i for $i = 1, \ldots, n$. The centre of mass of the cluster is defined as $\bar{r} = \frac{1}{n}\sum_{i=1}^{n} r_i$. Then the radius of gyration (R_g) is the mean square distance from the centre of mass of the cluster, $R^2 = \frac{1}{n}\sum_{i=1}^{n}(r_i - \bar{r})^2$. Above the threshold, the correlation length represents the size of the "holes" in the percolating cluster.

The correlation length also follows a power law of the occupancy probability near the threshold (Stauffer and Aharony, 1992),

$$\xi(p) \propto |p - p_c|^{-\nu}, \tag{2.3}$$

where ν is the correlation length exponent. The negative exponent shows that the cluster size diverges at the threshold. This is the emergence of the percolating cluster. This scaling law describes how the typical size of finite clusters increases, diverges at the threshold and then decreases again. It should be noticed that the correlation length ξ is related to the two-point correlation function $g(r)$, which

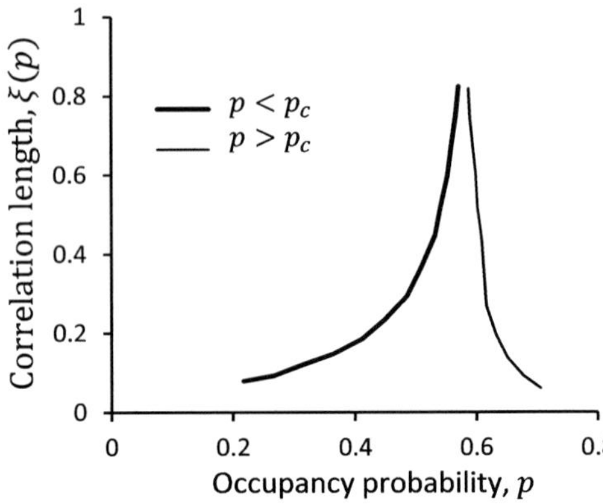

Figure 2.4. Divergence in correlation length $\xi(p)$ of infinite-size systems near the threshold.

is the probability that two sites separated by a distance r are in the same cluster, and is given by

$$g(r) \propto r^{-2\beta/\nu} e^{-r/\xi}. \tag{2.4}$$

This behaviour of the correlation length close to the threshold for an infinite-size system is shown in Figure 2.4. The reason for the decrease of correlation length above the threshold is that as p increases further, the finite clusters get absorbed into the infinite cluster and so the typical size of the finite clusters goes down.

2.2.3 *Cluster size*

There are various clusters of different sizes. Another measure of size is the number of sites (s) in the cluster. The number of clusters of a given size follows a power law at the percolation threshold, $n_s \sim s^{-\tau}$, where τ is another critical exponent. The mean cluster size, S, is another percolation quantity that has a simple power law dependency on the occupancy probability,

$$S(p) \propto (p - p_c)^{-\gamma}, \tag{2.5}$$

where γ is called the mean cluster size exponent.

2.2.4 *Backbone fraction*

So far, we have investigated the scaling properties of various measures of the size of the clusters. We now investigate more their structure. If we look at Figure 2.5, we can divide the sites in the percolating cluster into two types. If we were to imagine a fluid flowing from left to right, or an electric current flowing in the same way, then the pale blue sites would carry none of the flow or current. These are called *dangling ends*. Instead all of the flow must traverse the dark blue sites which make up the *backbone*.

As expected, the backbone $B(p)$ has a simple power law,

$$B(p) \propto (p - p_c)^{\beta_B}, \tag{2.6}$$

where β_B is called the backbone exponent. This is very important quantity in studying the dynamic behaviour of percolating systems because the tortuosity of the transport paths is controlled by the structure of the backbone. We will discuss this in more detail in Chapter 6.

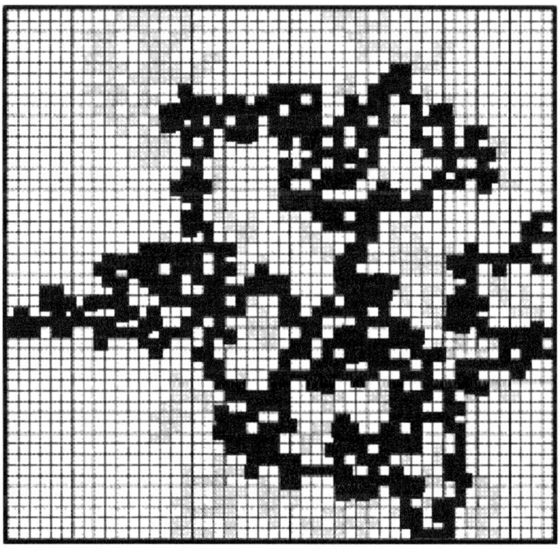

Figure 2.5. The percolating cluster on a lattice (50 × 50) of sites connected from left to right at the threshold, showing the backbone (dark blue) and dangling ends (pale blue).

2.2.5 *Effective permeability or conductivity*

We shall assume that the occupied sites all have the same permeability equal to unity and unoccupied sites are impermeable (equally we could discuss electrical conductivity where the occupied sites are conductors and the unoccupied ones are insulators). The effective permeability of the system (K_{eff}) is that single value permeability for the whole system that gives the same flow rate for a given pressure drop. Effective permeability is also referred to as the equivalent permeability of the upscaled region based on Darcy's law. Darcy's law is a linear force flux relationship for flow in porous media (like Ohm's law for electrical conductivity or Fourier's law for thermal transport) that states that the flow rate is proportional to the pressure gradient and inversely proportional to the viscosity of the fluid, $v = -\frac{K}{\eta}\nabla p$ (Renard and de Marsily, 1997). The constant of proportionality (K) is the permeability and has dimensions of an area.

In a percolating system, as not all the occupied sites contribute to the flow, we might expect a difference between the percolation probability and the conductivity of the system. The effective permeability $K_{\text{eff}}(p)$ has a simple power law dependency,

$$K_{\text{eff}}(p) \propto (p - p_c)^{\mu}, \tag{2.7}$$

where μ is called the conductivity exponent. A comparison between Eqs. (2.1) and (2.7) is shown schematically in Figure 2.6. Above the threshold, the connectivity increases rapidly whereas the effective permeability increases relatively more slowly. This is because the *dangling ends* do not contribute to the flow.

2.3 Universality in percolation theory

The critical exponents β, ν, γ, β_B, μ described in scaling laws above are found to be universal. By universality, we mean that the values of these exponents are independent of the kind of lattice or whether it is site or bond percolation. It only depends on the dimensionality of space (i.e., 2D or 3D). This is also true for continuum percolation systems. It is also found to hold for correlated percolation as long as

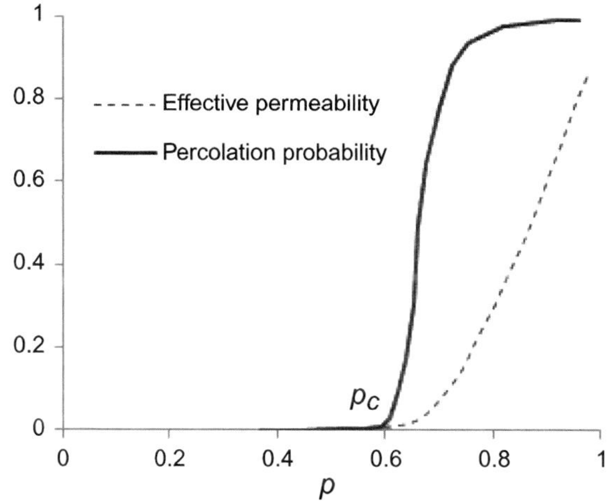

Figure 2.6. Illustration of how the percolation probability $P(p)$ and the effective permeability of infinite-size systems increase near the threshold.

the correlation length is small compared to the system size (which means that finite-range correlations do not change the universality). However, for some models with very long correlation length or objects with power law size distributions, the exponents may change (Weinrib and Halperin, 1983; Prakash *et al.*, 1992; Sahimi and Mukhopadhyay, 1996; Masihi and King, 2007, 2008; Schmittbuhl *et al.*, 1993; Sadeghnejad *et al.*, 2013).

Clearly this is a very powerful concept as it enables us to study and understand the behaviour of a very wide range of systems without needing to worry so much about the fine scale details. As emphasised by Berkowitz and Balberg (1993), the principle of universality is a result of the fact that the percolation transition (transition from disconnected to connected paths) is a second-order phase transition similar to the Curie point for magnets. The generally accepted literature values for these exponents for 2D and 3D systems are given in Table 2.2.

2.4 Finite-size effects

The scaling behaviour described so far is very simple, but strictly only applies to infinitely large systems. However, in applications,

Table 2.2. Selected percolation exponents in 2D and 3D (Stauffer and Aharony, 1992; Berkowitz and Balberg, 1993; Sahimi, 1994; Hunt and Ewing, 2005; Vogel *et al.*, 2010).

	2D		3D
Exponent	**Exact**	**Approx.**	**Approx.**
β	5/36	0.14	0.41
ν	4/3	1.33	0.88
γ	43/18	2.39	1.82
τ	187/91	2.055	2.18
μ	—	1.31	2
β_B	—	0.4758	1.05

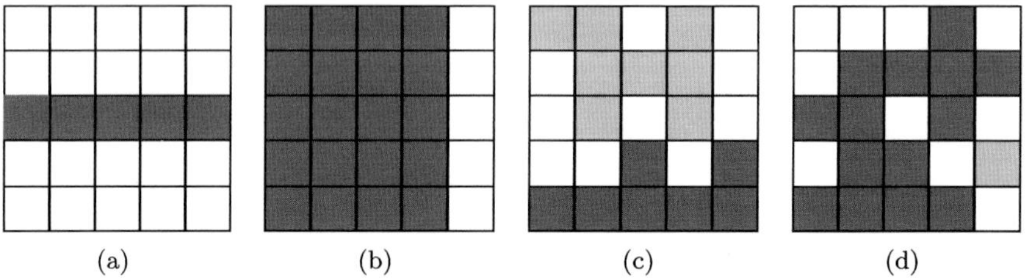

Figure 2.7. Illustration of four different realisations on a 5×5 grid showing that the connectivity from left to right may occur at (a) $p = 0.2$ which is much less than the infinite threshold, (b) $p = 0.8$ which is much higher than the infinite threshold or (c) and (d) $p = 0.6$ emphasising that at the same occupancy probability, there may be different values for the connectivity ($P = 0.46$ and 0.93). (a) $p = 0.2$, $P = 1$, (b) $p = 0.8$, $P = 0$, (c) $p = 0.6$, $P = 0.46$ and (d) $p = 0.6$, $P = 0.93$.

real systems are only finite in size. The finite boundaries modify the behaviour which we now need to investigate. We use a lattice of size $L \times L$ to model a finite system. We then generate different realisations by occupying sites with probability p as before. It is no longer possible to define percolation in terms of a single infinite cluster because different realisations can have very different connectivity for the same occupancy, as shown in Figure 2.7. In particular, Figures 2.7(c) and 2.7(d) have the same occupancy but very different spanning clusters. Also Figure 2.7(a) has a much lower occupancy, but there is a connected cluster, whereas Figure 2.7(b), at a much higher occupancy, does not. Note that in all these cases, we are only concerned with horizontal connectivity; this issue will be addressed later.

Whilst we do not have the same critical threshold behaviour as before, we expect that overall as the occupancy fraction increases, so does the fraction of occupied sites in a spanning cluster. We define the percolation probability $P(p, L)$ to be the fraction of occupied sites belonging to the spanning cluster. This is the finite-size analogue of $P(p)$ used in the infinite system. To predict the connectivity of these systems, we make use of what is known as *finite-size scaling*. It should be noted that finite size refers to finite but very large systems.

If we plot the connected fractions $P(p, L)$ as a function of occupancy probability p for a large number of realisations of a particular system size, we get a scatter of points (Figure 2.8). We see that generally the plots have similar behaviour (a sharp change from no connection to high connectivity) but that the threshold is "smeared out". Also the amount of this smearing is reduced as the system size gets larger. This is because the probability of finding configurations like Figure 2.7(a) gets smaller as the system gets larger.

Instead of working with the whole scatter, we use the average value $P(p, L)$ and standard deviation $\Delta(p, L)$ of connectivity obtained over all realisations. Note that the averaging is done at a fixed occupancy probability and system size; it is an ensemble

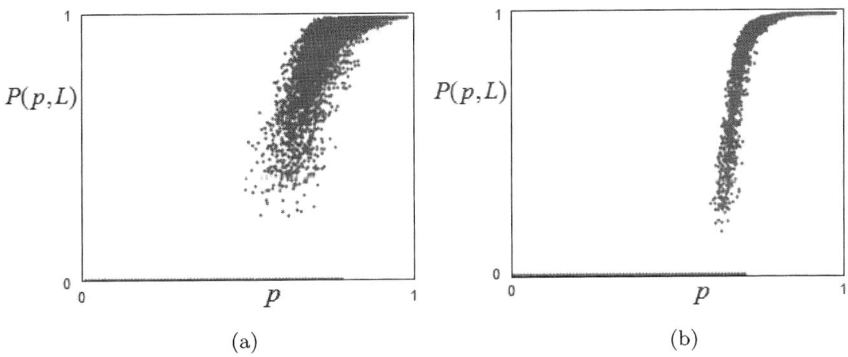

Figure 2.8. The typical scatter plots of percolation probability versus occupancy in two system sizes (a) $L = 16$ and (b) 64.

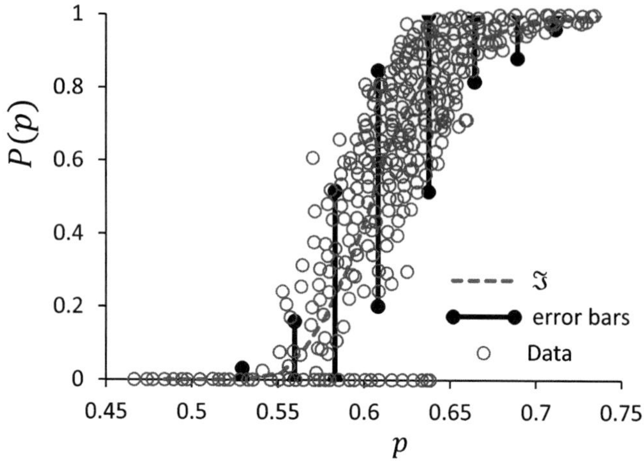

Figure 2.9. A typical scatter obtained for connectivity of a finite-size system. The curve and the lines represent the average and the error bars respectively determined over all realisations at the same occupancy probability.

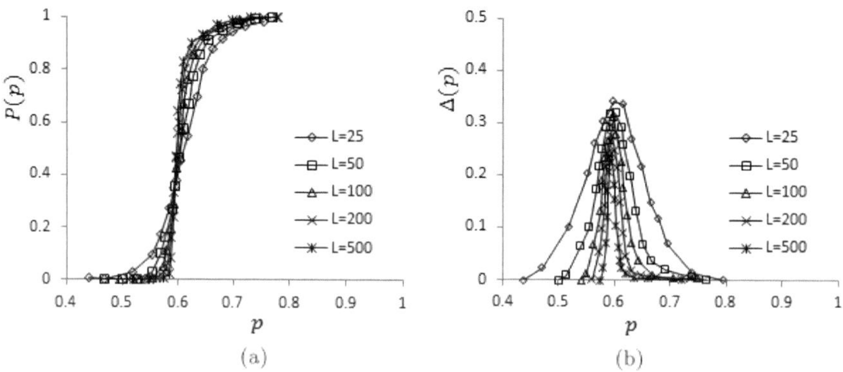

Figure 2.10. Finite-size effects on the percolation transition on (a) mean connectivity and (b) standard deviation in connectivity.

average over the realisations. The curve and error bars are shown in Figure 2.9. We will not distinguish in notation between $P(p, L)$ and its average.

We plot the average connectivity $P(p, L)$ against occupancy probability for different system sizes in Figure 2.10. As said before, it is clear that the sharp transition found in the infinite system is now smeared out. As the system size increases, the amount of smearing of the percolation transition decreases. In Figure 2.10, we also plot the standard deviation $\Delta(p, L)$ of connectivity (the error bars shown in Figure 2.9). We find a typical bell-shaped curve. Again as the system size increases, this curve becomes narrower and its peak becomes lower.

Finite-size scaling within percolation theory (which is linked with renormalisation scaling, e.g., Wilson (1975)) is a way of relating all of the connectivity curves for different system sizes to a single master curve. This scaling is based on the principle that all the properties depend only on the ratio of the system size to the correlation length L/ξ. This is because as the system size becomes large compared with the correlation length, it would behave like an infinite system, whereas when these lengths are similar, the effects of the boundaries become important, so this is the dimensionless length that controls the behaviour of the system. So we look for functions of the form

$$P(p, L) = F[L/\xi], \tag{2.8}$$

where F is some function. Given the dependence of the correlation length on the occupancy probability given by Eq. (2.3), we can rewrite this as

$$P(p, L) = F[(p - p_c)L^{1/\nu}]. \tag{2.9}$$

We now consider an arbitrary change of the system size by $L \rightarrow \kappa L$. We do not expect this transformation to alter the percolation behaviour of the system, i.e., $P(p, kL) \rightarrow c(k)P(p, L)$. The only function that satisfies this property is the power law,

$$P(p, L) = L^A \Im[(p - p_c)L^{1/\nu}], \tag{2.10}$$

where $\Im$ is a function and A is a power to be determined. As the system size gets larger ($L \rightarrow \infty$), we must recover the infinite system size behaviour given by Eq. (2.1). So $\Im[z] \rightarrow z^\beta$ as $z \rightarrow \infty$. To cancel the L dependence, we must have $A = -\beta/\nu$. Hence the finite-size scaling for the mean connectivity becomes

$$P(p, L) = L^{-\beta/\nu} \Im[(p - p_c)L^{1/\nu}], \tag{2.11}$$

where $\Im$ is a master function for the mean connected fraction. This needs to be determined numerically. This means that if we plot $L^{\beta/\nu}\ P(p, L)$ against $(p - p_c)^{1/\nu}$, then all the curves for different systems sizes L should collapse onto the same master curve. This is shown in Figure 2.11.

We observe that well above the critical region ($p \gg p_c$), the scaling collapse breaks down (as highlighted on Figure 2.11). However, over much of the regime, this works very well. The power of this result is that it shows that the mean connectivity for any system can be calculated directly from this master curve and then "undoing" the scaling transformation. So we do not need to do any more expensive simulations but just use the curve and simple algebra.

We can use similar arguments to show that the mean cluster size which has the finite-size scaling,

$$S(p, L) = L^{\gamma/\nu} R[(p - p_c^{\infty})L^{1/\nu}], \tag{2.12}$$

where R is some function. We use the relationship between the critical exponents, $d - \gamma/\nu = 2\beta/\nu$ (Stauffer and Aharony, 1992) and the

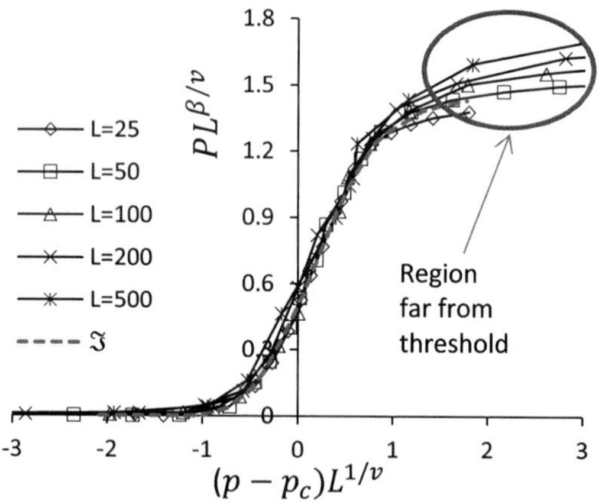

Figure 2.11. Illustration of finite-size-scaling transformations for the average connectivity.

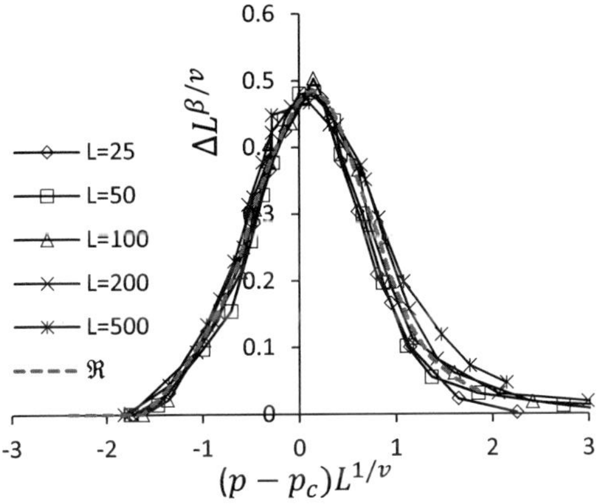

Figure 2.12. Illustration of finite-size scaling transformations for the standard deviation of the connectivity.

fact that the variance in connectivity $\Delta^2(p, L)$ is related to the mean cluster size $S(p, L)$ through

$$\Delta^2(p, L) \propto S(p, L)/L^d, \tag{2.13}$$

in d dimensions to give the scaling for the standard deviation of connectivity,

$$\Delta(p, L) = L^{-\beta/\nu}\Re[(p - p_c^\infty)L^{1/\nu}], \tag{2.14}$$

where R is the master function for the standard deviation of connectivity. This data collapse is shown in Figure 2.12.

We can now use these scaling laws to great effect because they enable us to predict the mean connectivity and its associated uncertainty for any system size very quickly, without performing any explicit realisations. We would expect to see similar finite-size scaling laws for other properties (like the backbone fraction, effective permeability and so on). We shall exploit these later in the book.

2.5 Calculation of percolation parameters

In order to study the percolation behaviour, we first need the numerical values of the percolation threshold p_c^∞, which depends on the underlying structure of the system, and the critical exponents (β, ν, γ, β_B, μ and so on) which are universal. These can be determined analytically only in some very specific cases. However, they can be estimated numerically by studying the finite-size effects in 2D or 3D. We will first discuss the 1D problem. This does not have reservoir applications, but it is the one case where nearly everything can be calculated analytically and gives insights into some of the mathematical arguments that could be used in higher dimensions. We then go on to discuss the numerical methods used to determine the percolation parameters. We note also that there is an extensive literature with some very accurate measurements from numerical experiments.

2.5.1 *Analytical method for 1D percolation*

There are specific percolation models for which we can determine the percolation threshold and the critical exponents analytically. These include percolation in 1D and the Bethe lattice (Fisher and Essam, 1961). Here, we shall concentrate on site percolation in 1D because it gives a good introduction to the mathematical tools used in percolation theory. We take a 1D grid as shown in Figure 2.13. As usual, the probability that a given site is occupied is p. If two occupied sites are next to each other, they are in the same cluster. The sites immediately either side of each cluster are empty.

Since all the sites are occupied randomly, the probability that neighbouring sites are occupied with two empty sites at either end can be written as

$$n_s = p^s(1-p)^2. \tag{2.15}$$

This is equivalent to the number of such s-clusters per lattice site. We note that the cluster numbers go exponentially to zero if the

Figure 2.13. Groups of occupied neighbouring sites forming clusters in 1D.

cluster size goes to infinity. In 1D, there is connection from left to right only if all sites are occupied as a single empty site would break up the connectivity. Hence, the percolation threshold in 1D becomes

$$p_c = 1. \tag{2.16}$$

So we cannot observe the behaviour above the threshold ($p > p_c$) in 1D. This is a very significant difference from all other dimensionalities.

We investigate the statistics of clusters in this 1D percolation problem. The probability that a site (occupied or not) belongs to a cluster of size s is $n_s s$. Every occupied site must belong to one cluster, so we have

$$\sum_s n_s s = p. \tag{2.17}$$

The probability that a cluster of an arbitrary occupied site contains exactly s sites is

$$w_s = n_s s \Big/ \sum\nolimits_s n_s S. \tag{2.18}$$

Then the mean cluster size S is

$$S = \sum w_s s = \sum \left(n_s s^2 \Big/ \sum\nolimits_s n_s s \right). \tag{2.19}$$

Using the relationship (2.17), it can be shown that (Stauffer and Aharony, 1992)

$$S = \frac{1+p}{1-p} \propto (p_c - p)^{-1}, \quad \text{for } p \to p_c. \tag{2.20}$$

This proves that the mean cluster size exponent γ is equal to 1 in 1D and shows that the mean cluster size diverges as p approaches p_c.

In 1D, the correlation function, which is the probability that two sites at distance r along the 1D grid are in the same finite cluster, becomes

$$g(r) = p^r = e^{r \ln p}, \tag{2.21}$$

as these two sites and all intermediate sites must be occupied. We can write this in the usual exponential decaying form with the correlation length:

$$g(r) = \exp(-r/\xi). \tag{2.22}$$

By comparing these, we find that

$$\xi = \frac{-1}{\ln p}. \tag{2.23}$$

Finally we expand the logarithm around the threshold $\ln p = \ln[p_c - (p_c - p)] \sim -(p_c - p)$ for small deviations away from the threshold. Then we can write

$$\xi = \frac{-1}{\ln(p)} \propto (p_c - p)^{-1}, \tag{2.24}$$

which shows that the critical exponent for the correlation length exponent in 1D is $\nu = 1$.

This gives the flavour of the kind of mathematical arguments used to derive the exponents and other percolation properties exactly in 1D. Unfortunately they are hard to translate into higher dimensions. The fundamental problem is that it is now no longer possible to write down the number of clusters of size s as in Eq. (2.15). This is because the clusters are no longer linear chains but can have very different configurations; these get increasingly complex as the clusters get larger. Consider a 2D problem. There are various cluster configurations (called lattice animals and denoted by g_{st}) for the same cluster size. Figure 2.14 shows the cluster configurations for a cluster of size $s = 4$.

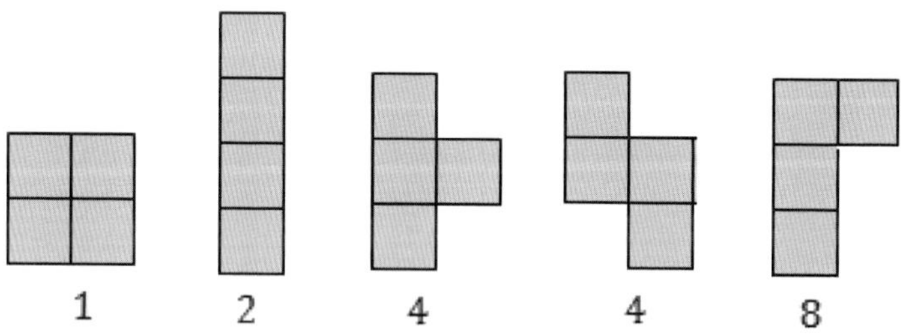

Figure 2.14. Cluster configurations for a cluster of size $s = 4$. Trivial reflections and rotations are not shown, but they are enumerated in the figures below each configuration.

The probability that a site is in a cluster of size $s = 4$ in all configurations (as shown in Figure 2.14) can be written as

$$n_4 = 2p^4(1-p)^{10} + 8p^4(1-p)^9 + 4p^4(1-p)^8 + 4p^4(1-p)^8 + p^4(1-p)^8. \quad (2.25)$$

The number of empty neighbours of a cluster is called the perimeter with a symbol t. The average number of s-clusters per lattice site becomes

$$n_s = \sum_t g_{st} p^s (1-p)^t. \quad (2.26)$$

However, there are 10^{13} configurations for clusters of size $s = 24$. This shows that it is impractical to find all clusters configurations. Hence we cannot find the percolation parameters analytically in higher dimensions, although this approach can lead to some useful approximations, but that is beyond the scope of this book. So we need to use numerical methods instead.

2.5.2 *Monte Carlo simulations for 2D and 3D percolation models*

2.5.2.1 *Percolation threshold*

In general, it is unclear how to define the threshold point of a finite system, as a spanning cluster may appear in one realisation and not in another at the same occupancy probability. Therefore, for a finite system, an apparent threshold $\tilde{p}_c(L)$ which depends on the system size can be introduced. There are several suggestions for the definition of this finite-size threshold (King, 1990; Galam and Mauger, 1996; Stauffer and Aharony, 1992; Berkowitz, 1995; Adler and Thovert, 1999, Harter, 2005):

- the point at which the connectivity is first non-zero. This gives an estimate for a single realisation.
- the intercept of the tangent to the mean connectivity curve taken from its inflexion. This gives an estimate for all realisations.
- the occupancy probability at which half of the realisations percolate. This is for all realisations.

- the point of intersection of all the curves of the mean probability of percolation obtained at different domain sizes.

It should be noticed that all these definitions have the same behaviour as the system size goes to infinity. We can use the apparent threshold, and finite-size scaling, to estimate the threshold for an infinite-size system (which from now on we denote by p_c^∞).

We first use the argument that the largest cluster spans when its size becomes comparable to the system size (or $\xi \sim L$), the finite-size dependency of the apparent threshold can be written as (Stauffer and Aharony, 1992)

$$\tilde{p}_c(L) - p_c^\infty \propto L^{-1/\nu}. \tag{2.27}$$

Hence, if we plot $\tilde{p}_c(L)$ against aspect ratio ω, we find a straight line whose intercept gives the infinite percolation threshold p_c^∞ and whose slope is the finite-size shift in the threshold.

The second method is based on the smearing of the sharp threshold found in finite-size systems. We define $\delta(L)$ to be the width of the transition zone between connected and unconnected zones. This can be considered as the standard deviation in the apparent threshold values observed over all realisations of the same system size. This also has the scaling

$$\delta(L) \propto L^{-1/\nu}, \tag{2.28}$$

which gives a useful way to estimate the infinite percolation threshold from

$$\tilde{p}_c(L) - p_c^\infty \propto \delta(L). \tag{2.29}$$

Therefore, without any knowledge of the exponent ν, a plot of $\tilde{p}_c(L)$ against $\delta(L)$ gives the percolation threshold by extrapolating to $\delta = 0$.

2.5.2.2 *Critical exponents*

From universality, it is unnecessary to determine the critical exponents as they are well documented. However, in some specific cases (e.g., where there are long-range correlations), we cannot rely on universality so we need to be able to estimate the exponents numerically. We introduce three methods here.

The first method uses the scaling laws for infinite systems. For example, to obtain an estimate of the connectivity exponent β, take a log–log plot of $P(p, L)$ against $(p - p_c^\infty)$ for a relatively large system. The gradient gives an estimate for the critical exponent β. A similar analysis can be performed to estimate the other critical exponents. A log–log plot of the correlation length ξ against $(p - p_c^\infty)$ for a large system gives an estimate of ν. The correlation length ξ is the radius of gyration of the largest cluster in the system,

$$\xi = \left[\frac{1}{n}\sum_{i=1}^{n}(r_i - \bar{r})^2\right]^{1/2}, \tag{2.30}$$

where $\bar{r}$ is the centre of the largest cluster and the r_i are the locations of sites in the cluster.

The second approach is to use finite-size scaling at the threshold. So from the scaling result (Eq. (2.11)) taken at the threshold, we see that

$$P(p_c, L) \propto L^{-\beta/\nu}. \tag{2.31}$$

We see that a plot of the connected fraction at the threshold against the system size gives us the ratio of exponents β/ν. So knowing one (from one of the methods just given) will give the other.

The third method is to use the width of the transition zone between connected and unconnected zones. We use the width measure $\delta(L)$ defined previously. A log–log plot of $\delta(L)$ against L gives an estimate for the correlation length exponent. We could also use p_f and p_{1-f} (which are the occupancy probabilities at which a fraction f and $1-f$ of all realisations percolate) as a measure of the transition zone. This has the scaling (Gawlinski and Stanley, 1981)

$$p_{1-f} - p_f \propto L^{-1/\nu}. \tag{2.32}$$

Using different values of fraction f gives different estimates which can be averaged to give a more reliable estimate of the critical exponent ν.

2.6 Anisotropy and percolation

So far, we have only investigated isotropic systems where the number of cells in each direction on the lattice is the same. Clearly for infinite-size systems, there are no boundaries, so the issue of isotropy or anisotropy doesn't arise. However, for finite-size systems, this is an issue, and we shall show that it modifies the finite-size scaling laws presented so far. There are a variety of ways that anisotropy could be introduced, but we shall restrict ourselves to the effect of using a rectangular grid instead of a square one. All other aspects of the model are the same. We shall see later that this is the most relevant extension for finite reservoirs with connecting flow units.

The first thing to notice is that for low values of the occupancy probability the clusters are small; the correlation length is small compared with the shortest length in the system. So the boundaries have no impact. But there will come a point where the correlation length becomes comparable to the shortest distance across the grid. This leads to the creation of an easy direction for connected paths which is in the short direction and a difficult direction which is along the long axis.

We use these ideas of "easy" and "difficult" directions for connectivity to define apparent percolation thresholds in these directions based on the occupied fraction at which 50% of realisations connect in each direction (Masihi *et al.*, 2006). Although we describe this kind of analysis on lattices, we expect that the same framework can be applied to other anisotropic systems.

We take an anisotropic grid of size $L_x \times L_y$ and define an aspect ratio $\omega = L_x/L_y$ to represent the anisotropy. The same algorithm as before is used to identify various clusters and their statistics. Figure 2.15 shows three configurations for anisotropic lattices of size $L_x = 200$ but at aspect ratios $\omega = 4$, 10 and 20.

Then as usual, we investigate the percolation probability as a function of the occupancy probability and system size. We also have to incorporate the effects of the aspect ratio. First we use a fixed aspect ratio ω. We shall assume that $\omega > 1$, which means that the short direction is y. Clearly there is symmetry between connectivity

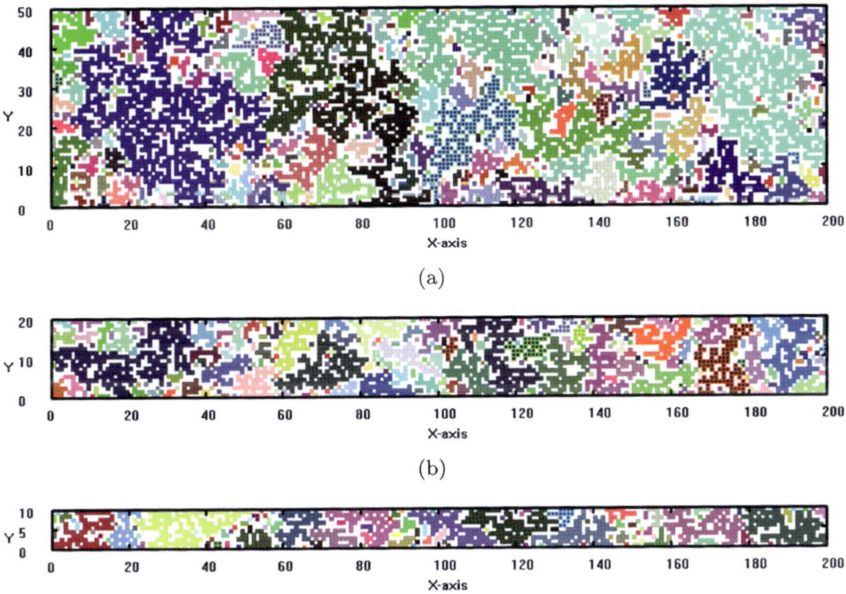

Figure 2.15. Three typical configurations for anisotropic lattices at a fixed system size $L_x = 200$ but at different aspect ratios. Note how vertical connectivity is easier than horizontal: (a) for $\omega = 4$ and $p = 0.545$, one vertically connected cluster, (b) for $\omega = 10$ and $p = 0.55$, three vertically connected clusters and (c) for $\omega = 20$ and $p = 0.57$, 10 vertically connected clusters.

in the x direction for aspect ratio ω and that in the y direction for aspect ratio $1/\omega$ (and vice versa transposing x and y). We shall exploit this symmetry later.

We should also note that we only study the impact of moderate aspect ratios. As the aspect ratio approaches $1/L_y$, the system becomes 1D and we would expect to see a crossover to 1D behaviour. This is clearly shown in Figure 2.15 in which the connectivity in the x direction at large aspect ratio is ruled by the number of vertical connected clusters. The transition from 2D to 1D behaviour is complicated, but we find that in real applications, the aspect ratio is at most of order 10. For this reason, we restrict our analysis to system sizes where the short dimension is greater than 10 and to aspect ratios less than 20. We first start by examining the impact of the anisotropy on the apparent percolation threshold.

2.6.1 *Apparent threshold for anisotropic systems*

We count the number of realisations which form clusters that connect from one side to the opposite side of the system in the two directions. The apparent threshold in the x direction is that occupancy probability at which 50% of realisations connecting in that direction (with a similar definition in the y direction). We make the *ansatz* that these scale with the system size as (Masihi *et al.*, 2006)

$$\tilde{p_c^i} = p_c^\infty + \Lambda_i(\omega)L_x^{-1/\nu}, \tag{2.33}$$

where i labels the coordinate direction x or y, p_c^∞ is the usual infinite system size percolation threshold (i.e., 0.59275 for site percolation) and Λ_i is simply the constant of proportionality which depends on the aspect ratio ω. For isotropic lattices ($\omega = 1$), this relation should reduce to Eq. (2.27).

Figure 2.16 shows a plot of the apparent threshold determined in the x and y directions for five different aspect ratios as a function of $L_x^{-1/\nu}$. This shows that the scaling in Eq. (2.33) is obeyed. These results also show that the separation of the apparent thresholds is symmetrically placed about the isotropic case ($\omega = 1$). This implies that, at least to a reasonable approximation,

$$\Lambda_x(\omega) = -\Lambda_y(\omega). \tag{2.34}$$

This symmetry allows us to define a single function for the coefficient Λ such that

$$\tilde{p_c^x} - p_c^\infty = \Lambda(\omega)L_x^{-1/\nu}, \tag{2.35a}$$

$$\tilde{p_c^y} - p_c^\infty = -\Lambda(\omega)L_x^{-1/\nu}. \tag{2.35b}$$

The other symmetry mentioned above arising from swapping x and y and inverting the aspect ratio allows us to show

$$\tilde{p_c^x}(\omega) = \tilde{p_c^y}(1/\omega). \tag{2.36}$$

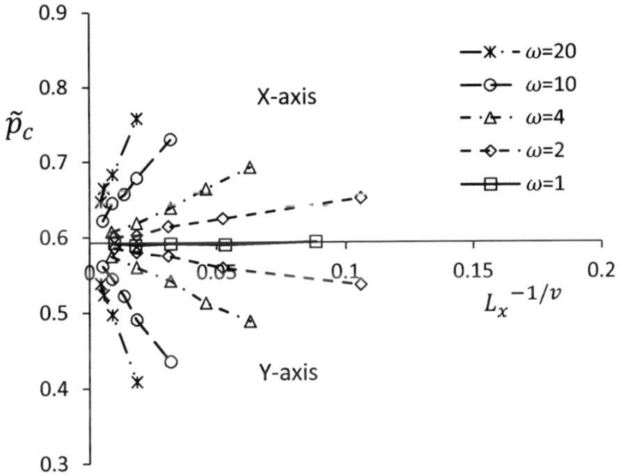

Figure 2.16. Plot of apparent threshold in x and y directions as a function of $L_x^{-1/\nu}$.

This implies that Λ has the following property:

$$\Lambda(\omega) = -\omega^{1/\nu}\Lambda(1/\omega). \tag{2.37}$$

Neglecting a very small finite-size shift in the threshold for the isotropic case (so assuming that $\Lambda(\omega = 1) \approx 0$), a simple scaling function that satisfies all these properties is

$$\Lambda(\omega) = c(\omega^{1/\nu} - 1). \tag{2.38}$$

Figure 2.17 shows a plot of the coefficient Λ against aspect ratio ω, which indicates that there is reasonable agreement with this conjecture. The coefficient $c \approx 0.922$ is found from numerical experiments. This value is found for site percolation on rectangular lattices. We expect the same formula to apply for other variants (such as continuum percolation) but with a different coefficient. We shall investigate this in subsequent chapters.

2.6.2 *Finite-size scaling laws for anisotropic systems*

We define the horizontal (P_x) and vertical (P_y) connected fractions as the fraction of occupied sites in the cluster that spans horizontally

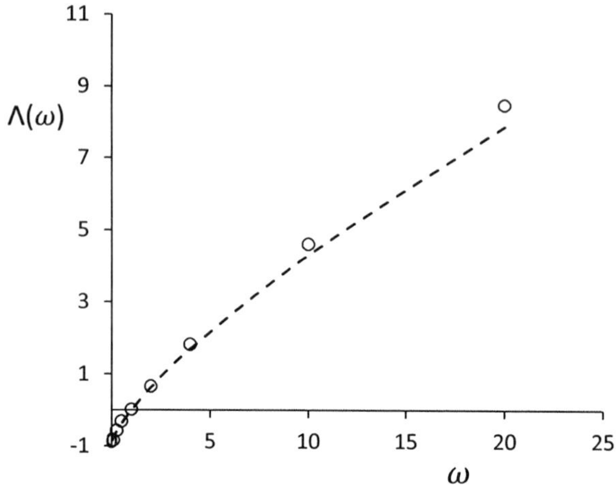

Figure 2.17. Shift in the apparent threshold Λ as a function of aspect ratio.

or vertically respectively. As for the isotropic case, plotting these for a set of realisations creates a scatter of points, and as before, we consider the average and standard deviation of this scatter. We expect the average connected fraction to obey a similar finite-size scaling law given as before (Eq. (2.11)).

In Figure 2.18(a), we plot $L_x^{\beta/\nu}P$ against $(p-p_c^\infty)L_x^{1/\nu}$ for a fixed value of the aspect ratio $\omega = 4$. Under this rescaling of the axes, the curves do collapse again, but now the horizontal and vertical curves are separated and displaced from the isotropic case (shown as a dashed line). In this case, we had used the infinite percolation threshold; if instead we had used the finite-size threshold corrected for anisotropy (from Eq. (2.35)) in Eq. (2.11), we get, to reasonable approximation, the collapse onto the single universal isotropic curve ($\Im$) as before (Figure 2.18(b)).

Hence, the appropriate finite-size scaling for anisotropic lattices becomes

$$P(p, L_x, \omega) = L_x^{-\beta/\nu}\Im[(p - \tilde{p}_c)L_x^{1/\nu}]. \tag{2.39}$$

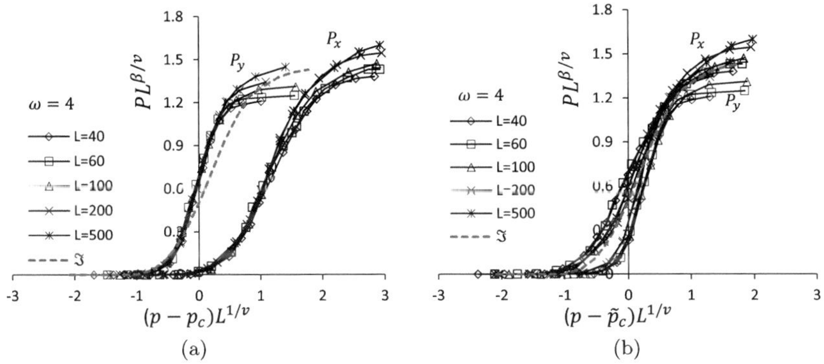

Figure 2.18. Finite-size scaling for the average connected fraction in horizontal P_x and vertical P_y directions for a fixed value of aspect ratio of 4 using (a) infinite threshold as usual and (b) the finite-size threshold given by Eq. (2.35).

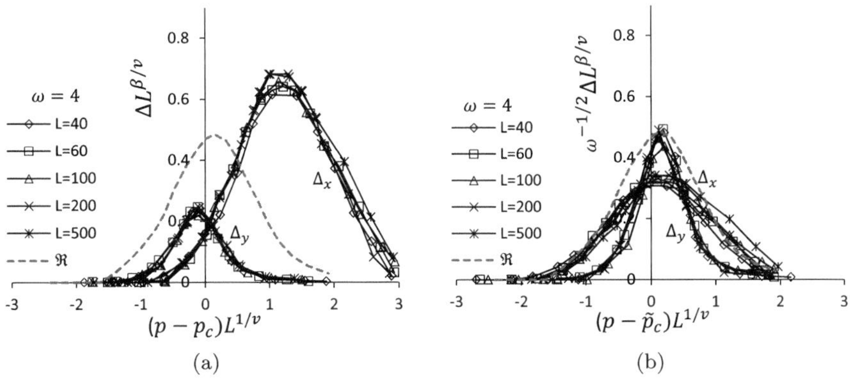

Figure 2.19. Finite-size scaling for the standard deviation of connectivity in the horizontal Δ_x and vertical Δ_y directions for a fixed value of aspect ratio of 4 using (a) infinite threshold and (b) the apparent threshold and modifying the amplitude according to Eq. (2.40).

Similarly the standard deviation of the connectivity results can be rescaled in a similar way. Here the anisotropic curves for the standard deviation of the connectivity are displaced from the isotropic case ($\Re$), but there is also a change in their magnitudes (Figure 2.19(a)). To scale the standard deviation curves to the same universal isotropic curve ($\Re$), we again use the finite-size threshold. We also use a change in magnitude of the error which can be accounted for by rescaling with the geometric mean length, $(L_x L_y)^{1/2}$. This arises from the dependency of the variance in connectivity on the mean cluster size S. We use $L_x L_y = \omega^{-1} L_x^2$ instead of L^2 in Eq. (2.13) for an anisotropic system. This introduces a pre-factor $\omega^{1/2}$ in the scaling law for the standard deviation of connectivity. Hence the finite-size scaling with L_x in the presence of anisotropy becomes

$$\Delta(p, L_x, \omega) = \omega^{1/2} L_x^{-\beta/\nu} \Re[(p - \tilde{p}_c) L_x^{1/\nu}]. \tag{2.40}$$

The result of data collapse for the standard deviation of the connectivity is shown in Figure 2.19(b). It can be seen that the use of the pre-factor $\omega^{1/2}$ improves the fit for the standard deviation of connectivity results although it is not perfect. However, these results enable us to use the same isotropic universal curves ($\Im$ and $\Re$) for predicting connectivity of anisotropic lattices which is a good enough approximation for engineering purposes.

2.7 Percolation and fractals

Finally we discuss the relationship between percolating clusters and fractals. Fractals are geometrical structures with a non-Euclidean dimension first discussed by Mandelbrot (1982). There is a very extensive literature on fractals which goes well beyond the scope of this book. We shall restrict ourselves to defining some salient features. We observe that percolation clusters are fractal and report their dimensions.

Fractals are mathematical entities which are invariant under length scale transformations (Falconer, 2003). By this, we mean that they look the same regardless of the scale at which they are observed (imagine zooming in or out on one). A classic example is the *Koch snowflake.* It is produced by taking a line of length l (Figure 2.20(a)). Then replace the middle third with two new line segments arranged to form a triangle (Figure 2.20(b)). This process is repeated for each of the four line segments thus formed (Figure 2.20(c) and further).

Generation stage	Shape	No. of segments	Size of each segment
(a)		1	l
(b)		4	$l/3$
(c)		16	$l/9$
(d)	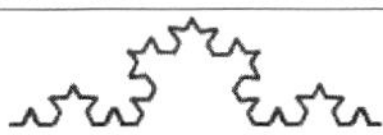	64	$l/27$

Figure 2.20. Generation of the Koch snowflake as a regular fractal object.

It should be clear that as the length scale is decreased by a factor 3, the number of line segments goes up by 4. One definition of the dimension of an object is to look at how the number of objects that makes it up scales with the linear size of the elements. Imagine a regular lattice with blocks of size l. In a line, the number of cells goes like l^1; for a square it is l^2, and for a cube l^3. So the dimension of the object is the power in the scaling $N \sim l^D$. Now consider the Koch snowflake and consider the ratio of number of segments to length of segments between say the second and third generations:

$$D = \frac{\ln(n_3/n_2)}{\ln(l_3/l_2)} = \frac{\ln(64/16)}{\ln(27/9)} = \frac{\ln 4}{\ln 3} \sim 1.26.$$

An object with a non-integer (i.e., non-Euclidean) dimension is a fractal.

The Koch snowflake is a regular fractal; however, many natural phenomena (such as rivers, coastlines, clouds, mountains) can be described as random fractals. Figure 2.21 shows a random fractal object formed by aggregating colloidal particles.

We will not discuss here the ways of estimating fractal dimensions of objects or go further into the subject. Instead we just note that at the percolation threshold the infinite cluster is found to have a fractal dimension D_c. It is also found that it is related to the critical exponents by what is known as a *hyperscaling relation* of the

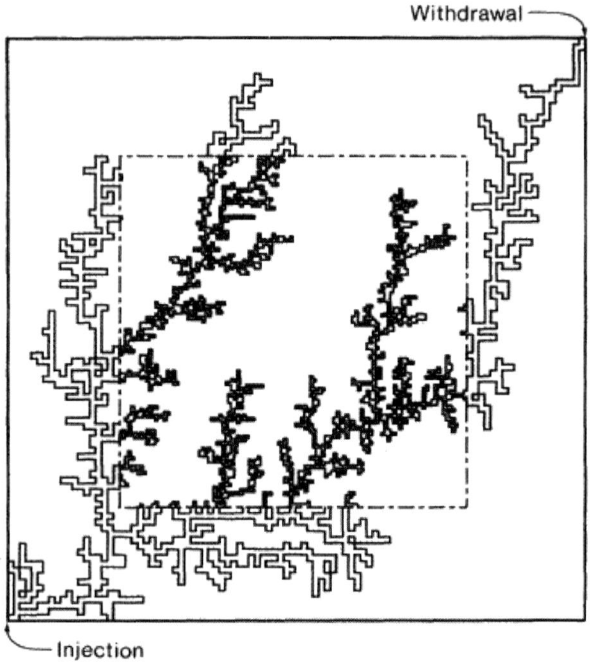

Figure 2.21. A random fractal formed by aggregating colloidal particles (also by injecting a very low viscosity fluid to displace a more viscous one in a porous medium) (Paterson, 1984).

Table 2.3. The fractal dimensions found in percolation (Sahimi, 1994; Stauffer and Aharony, 1992; Hunt and Ewing, 2005; Grassberger, 1999).

Fractal dimension	2D	3D
D_c	91/48	2.53
D_B	1.6432	1.87

percolating cluster (Stauffer and Aharony, 1992):

$$D_c = d - \beta/\nu. \tag{2.41}$$

Here d is the Euclidean dimension of the underlying grid (two or three).

There is a similar relationship for the fractal dimension of the backbone:

$$D_B = d - \beta_B/\nu. \tag{2.42}$$

Finally we note the values of these dimensions in 2D and 3D as given in Table 2.3.

2.8 Problems

2.1. Why is p_c not well defined for a finite-size lattice?

2.2. Starting with the definition of the correlation function for 1D percolation problems, prove that the correlation length $\xi(p)$ follows the scaling $\xi \propto (p_c - p)^{-1}$ near the threshold.

2.3. Prove that in 1D percolation problems, the mean cluster size $S(p)$ follows the scaling $S \propto (p_c - p)^{-1}$ near the threshold.

2.4. A simple lattice model with which one can analytically analyse the percolation properties is called the Bethe lattice which has no closed loop in the structure. In this lattice, each site has z neighbouring sites (e.g., see figure below for the Bathe lattice with $z = 3$). Also the fraction of sites on the boundary or surface sites is $\frac{z-2}{z-1}$ (e.g., three sites on the circle shown in figure below for the first generation). Prove that the percolation threshold is $p_c = \frac{1}{z-1}$ and the critical exponent for the percolation probability is $\beta = 1$ and the critical exponent of the mean cluster size is $\gamma = 1$.

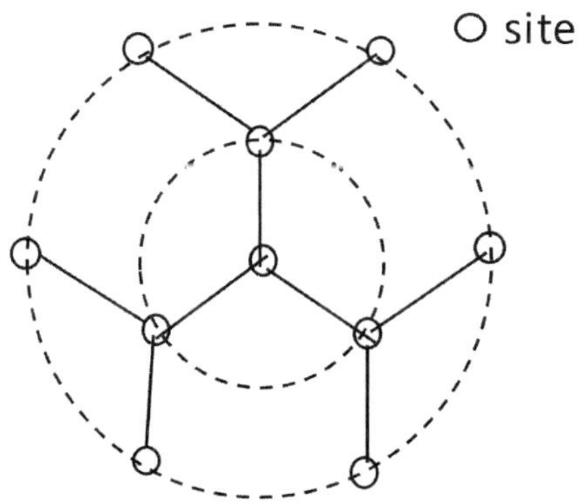

2.5. Let us assume the site percolation model at the threshold representing a porous medium where the oil resides in occupied sites. Use the fractal properties of percolating cluster to estimate the amount of oil at the field scale of $2,000 \times 2,000 \times 2,000\,\text{ft}^3$ based on the information you have already obtained from a rock sample of size $0.1 \times 0.1 \times 0.1\,\text{ft}^3$.

2.6. Consider a site percolation problem with L as the system size, N as the number of occupied sites in the system, n as the average number of occupied sites in the percolating cluster over all realisations and Δ as the standard deviation in the number of occupied sites in percolating cluster over all realisations. The statistics from a Monte Carlo simulation of three different system sizes $L = 10$, 50 and 200 are given in table below.

$L = 10$			$L = 50$			$L = 200$		
N	n	Δ	N	n	Δ	N	n	Δ
1	0	0	5	0	0	5	0	0
31	0	0	465	0	0	6,445	0	0
42	0.25	2.511	1,255	0	0	13,805	0	0
48	1.7925	6.821	1,295	1.192	18.826	22,505	0	0
52	5.1875	11.803	1,385	34.779	140.873	22,535	8.081	161.620
54	9.7625	16.053	1,425	154.44	294.146	22,995	371.407	1,556.04
57	17.0575	20.516	1,445	246.708	366.354	23,265	1,133.06	3,080.14
60	27.8675	23.020	1,465	372.638	434.566	23,465	2,643.26	4,759.74
62	35.09	23.464	1,485	524.917	476.937	23,575	4,348.65	5,727.149
63	39.7675	22.686	1,505	698.032	484.293	23,745	7,532.93	6,646.816
67	52.985	19.068	1,525	880.748	451.480	23,915	10,528.1	6,771.410
71	64.2125	11.952	1,545	1,033.34	405.494	24,065	13,129.9	6,413.215
73	68.0875	9.891	1,575	1,222.13	313.363	24,255	16,077.3	5,178.573
77	74.8975	5.970	1,605	1,376.9	199.136	24,545	19,131.9	3,258.930
84	83.39	4.303	1,645	1,500.63	67.572	24,775	20,735.1	1,512.445
96	95.995	0.0706	1,735	1,668.88	33.996	24,995	21,790.1	706.662
97	97	0	1,845	1,818.45	15.917	25,405	23,058.2	374.017
98	98	0	2,045	2,040.45	3.485	26,785	25,740.2	131.840
99	99	0	2,235	2,234.8	1.154	30,005	29,791.3	29.620
100	100	0	2,475	2,475.5	0.0350	36,425	36,422.4	1.8463

The statistics for the number of occupied sites in the system at the threshold point (N_c), the average number of occupied sites in the percolating cluster over all realisations at the threshold point (n_c) of three different system sizes $L = 10$, 50 and 200 are given in table below.

$L = 10$				$L = 50$				$L = 200$			
N_c	n_c	N_c	n_c	N_c	n_c	N_c	n_c	N_c	n_c	N_c	n_c
51	29	49	29	1,448	943	1,508	758	22,947	4,794	22,528	3,231
68	55	61	54	1,571	1,252	1,594	1,296	23,438	8,701	24,210	15,918
51	37	68	44	1,388	263	1,470	465	24,111	11,531	23,706	8,837
60	50	59	25	1,380	798	1,413	351	23,527	10,393	23,378	11,716
59	50	63	53	1,420	406	1,470	938	23,115	12,090	23,689	4,890
70	33	63	48	1,526	1,424	1,553	777	23,858	9,891	23,911	13,596
47	26	60	45	1,576	1, 341	1,406	487	23,697	10,728	23,341	6,106
51	25	56	43	1,392	703	1,376	611	24,200	15,744	24,139	15,661
54	37	48	20	1,482	887	1,557	1,108	23,505	14,615	23,552	9,685
62	50	59	45	1,453	528	1,479	660	23,113	7,219	23,043	6,115
49	27	67	52	1,462	693	1,507	1,192	23,801	14,469	23,853	15,829
59	54	54	30	1,595	1,353	1,523	1,133	22,884	6,283	24,105	17,070
62	51	52	33	1,426	585	1,399	728	23,552	6,907	23,620	10,698
49	28	69	60	1,555	1,023	1,528	934	24,763	17,342	23,244	5,533
61	49	57	46	1,336	523	1,510	789	23,845	8,691	23,191	7,725
65	38	66	64	1,412	531	1,530	900	23,995	15,726	23,556	7,954
69	66	60	39	1,389	630	1,627	1,555	23,671	13,864	23,861	9,657
67	57	59	51	1,448	796	1,524	920	23,283	5,870	24,695	17,446
53	36	36	22	1,414	485	1,508	887	23,733	15,497	23,898	12,206
59	27	54	34	1,577	1,462	1,500	799	22,958	6,008	24,217	14,136
55	42	54	19	1,514	853	1,511	485	23,022	3,869	23,495	7,913
66	61	66	60	1,410	504	1,459	574	23,681	11,046	23,904	11,904
55	43	57	31	1,550	1,153	1,505	1,112	24,129	18,320	22,961	4,961
60	17	51	24	1,550	1,044	1,417	594	23,844	10,829	24,156	16,550
63	63	62	58	1,467	881	1,480	973	23,404	11,127	23,912	9,125
63	41					1,522	1,141			23,529	6,429

(a) Estimate the critical exponents (β, ν) and the infinite percolation threshold.

(b) Plot the mean connectivity and standard deviation of connectivity against the occupancy probability p at three different system sizes.

(c) Show how to relate different curves to one single universal curve by using finite-size scaling transformations (Eqs. (2.11) and (2.14)).

References

Adler, P.M. and Thovert, J.F., *Fractures and Fracture Networks*. Kluwer Academic Publishers, London, pp. 163–218, 1999.

Belayneh, M., Masihi, M., Matthäi, S.K. and King, P.R., Prediction of vein connectivity using the percolation approach: model test with field data. *Journal of Geophysical Engineering*, 3: 219, 2006.

Berkowitz, B., Analysis of fracture network connectivity using percolation theory. *Mathematical Geology*, 27(4): 467–483, 1995.

Berkowitz, B. and Balberg, I., Percolation theory and its application to ground water hydrology. *Water Resources Research*, 29(4): 775–794, 1993.

Broadbent, S.R. and Hammersley, J.M., Percolation processes: 1. Crystal and mazes. *Proceedings of the Cambridge Philosophical Society*, 53: 629–641, 1957.

Christensen, K., Flyvbjerg, H. and Olami, Z., Self-organized critical forest-fire model: mean-field theory and simulation results in 1 to 6 dimensions. *Physical Review Letters*, 71: 2737, 1993.

Coleman, J.N., Curran, S., Dalton, A.B., Davey, A.P., *et al.*, Percolation-dominated conductivity in a conjugated-polymer-carbon-nanotube composite. *Physical Review B*, 58(12): R7492–R7495, 1998.

Falconer, K., *Fractal Geometry: Mathematical Foundations and Applications, Second Edition*. John Wiley & Sons, New York, 2003.

Fisher, M.E. and Essam, J.W., Some cluster size and percolation problems. *Journal of Mathematical Physics*, 2: 609–619, 1961.

Galam, S. and Mauger, A., Universal formulas for percolation thresholds. *Physical Review E*, 53: 2177, 1996.

Gawlinski, E.T. and Stanley, H.E., Continuum percolation in two dimensions: Monte Carlo tests of scaling and universality for non-interacting discs. *Journal of Physics A: Mathematical and General*, 14: L291–L299, 1981.

Glass, R.J., Conrad, S.H., Yarrington, L., Gravity-destabilized nonwetting phase invasion in macroheterogeneous porous media: near-pore-scale macro modified invasion percolation simulation of experiments. *Water Resource Research*, 37(5): 1197–1207, 2001.

Golden, K.M., Murphy, N.B. and Cherkaev, E., Spectral analysis and connectivity of porous microstructures in bone. *Journal of Biomechanics*, 44: 337–344, 2011.

Grassberger, P., Conductivity exponent and backbone dimension in 2D percolation. *Physica A: Statistical Mechanics and its Applications*, 262: 251–63, 1999.

Harter, T., Finite size scaling analysis of percolation in three dimensional correlated binary Markov chain random fields. *Physical Review E*, 72: 026120.1–026120.7, 2005.

Havlin, S. and ben-Avraham, D., Diffusion in porous media. *Advances in Physics*, 36(6): 695–798, 1987.

Hoshen, J. and Kopelman, R., Percolation and cluster distribution. I. Cluster multiple labelling technique and critical concentration algorithm. *Physical Review B*, 14(8): 3438–3445, 1976.

Hunt, A.G. and Ewing, R., *Percolation Theory for Flow in Porous Media.* Lecture Notes in Physics, vol. 771. Springer, Berlin, 2005.
King, P.R., The connectivity and conductivity of overlapping sandbodies. In: Buller, A.T. *et al.* (eds.), *North Sea Oil and Gas Reservoirs — II*. Graham & Trotman, London, pp. 353–361, 1990.
Lagorio, C., Migueles, M.V., Braunstein, L.A., López, E. and Macri, P.A., Effects of epidemic threshold definition on disease spread statistics. *Physica A: Statistical Mechanics and its Applications*, 388(5): 755–763, 2009.
Mandelbrot, B.B., *The Fractal Geometry of Nature.* W.H. Freeman, New York, pp. 468, 1982.
Masihi, M., King, P.R. and Nurafza, P., The effect of anisotropy on finite size scaling in percolation theory. *Physical Review E*, 74: 042102, 2006.
Masihi, M. and King, P.R., A correlated fracture network: modeling and percolation properties. *Water Resources Research Journal*, 43, doi: 10.1029/2006WR005331, 2007.
Masihi, M., King, P.R. and Nurafza, P., Fast estimation of connectivity in fractured reservoirs using percolation theory. *SPE Journal*, 12(2): 167–178, 2007.
Masihi, M. and King, P.R., Connectivity prediction in fractured reservoirs with variable fracture size: analysis and validation. *SPE Journal*, 13(1): 88–98, 2008.
Moore, C. and Newman, M.E.J., Exact solution of site and bond percolation on small-world networks. *Physical Review E*, 62: 7059, 2000.
Mourhatch, R., Tsotsis, T.T. and Sahimi, M., Determination of the true pore size distribution by flow permporometry experiments: an invasion percolation model. *Journal of Membrane Science*, 367: 55–62, 2011.
Mourzenko, V.V., Thovert, J.F. and Adler, P.M., Percolation of three dimensional fracture networks with power law size distribution. *Physical Review E*, 72: 036103.1–036103.14, 2005.
Paterson, L., Diffusion-limited aggregation and two-fluid displacements in porous media. *Physical Review Letters*, 52(18): 1621–1624, 1984.
Prakash, S., Havlin, S., Schwartz, M. and Stanley, H.E., Structural and dynamical properties of long-range correlated percolation. *Physical Review A*, 46: R1724, 1992.
Renard, P.H. and de Marsily, G., Calculating equivalent permeability: a review. *Advances in Water Resources*, 20(5–6): 253–278, 1997.
Sadeghnejad, S., Masihi, M. and King, P.R., Dependency of percolation critical exponents on the exponent of power law size distribution. *Physica A*, 392: 6189–6197, 2013.
Sahimi, M., *Applications of Percolation Theory*. Taylor and Francis, London, pp. 258, 1994.
Sahimi, M. and Mukhopadhyay, S., Scaling properties of a percolation model with long range correlations. *Physical Review E*, 54: 3870–3880, 1996.
Schmittbuhl, J., Vilotte, J.P. and Roux, S., Percolation through self-affine surfaces. *Journal of Physics A: Mathematical and General*, 26: 6115, 1993.

Selyakov, V.I. and Kadet, V.V., *Percolation Models for Transport in Porous Media with Applications to Reservoir Engineering*. Kluwer Academic Publishers, London, pp. 241, 1996.

Stauffer, D. and Aharony, A., *Introduction to Percolation Theory*. Taylor and Francis, London, pp. 181, 1992.

Vogel, E.E., Lebrecht, W. and Valdés, J.F., Bond percolation for homogeneous two-dimensional lattices. *Physica A: Statistical Mechanics and its Applications*, 389: 1512–1520, 2010.

Weinrib, A. and Halperin, B., Critical phenomena in systems with long range correlated quenched disorder. *Physical Review B*, 27: 413, 1983.

Wilson, K.G., Renormalization group methods. *Advances in Mathematics*, 16(2): 170–186, 1975.

Zhang, G., Xia, Y., Wang, H. *et al.*, A percolation model of thermal conductivity for filled polymer composites. *Journal of Composite Materials*, 44(8): 963–970, 2009.

Chapter 3

Continuum percolation for geological models

So far, we have described percolation theory in terms of a simple 2D lattice. In this chapter, we extend these ideas to be relevant for realistic geological models. In particular, we use continuum percolation to represent a number of geological environments, as described in Chapter 1. We shall focus on two environments: overlapping sandbodies and fracture networks. We briefly describe the basic percolation quantities in 2D which are of most importance to reservoir engineering. In particular, we shall concentrate on the connectivity of these models. The other percolation properties such as the backbone fraction, dangling ends and conductivity will be described in Chapter 6. We first describe continuum percolation which is the basis of these.

Although percolation theory was initially developed for lattices, there are many systems not based on a lattice (Stauffer and Aharony, 1992; Stell, 1996). In continuum percolation, geometrical objects are placed randomly and independently (with Poisson statistics) in a region. This is very similar to the use of object-based models in reservoir engineering (Haldorsen *et al.*, 1988; Sahimi, 1994; Berkowitz, 1995; Baker *et al.*, 2002; Consiglio *et al.*, 2004).

Firstly, the objects are not restricted to points on a fixed lattice which means that there is no maximal concentration of objects (in lattice-based percolation, the upper bound of the occupancy probability is one). In this we are speaking mathematically, there are constraints on the number of geological objects one might find.

Secondly, the objects can be of any shape with variable length and/or orientation (Berkowitz, 1995). Again geologically there is a very wide range of possible shapes of flow units, but we usually model them with just a range of simple shapes (such as squares, rectangles, discs or ellipses in 2D). We again assume that the whole reservoir is just made up of two types of unit, the objects and the background, one of which is permeable and the other impermeable. This is the situation described in the outcrop images in Figure 1.5 from Chapter 1. Which is which depends on whether the objects represent flow units or flow barriers. We shall consider the more general case of variable permeability flow units in later chapters.

In order to make use of the results of standard percolation, we have to define the appropriate equivalent terms to the effective length scale of the system L and the percolation quantities such as the occupancy probability p and the percolation probability $P(p)$ and so on. However, from the principle of universality, we can use the same scaling laws with the same numerical values of the critical exponents as in lattice percolation (King, 1990; Berkowitz, 1995).

In continuum percolation, we get the same threshold behaviour arising from clusters growing, merging and connecting across the system. We expect the geometric properties to satisfy universality and share the same critical exponents. However, it is found that the exponents of some transport properties may be different from those in discrete networks (Sahimi, 1994; Hunt and Ewing, 2005). We will discuss this later.

Finally, as the percolation threshold depends on the shape of the objects, it has to be found by direct numerical simulation. Fortunately it is found that for many simple convex shapes used in object modelling, the threshold values are all very similar, so we can comfortably use a single value in most cases. There are a number of studies applying percolation theory to continuum systems that check the universality and determine the percolation threshold and critical exponents (Gawlinski and Stanley, 1981; Xia and Thorpe, 1988; Lee and Torquato, 1990; King, 1990; Berkowitz, 1995; Choi *et al.*, 1995; Garboczi *et al.*, 1995; Lin and Hu, 1998; Lorenz and Ziff, 2001; Baker *et al.*, 2002).

3.1 Simple sandbody model

3.1.1 *Isotropic overlapping sandbody model in 2D*

To study the percolation properties of overlapping sandbody systems, we first need to construct a model by using an object-based technique. It is assumed that sandbodies can be represented by squares (circles or ellipses) in 2D. These objects are distributed randomly and independently in an impermeable background. The bodies can overlap each other to make a geometrically complex pattern. Furthermore, not only are the bodies assumed to be entirely permeable, but also it is assumed that there is a perfect hydraulic contact between them. Finally we assume that the background is entirely impermeable and so contributes nothing to the flow. Real sandbodies are more complicated than this, but this can be used to establish the principles of continuum percolation.

For simplicity, we consider an arbitrary square region of size $X \times X$ within which a number of squares of side size a are placed independently. So the centre point of each body is assigned randomly, and the bodies are allowed to overlap. The effective size of the system is defined by the dimensionless length $L = X/a$. The net-to-gross ratio is the area fraction of permeable sand which is the same as the probability that a point chosen at random lies within a sand. So it is equivalent to the occupancy probability p of lattice percolation theory. Note that overlaps are only counted once, so the sand area is not the same as the number of bodies multiplied by the area of each body.

As in lattice percolation problem, we observe that as the net-to-gross ratio increases, clusters of overlapping sandbodies are formed that increase in size until at a critical value one spanning cluster is formed. The connectivity $P(p, L)$, is defined as the probability that a point in the sandbodies belongs to this percolating cluster connecting the two sides. This connected sand fraction is the same

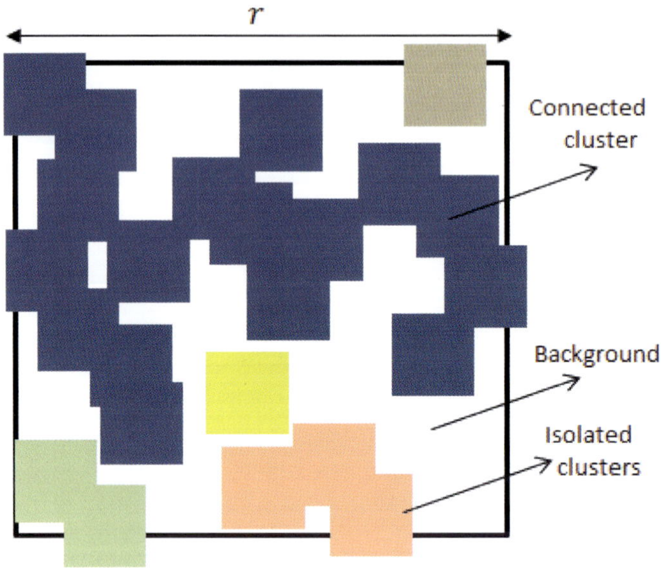

Figure 3.1. Schematic of an overlapping sandbody model made of squares showing a connected cluster (from left to right) in blue and isolated clusters with different colours.

as the fraction of the total sands belonging to this connecting cluster. It is a function of net-to-gross ratio, p, and the dimensionless size of the system, L. Also we want to know the probability that two wells, separated by a Euclidean distance r in a reservoir of size L, are connected (see Figure 3.1 in which wells are placed on the left and right boundaries).

3.1.2 *Computational algorithm*

We need an appropriate computational algorithm to generate a large number of realisations. The region is divided into a coarser grid, each block having the same size as the sandbodies. Any sandbodies whose centre lies within the same block must overlap. Sandbodies whose centres do not lie within the eight blocks surrounding the current one cannot overlap (Figure 3.2).

Because the sands are distributed independently, the squares have a Poisson probability distribution to occupy one of these blocks.

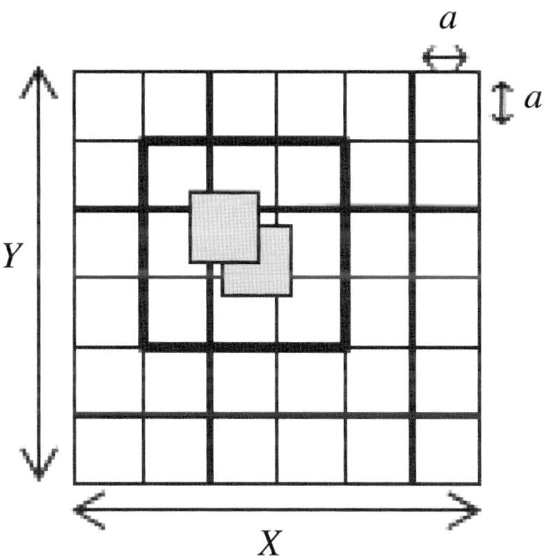

Figure 3.2. Coarse grid used for the sandbody model.

The probability P_k that a block is occupied by k sandbodies is

$$P_k = x^k e^{-x}/k! \quad \text{with} \quad x = Na^2/A^2, \tag{3.1}$$

where N is the total number of sandbodies and $A = X \times X$. We generate the sandbodies by sweeping through the blocks sequentially. In a given block, generate a random number y in the range 0–1; then generate the cumulative of the Poisson distribution, $C_i = \sum_0^k P_i$. The number of sandbodies to be located within this block is k such that $C_k < y < C_{k+1}$. These k sandbodies are then distributed randomly within the block.

The algorithm to generate the realisations and to calculate the net-to-gross ratio and connected sand fraction is shown in Figure 3.3.

As each object is placed, it is given an index number which represents the cluster that the body belongs to. A check is made to see if it intersects with previously placed bodies or not. If two bodies intersect each other, then they get the same index. Since it is possible for a body to overlap with more than one other body connecting existing clusters, the two sets of cluster indices must now

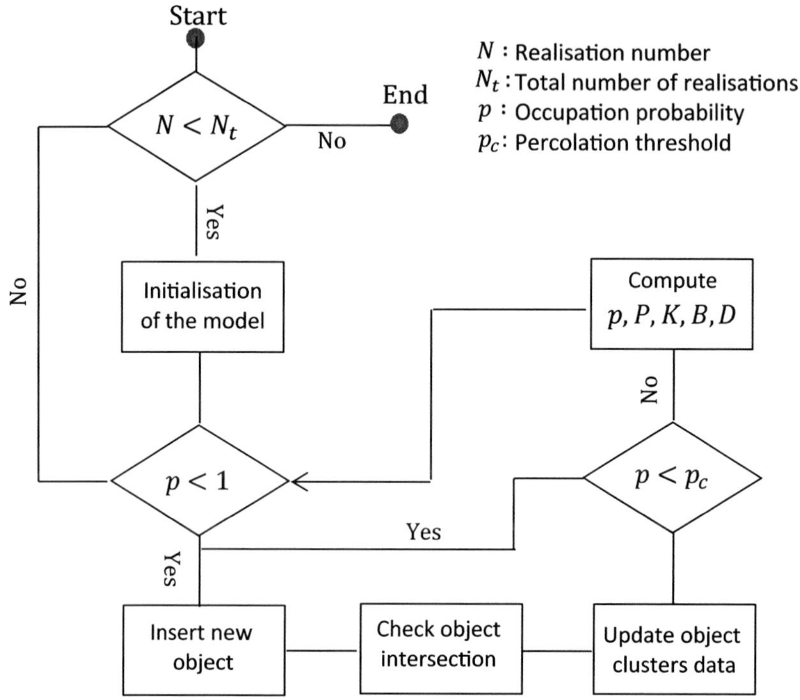

Figure 3.3. Flowchart of the algorithm used to generate the realisations and do the computations.

be set the same to make one index for the whole cluster. Then a new square is placed and the whole procedure repeated until a preset number of bodies, N, have been placed and checked.

The check for intersection of two bodies is straightforward for simple geometrical objects such as aligned squares, rectangles or circles. The criterion is that if the distance between the centres of the two objects is greater than the sum of their half lengths (or radii). This is more complex for other geometrical shapes or non-aligned ellipses or rectangles (Baron, 1972; Baker *et al.*, 2002).

Once all the bodies are placed and the clusters identified, the algorithm computes the net-to-gross ratio and finds the percolating clusters in the X or Y direction and their total occupied fraction. The percolating clusters are identified by checking bodies connected

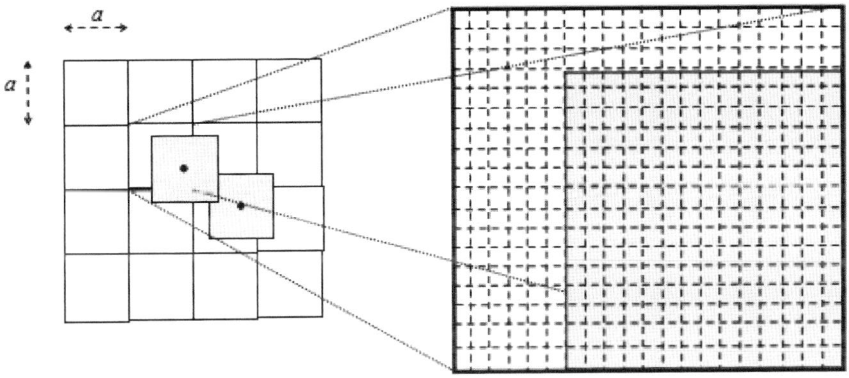

Figure 3.4. Fine grid used to calculate the net-to-gross ratio and the connected sand fraction.

to the both sides of the region to find if they have the same index number.

The net-to-gross ratio is calculated by placing a fine grid over the system (Figure 3.4). Each point of this grid is checked to see if it lies in a sandbody or not. The fraction of points that do is a reasonable approximation to the net-to-gross ratio if the grid is fine enough. This calculation of p ensures that the overlap area is not counted twice because if a fine-grid node lies within one or more bodies, it is only counted only once.

It is clear that the finer this grid is, the more accurate the result is. However, finer grids increase the computational time. So, there is a trade-off between the accuracy and the CPU time. The optimum point can be evaluated by plotting apparent threshold against the fine-to-coarse-grid ratio as the point at which there is no change in this estimate (Figure 3.5).

The total number of fine-grid points gives the total area of the sand. The net-to-gross ratio is then the ratio of this area to the total area of the region. If there is a percolating cluster, the same fine grid can be used to count the nodes which lie within the bodies belonging to this percolating cluster. This gives the total area of the connected sand. The total fraction of percolating clusters is then the ratio of this area to the total area of the sand. This sand fraction (P) connecting opposite sides of the system is important in reservoir engineering

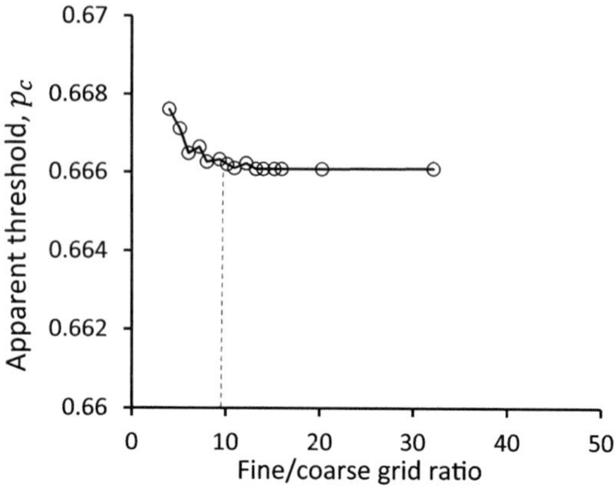

Figure 3.5. Sensitivity analysis on the calculation of apparent threshold with respect to the fine-to-coarse-grid ratio.

and hydrology (Larue and Hovadik, 2006; Wen, 2006; Knudby and Carrera, 2005).

As with the lattice case, we run this multiple times to generate different realisations of the sandbodies. We find the same critical behaviour of growing and merging clusters growing until one spans the region at the percolation threshold. Two realisations of this simplified sandbody model below and just above the threshold are shown in Figure 3.6.

The apparent threshold, $\tilde{p}_c(L)$, is the sand fraction at which half of the realisations percolate. The total number of realisations required to obtain reliable statistics is dependent on the effective system size. We need to take a sufficiently large number of realisations to get a stable estimate of the threshold, but again this can increase computer run time, so there is a compromise. A plot of the estimate of the percolation threshold against the number of realisations is shown in Figure 3.7.

Moreover, the fine grid placed on the system has to be fine enough to accurate estimates of the threshold. This is especially true for more complex sandbody shapes. This algorithm remains essentially unchanged when applied to more complex models as discussed in Chapter 4.

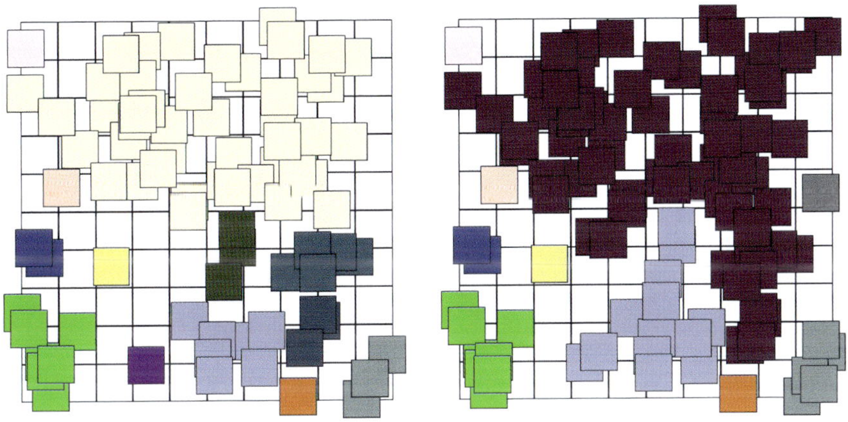

(a) (b)

Figure 3.6. Two realisations of isotropic sandbodies represented by squares (a) below the threshold ($N = 80$, $p = 0.54$, $P = 0$) and (b) just above the threshold ($N = 100$, $p = 0.60$, $P = 0.6$) for the effective system size of $L = 10$.

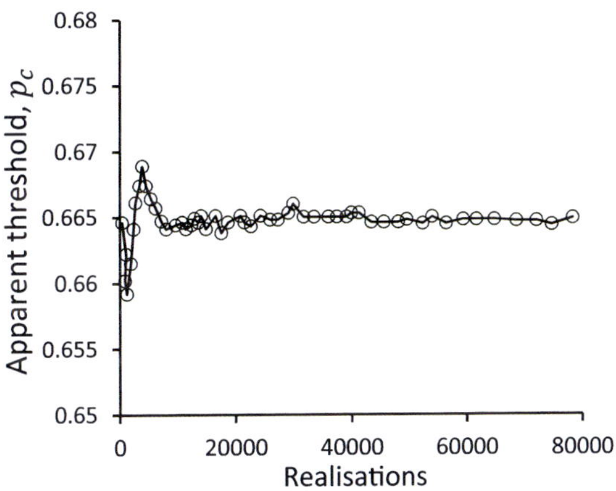

Figure 3.7. Illustration for choosing the adequate number of realisations to be statistically sufficient.

3.1.3 *Connectivity*

We observe that continuum percolation shows similar behaviour to lattice percolation and so we expect to be able to apply the same scaling laws to predict connectivity and so on. To do this, we need to determine the various percolation parameters. From universality, we expect that the exponents to be unchanged and independent of the body size and shape. They only depend on the dimensionality of the system. However, the percolation threshold p_c does depend on the details of the system. Demonstrations that universality does apply can be found elsewhere (Baker *et al.*, 2002; Lee and Torquato, 1990; Lin and Hu, 1998; Lorenz and Ziff, 2001; Xia and Thrope, 1988; Masihi and King 2007b; Sadeghnejad *et al.*, 2010, 2013b).

We find the threshold for the aligned squares percolation from the scale dependency of the apparent threshold to the error observed on the threshold (Eqs. (2.27) and (2.29)) and by plotting the estimated value of the apparent threshold against system size L (Figure 3.8). We find that $p_c^{\infty} = 0.667 \pm 0.004$.

It is found that this value of percolation threshold for aligned squares is very similar to that found for overlapping discs (Baker *et al.*, 2002). The numerical values of the percolation threshold for randomly placed constant-size sandbodies of various shapes is presented in Table 3.1. As before the threshold values of 3D systems are much lower than the thresholds obtained from 2D models. It

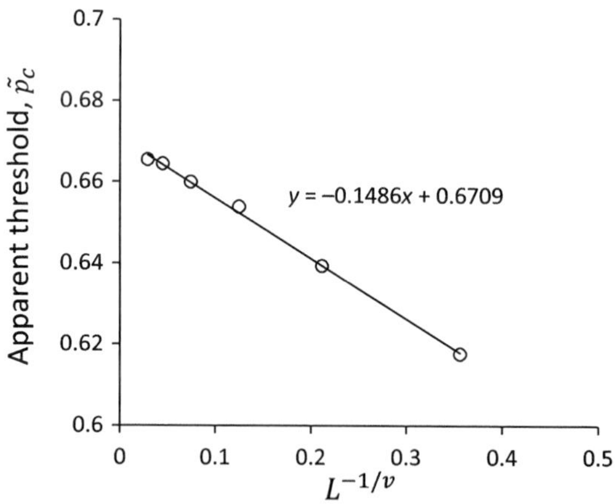

Figure 3.8. Extrapolation of finite-size apparent threshold to find the infinite threshold for continuum percolation using aligned squares.

Table 3.1. Thresholds of various overlapping sandbody shapes (Xia and Thrope, 1988; King, 1990; Garboczi *et al.*, 1991; Quintanilla *et al.*, 2000; Lorenz and Ziff, 2001; Baker *et al.*, 2002; Sadeghnejad *et al.*, 2013a, 2013b).

	Object shape	**Threshold value**
2D	Discs	0.676 ± 0.003
	Ellipses $(a/b = 2)$	0.635 ± 0.005
	Aligned squares	0.667 ± 0.004
	Randomly oriented squares	0.625 ± 0.002
3D	Spheres	0.289 ± 0.005
	Ellipsoids $(a/b = 1.33)$	0.286 ± 0.02
	Aligned cubes	0.277 ± 0.002
	Randomly oriented cubes	0.216 ± 0.002

also shows that the actual shape of the objects used to represent the geological sandbodies is not that crucial.

In principal, it is unnecessary to determine the critical exponents as they are universal. Previous estimates of the exponents were found to be close to the lattice percolation values given in Table 2.2 (King, 1990; Koudine *et al.*, 1998; Adler and Thovert, 1999; Nurafza *et al.*, 2006). As an example of how this is done, to estimate the exponent β, one can take a log–log plot of the mean connected fraction P against $(p - p_c)$ for a very large system. Figure 3.9(a) shows such a plot; the slope gives the exponent β. The exponent ν can be found from a log–log plot of the variance in probability distribution that the system percolates, at a given net-to-gross ratio, against system size L, as shown in Figure 3.9(b). The estimated values of the exponents are $\beta = 0.139$ and $\nu = 1.33$ which are in very good agreement with the literature values for lattice percolation in Table 2.2. From now on, we shall assume universality and use literature values for all other exponents.

We can now proceed to examine the finite-size scaling for the connected sand fraction. We find the same finite-size scatter as before with a rapid increase in the connectivity above the percolation threshold. Again we look at the average connected fraction $(P(p, L))$ and its standard deviation $(\Delta(p, L))$. These are shown in Figure 3.10

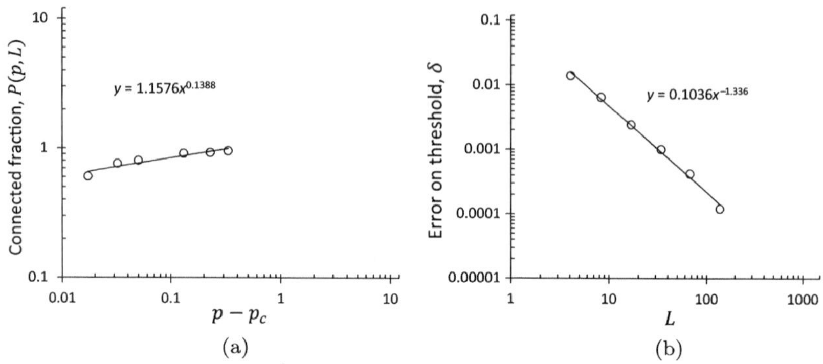

Figure 3.9. Plots used to estimate the critical exponents (a) β and (b) ν for the square sandbodies model in 2D.

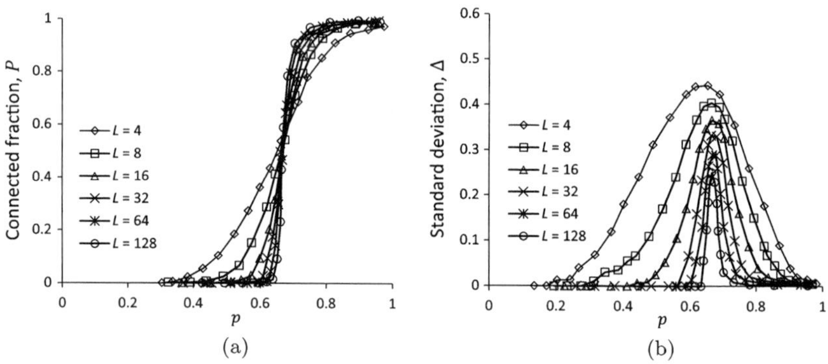

Figure 3.10. (a) Mean connected sand fraction and (b) its standard deviation for various system sizes of the square sandbodies model in 2D.

(we recall here that the finite-size length L is the dimensionless length X/a).

We apply the same finite-size scaling laws as before (Eqs. (2.11) and (2.14)) and get the same data collapse, as shown in Figures 3.11 and 3.12. This gives the master curves required for continuum percolation.

We reiterate that percolation scaling is only valid in close proximity to the percolation threshold and there will be deviations far above the threshold. In that regime, other approaches such as effective

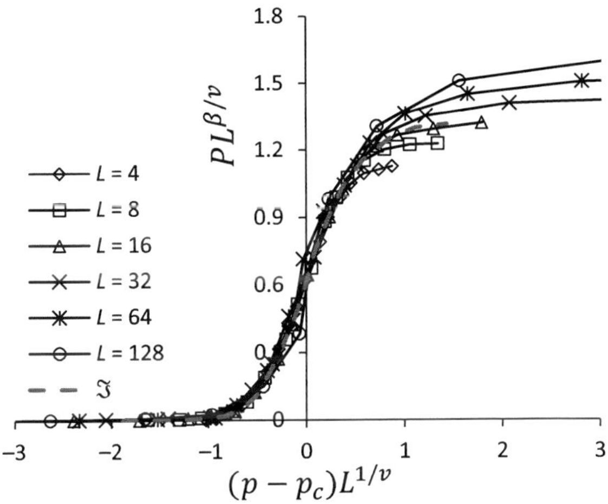

Figure 3.11. Rescaling the mean connected sand fraction by using scaling law (Eq. (2.11)) to get the mean connectivity master curve for the square sandbodies model in 2D.

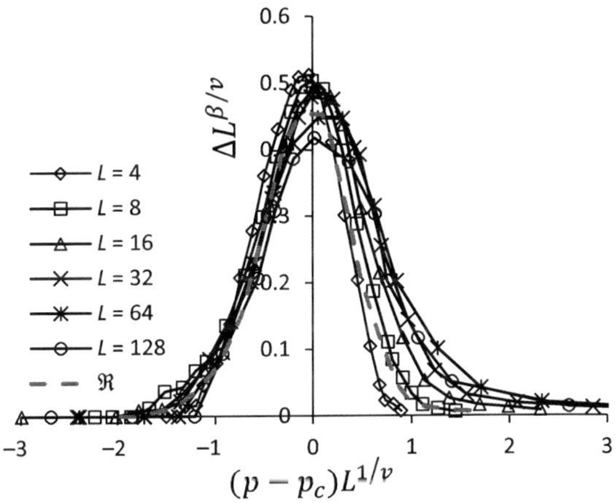

Figure 3.12. Rescaling the standard deviation in connectivity by using scaling law (Eq. (2.14)) for the square sandbodies model in 2D.

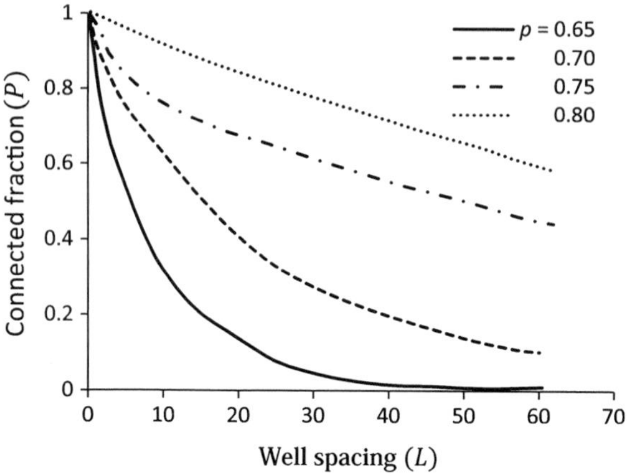

Figure 3.13. The connected sand fraction as a function of well spacing at various net-to-gross ratios for a given reservoir size.

medium theory can be applied (Kirkpatrick, 1973; Mukhopadhyay and Sahimi, 2000; Adler and Berkowitz, 2000; Torquato, 2002; Knudby *et al.*, 2006; Savik *et al.*, 2013). This will be discussed in more depth in Chapter 7.

These master curves also allow us to calculate the mean connected sand fraction as a function of well spacing at various net-to-gross ratios (Figure 3.13). The product of the connected sand fraction (P) and the volume of hydrocarbon initially in place gives the connected volume of hydrocarbons between two wells. So, the connected sand fraction curves in Figure 3.13 can be used to determine the fraction of the original oil in place that is in contact with the two wells. This is not the same as the recoverable oil from these wells as not all parts of the connected sand cluster are accessible. We will discuss this further in Chapter 6.

3.2 Simple fracture network model

We now discuss a different kind of continuum percolation model. The concept is the same in that geometrical objects are placed randomly in space. In this case, the objects are not squares but line segments. The rationale for this is that the line segments represent faults or fractures (we shall use these terms synonymously although from geological and geomechanical points of view they are very different). We recognise that real fractures are not placed independently, the geomechanics of their formation leads to complex patterns. But we use this model as a basic starting point in the same way as we represent geologically complex sand units with simple squares.

All hydrocarbon reservoirs are naturally fractured. Generally most of the volume of the fluids occupies the porous matrix. However, the fractures act as the main conduits for flow. This is another system where the connectivity of the objects (fractures) is key to understanding the flow properties of the whole system (Wilke *et al.*, 1985; Adler and Thovert, 1999; Berkowitz, 2002; Belayneh *et al.*, 2006; Tóth and Vass, 2011).

We represent fractures with line segments in 2D and squares (rectangles or discs) in 3D. We assume that they are the only pathways for the flow. The conductive network of fractures is embedded in an impermeable background. In reality, the situation is far more complicated than this basic model. In some cases, the matrix may have a non-negligible permeability compared to the permeability of the fracture (Saidi, 1987; Taylor *et al.*, 1999; Matthäi and Belayneh, 2004). Moreover fractures may be in physical contact but not in hydraulic contact because of a variety of mechanical or geochemical processes which leads to their closure or infilling (Bear *et al.*, 1993). At this stage, we neglect these details and more realistic models will be described in Chapter 5.

3.2.1 *Simple fracture network model in 2D*

We take line segments (which represent the fractures in 2D) distributed randomly within a squared region of size $L \times L$. All the lines have the same length l with orientations that are distributed uniformly on the interval $-\pi/2 < \theta < \pi/2$. The centre point of the line is taken as the centre of the fracture. The problem of what fraction of fractures connects opposite boundaries is a continuum percolation problem. First we need to choose an analogue for the occupancy probability p. This is not obvious; it cannot be the area fraction of the line segments as they have no area, but clearly it must be related to their number density. It is clear that the number density alone is not an appropriate parameter to use as it makes no distinction between a small and large fractures. Whilst in the simple model all fractures have the same length, we will want to generalise this later, so it is useful to have a measure that also generalises. These measures of percolation threshold may be related to each other, depending on the shape, orientation, and size variation of bodies, which will be discussed in Chapter 5. The most common choices used in the literature are as follows.

The first is the dimensionless number density defined as

$$\rho = Nl^2/4L^2, \tag{3.2}$$

which is the number density (N/L^2) multiplied by the effective area "swept" by the fracture (Bour and Davy, 1997). By the area swept, we mean that, since the fracture can take any orientation, a very large number of realisations centred at the same point would look like a disc of radius $l/2$. Hence, provided the centres of two fractures are more than a distance l apart, they cannot intersect regardless of their actual orientation. Clearly this argument becomes more involved if the fractures do not take all orientations with equal probability. This will be discussed later.

The second is the average number of intersections per fracture, I (Robinson, 1983; Balberg *et al.*, 1984; Berkowitz, 1995).

The third is the excluded area fraction, ρ_{ex}, which is the fraction of excluded area (a_{ex}) to the total area of the system, A. By definition, excluded area is the area around an object in which the centre of another object must lie in order for them to overlap (Balberg *et al.*, 1984; Adler and Thovert, 1999; Masihi *et al.*, 2007).

The fourth is the probability that a point chosen at random is in the effective area of a fracture defined by one-fourth of its average excluded area, i.e.,

$$p = 1 - e^{-N\langle a_{\text{ex}}\rangle/4L^2}. \tag{3.3}$$

This is based on the assumption that the fractures have a Poisson distribution, N is the number of fractures in the region, L is the system size and $\langle a_{\text{ex}}\rangle$ is the average excluded area over the

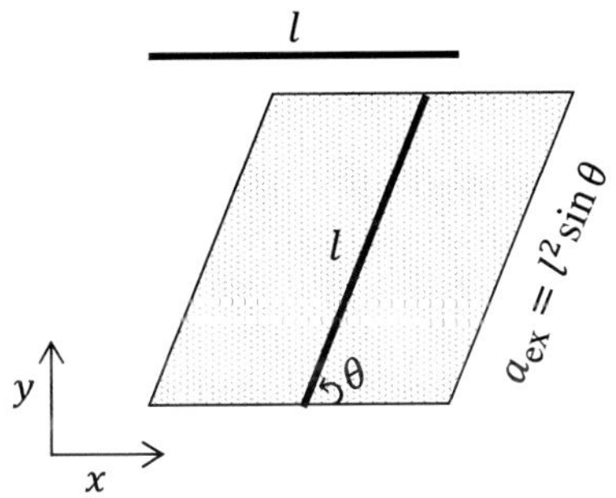

Figure 3.14. Excluded area for two fractures.

distribution of the fracture orientations. The derivation of this equation will be given in Chapter 4.

The excluded area is that region around each fracture where the centre of another fracture must not lie if they are to intersect. The excluded area for two fractures with a relative orientation of θ is shown in Figure 3.14 and is found to be $l^2 \sin\theta$.

We then average over the orientation distribution. If this is uniform on the range $-\theta_0 < \theta < \theta_0$, the mean excluded area is

$$\langle a_{\text{ex}} \rangle = \frac{l^2}{\theta_0}\left[1 - \frac{\sin\theta_0 \cos\theta_0}{\theta_0}\right]. \tag{3.4}$$

For the isotropic case where $\theta_0 = \frac{\pi}{2}$, the mean excluded area becomes

$$\langle a_{\text{ex}} \rangle = 2l^2/\pi. \tag{3.5}$$

Using this mean excluded area in Eq. (3.3) will give the percolation parameter p.

Having defined the analogue to the occupancy probability, we now need to define the equivalent of the connected fraction P. This is easier; it is just the number of fractures in the spanning cluster (n_s) divided by the total number of fractures (N):

$$P = \frac{n_s}{N}. \tag{3.6}$$

We first need to check whether this model belongs to the general class of continuum percolation. There are many publications which confirm that it does (Balberg *et al.*, 1983; Berkowitz, 1995; Koudine *et al.*, 1998; Adler and Thovert, 1999; Masihi and King 2007a), and so we can use the conventional percolation exponents (Balberg *et al.*, 1983; Robinson, 1984; Charlaix, 1986; Bour and Davy, 1997, 1998; Adler and Thovert, 1999; Masihi *et al.*, 2007). However, we still need to determine the threshold for it.

3.2.2 *Computational algorithm*

To estimate the threshold, we need an algorithm to generate realisations of the simple model. We generate the Cartesian coordinates for the centre points of fractures at random and select an orientation from the distribution. As with the simple square sandbody model, we use an index number to identify the fracture and which cluster it lies in. This labelling is the same as used previously but it is slightly more complicated to determine if two fractures intersect or not.

We define a vector joining the centres of two fractures with orientation α and magnitude r_{ij} as shown in Figure 3.15.

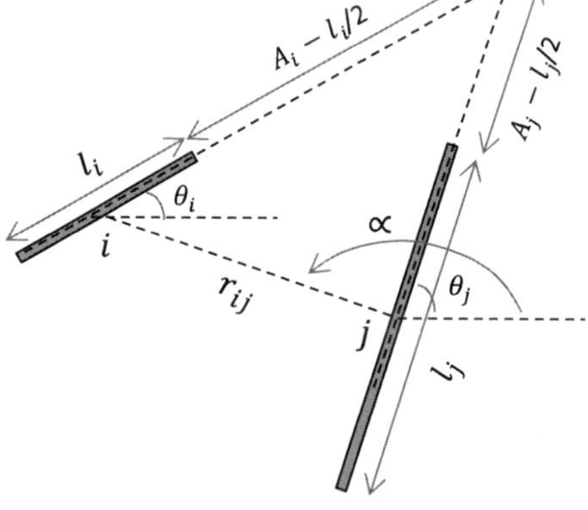

Figure 3.15. Intersection criterion for two fractures.

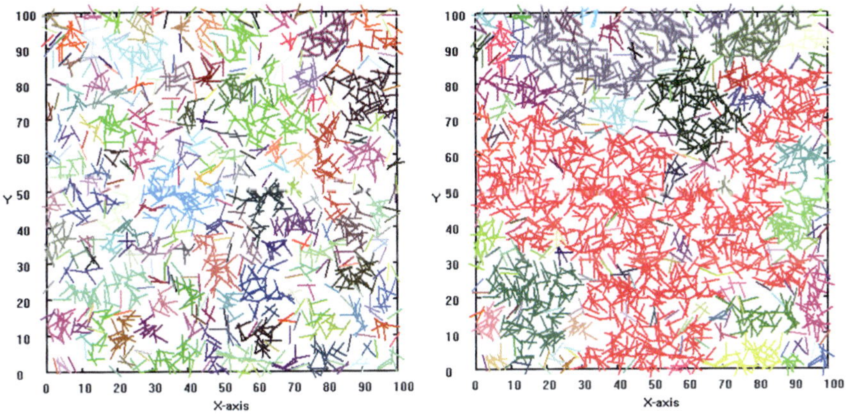

(a) (b)

Figure 3.16. Two realisations of fracture networks (a) below the threshold ($N = 1{,}500$ and $p = 0.45$) and (b) above the threshold ($N = 2{,}066$ and $p = 0.56$) with a spanning cluster (in red) consisting of 897 fractures.

We then calculate the distance from the centres of the fractures to the point where the continuation of the line segments would cross (A_i and A_j for the two fractures i and j). These are as follows:

$$A_i = r_{ij}|\sin(\propto - \theta_j)/\sin(\theta_i - \theta_j)|, \tag{3.7a}$$

$$A_j = r_{ij}|\sin(\propto - \theta_i)/\sin(\theta_i - \theta_j)|. \tag{3.7b}$$

Two fractures intersect if both $A_i \leq l_i/2$ and $A_j \leq l_j/2$. If the intersection point is in the region of interest, then the two fractures are assigned a common cluster number. Each time a fracture is placed, it can (1) form a new cluster (with no intersection), (2) extend an existing cluster or (3) unite two or more existing clusters. Two realisations, below and above the threshold, are shown in Figure 3.16; the different colours label different clusters.

3.2.3 *Connectivity*

As before, we first concentrate on the connectivity of the fracture networks. We find the behaviour very similar to that in the other percolation models with a rapid rise in connectivity and the same sort of finite-size corrections (Figure 3.17).

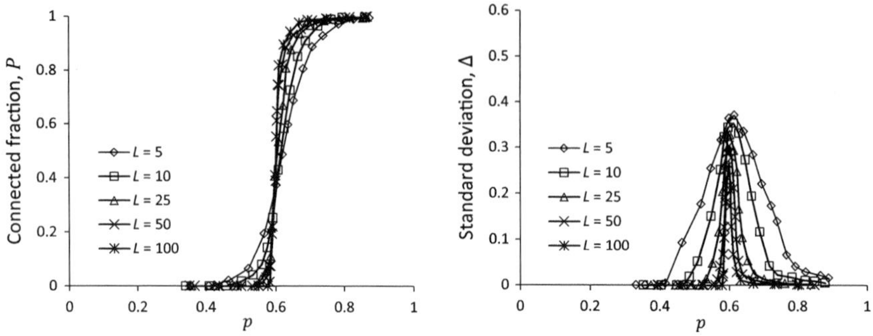

(a) (b)

Figure 3.17. Finite-size effects for (a) the mean connectivity and (b) its associated uncertainty in fracture networks (Masihi *et al.*, 2007).

We expect to be able to use the same finite-size scaling to get the collapse of all these curves onto a common master curve. To do this, we need the percolation threshold, which will depend on the measure of occupancy used. For constant-length randomly oriented fractures and depending on various definitions, it is found to be $\rho_c \approx 1.41$, $I_c \approx 3.6$, $\rho_{\text{ex}} \approx 3.6$ and $p_c \approx 0.592$ (Robinson, 1983; Balberg, 1985; Berkowitz, 1995; Bour and Davy, 1997; Adler and Thovert, 1999; Masihi *et al.*, 2007).

As before, we can use finite-size scaling to extrapolate the infinite threshold from finite-size estimates and find $p_c^\infty = 0.59$ (Figure 3.18).

In Figure 3.19, we show the estimation of the connectivity exponent by looking at the connected fraction near the threshold and find $\beta = 0.13 \pm 0.04$.

And similarly in Figure 3.20, we estimate the correlation exponent to be $\nu = 1.11$.

We can now rescale the connectivity plots to determine the master curves for the mean and standard deviation in connectivity (Figure 3.21).

Now from these two connectivity master curves, one can reconstruct the previous connectivity curves (or construct the connectivity curves for any other system size) without performing any explicit realisations.

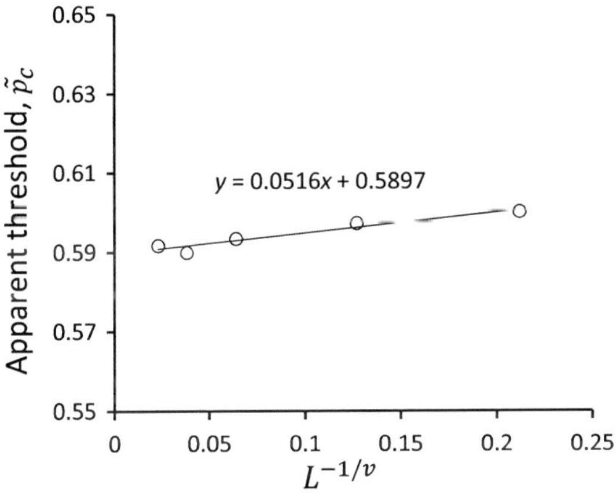

Figure 3.18. Estimate of the infinite threshold from finite-size extrapolation.

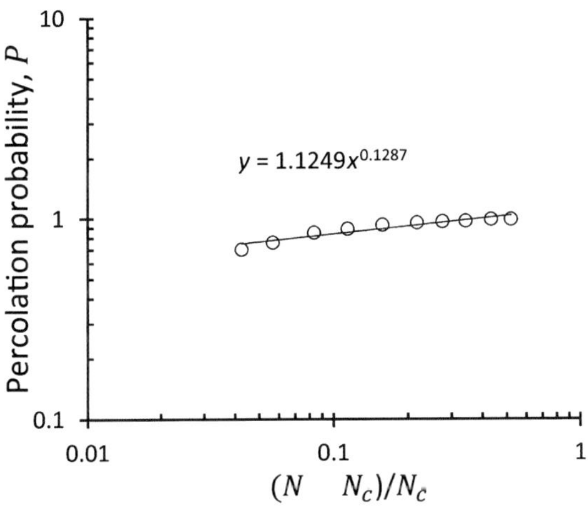

Figure 3.19. Plot of the percolation probability versus $(N - N_c)/N_c$ for a large fracture system ($L = 100$).

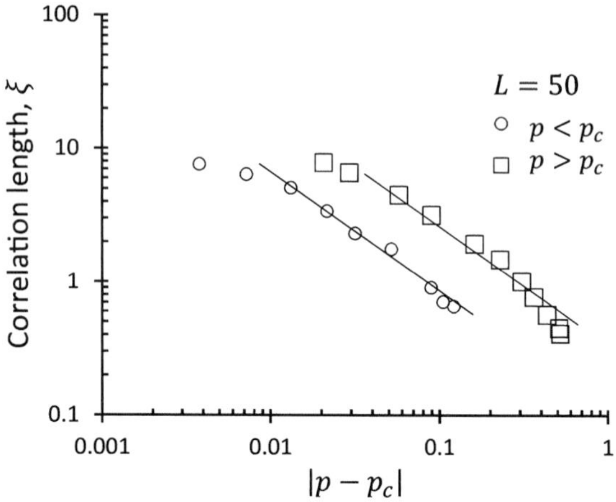

Figure 3.20. Plot of the correlation length against $|p - p_c|$ of fracture system for a large system size $L = 50$.

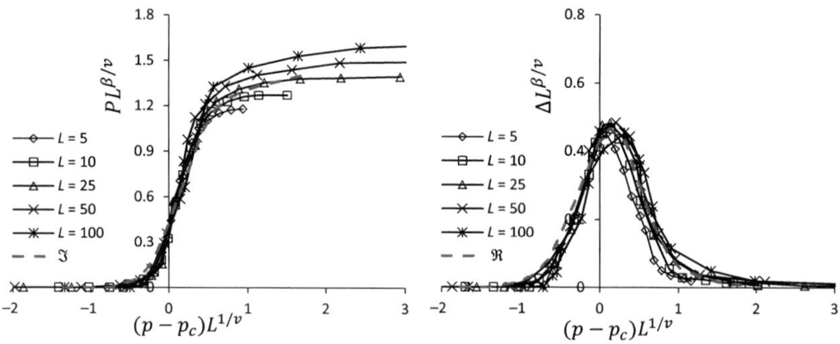

(a) (b)

Figure 3.21. Illustration of the finite-size scaling law (Eqs. (2.11) and (2.14)) for (a) the mean and (b) the standard deviation of connectivity results of fracture networks.

3.3 Problems

3.1. Consider a typical cross-sectional reservoir model of 50 m thickness and interwell spacing 1 km. Sandbodies of 5 m thickness and 100 m length are distributed randomly within this reservoir. Using the connectivity master curves (Figures 3.11 and 3.12), estimate the mean connected sand fraction and associated error between two wells for net-to-gross ratio of 0.75.

3.2. Consider the reservoir data in problem 3.1. Plot the average sand fraction as a function of well spacing for net-to-gross ratio of 0.6, 0.7, 0.8 by inverting the scaling equations.

3.3. Consider an isotropic overlapping sandbody model in 2D made by circles representing sands of diameter l that are distributed randomly in the square medium of side size L. How do you compare the connectivity of this simple sandbody model with the connectivity of the overlapping sandbody model discussed in Section 3.1 where sandbodies represented by squares?

3.4. Consider a randomly oriented fracture network model in 2D made by fractures of size $l = 1$ distributed randomly in the square medium of side size $L = 200$. Use the two connectivity master curves ($\Im$ and $\Re$) shown in Figure 3.21 and construct the mean and standard deviation connectivity curves (P and Δ versus p) for this model.

3.5. Consider a horizontal–vertical fracture network model in 2D made by fractures of size l with the same density of horizontal and vertical fractures that are distributed randomly in the square medium of side size L. How do you compare the connectivity of this fracture network model with the connectivity of the randomly oriented fracture network model discussed in this chapter?

3.6. Instead of occupancy probability p used in lattice percolation, four measures, namely the dimensionless density (Eq. (3.2)), the average number of intersections per fracture I, the excluded area fraction of fractures and the probability density function (Eq. (3.3)), have been introduced for fracture systems in Section 3.2. Investigate the possible relationship between these measures.

References

Adler, P.M. and Thovert, J.F., *Fractures and Fracture Networks*. Kluwer Academic Publishers, London, 1999.

Adler, P.M. and Berkowitz, B., Effective medium analysis of random lattices. *Transport in Porous Media*, 40: 145–151, 2000.

Baker, D.R., Paul, G., Sreenivasan, S. *et al.*, Continuum percolation threshold for interpenetrating squares and cubes. *Physical Review E*, 66: 0546136, 2002.

Balberg, I., Universal percolation threshold limits in the continuum. *Physical Review B*, 31(6): 4053–4055, 1985.

Balberg, I., Binenbaum, N. and Anderson, C.H., Critical behaviour of the two dimensional sticks system. *Physical Review Letters B*, 51(18): 1605–1608, 1983.

Balberg, I., Anderson, C.H., Alexander, S. and Wagner, N., Excluded volume and its relation to the onset of percolation. *Physical Review B*, 30(7): 3933–3943, 1984.

Baron, J.V., Phase transition of the classical hard ellipse systems. *Journal of Chemical Physics*, 56(10): 4730, 1972.

Bear, J., Tsang, C.F. and de Marsily, G., *Flow and Contaminant Transport in Fractured Rock*. Academic Press, San Diego, CA, pp. 169–231, 1993.

Belayneh, M., Masihi, M., Matthäi, S.K. and King, P.R., Prediction of vein connectivity using the percolation approach: model test with field data. *Journal of Geophysical Engineering*, 3: 219–229, doi: 10.1088/1742-2132/3/3/003, 2006.

Berkowitz, B., Analysis of fracture network connectivity using percolation theory. *Mathematical Geology*, 27(4): 467–483, 1995.

Berkowitz, B., Characterizing flow and transport in fractured geological media: a review. *Advances in Water Resources*, 25: 861–884, doi: 10.1016/S0309-1708(02)00042-8, 2002.

Bour, O. and Davy, P., Connectivity of random fault networks following a power law fault length distribution. *Water Resources Research*, 33(7): 1567–1583, 1997.

Bour, O. and Davy, P., On the connectivity of three dimensional fault networks. *Water Resources Research*, 34(10): 2611–2622, 1998.

Charlaix, E., Percolation threshold of random array of discs: a numerical simulation. *Journal of Physics A: Mathematical and General*, 19(9): L533–L536, 1986.

Choi, H.S., Talbot, J., Tarjus, G. and Viot, P., Percolation and structural properties of particle deposits. *Physical Review E*, 51(2): 1353–1363, 1995.

Consiglio, R., Zouain, R.N.A., Baker, D.R., Paul, G. and Stanley, H.E., Symmetry of the continuum percolation threshold in systems of two different size objects. *Physica A*, 343: 343–347, doi: 10.1016/j.physa.2004.05.051, 2004.

Garboczi, E.J., Thorpe, M.F., de Vries, M.S. and Day, A.R., Universal conductivity curve for a plane containing random holes. *Physical Review A*, 43: 6473, 1991.

Garboczi, E.J., Snyder, K.A., Douglas, J.F. and Thorpe, M.F., Geometrical percolation threshold of overlapping ellipsoids. *Physical Review E*, 52(1): 819–827, 1995.

Gawlinski, E.T. and Stanley, H.E., Continuum percolation in two dimensions: Monte Carlo tests of scaling and universality for non-interacting discs. *Journal of Physics A: Mathematical and General*, 14: L291–L299, 1981.

Haldorsen, H.H., Brand, P.J. and Macdonald, C.J., Review of the stochastic nature of reservoirs. In: Edwards, S. and King, P.R. (eds.), *Mathematics in Oil Production*. Clarendon Press, Oxford, pp. 109–209, 1988.

Hunt, A.G. and Ewing, R., *Percolation Theory for Flow in Porous Media*. Lecture Notes in Physics, vol. 771. Springer, Berlin, 2005.

King, P.R., The connectivity and conductivity of overlapping sandbodies. In: Buller, A.T. *et al.* (eds.), *North Sea Oil and Gas Reservoirs — II*. Graham & Trotman, London, pp. 353–362, 1990.

Kirkpatrick, S., Percolation and conduction. *Reviews of Modern Physics*, 45: 574, 1973.

Knudby, C. and Carrera, J., On the relationship between indicators of geostatistical, flow and transport connectivity. *Advances in Water Resources*, 28: 405–421, 2005.

Knudby, C., Carrera, J., Bumgardner, J.D. *et al.*, Binary upscaling: the role of connectivity and a new formula. *Advances in Water Resources*, 29: 590–604, doi: 10.1016/j.advwatres.2005.07.002, 2006.

Koudine, N., Garcia, R.G., Thovert, J.F. and Adler, P.M., Permeability of three dimensional fracture networks. *Physical Review E*, 57(4): 4466–4479, 1998.

Larue, D.K. and Hovadik, J., Connectivity of channelized and other reservoirs. In: AAPG Annual Convention, Houston, TX, USA, 2006.

Lee, S.B. and Torquato, S., Monte Carlo study of correlated continuum percolation: universality and percolation thresholds. *Physical Review A*, 41: 5338–5344, 1990.

Lin, C.Y. and Hu, C.K., Universal finite size scaling functions for percolation on three dimensional lattices. *Physical Review E*, 58: 1521–1527, 1998.

Lorenz, C.D. and Ziff, R.M., Precise determination of the critical percolation threshold for the three dimensional Swiss cheese model using a growth algorithm. *Journal of Chemical Physics*, 114: 3659–3661, doi: 10.1063/1.1338506, 2001.

Masihi, M. and King, P.R., A correlated fracture network: modelling and percolation properties. *Water Resources Research Journal*, 43, doi: 10.1029/2006WR005331, 2007a.

Masihi, M. and King, P.R., Connectivity of spatially correlated fractures: simulation and field studies. Paper SPE 107132, Proceedings of the SPE Europec, London, UK, 11–14 June, 2007b.

Masihi, M., King, P.R. and Nurafza, P., Fast estimation of connectivity in fractured reservoirs using percolation theory. *SPE Journal*, 12(2): 167–178, 2007.

Matthäi, S.K. and Belayneh, M., Fluid flow partitioning between fractures and a permeable rock matrix. *Geophysical Research Letters*, 31: L07602.1–L07602.5, 2004.

Mukhopadhyay, S. and Sahimi, M., Calculation of the effective permeabilities of field-scale porous media. *Chemical Engineering Science*, 55: 4495–4513, doi: 10.1016/S0009-2509(00)00098-1, 2000.

Nurafza, P., King, P.R. and Masihi, M., Facies connectivity modelling: analysis and field study. Paper SPE 100333, Proceedings of the SPE Europec,Vienna, Austria, 12–15 June, 2006.

Quintanilla, J., Torquato, S. and Ziff, R.M., Efficient measurement of the percolation threshold for fully penetrable discs. *Journal of Physics A: Mathematical and General*, 33: L399–L407, doi: 10.1088/0305-4470/33/42/104, 2000.

Robinson, P.C., Connectivity of fracture systems-a percolation theory approach. *Journal of Physics A: Mathematical and General*, 16: 605–614, 1983.

Robinson, P.C., Numerical calculations of critical densities for lines and planes. *Journal of Physics A: Mathematical and General*, 17(14): 2823–2830, 1984.

Sadeghnejad, S., Masihi, M., King, P.R. *et al.*, Effect of anisotropy on the scaling of connectivity and conductivity in continuum percolation theory. *Physical Review E*, 81: 061119, doi: 10.1103/PhysRevE.81.061119, 2010.

Sadeghnejad, S., Masihi, M., Pishvaie, M. and King, P.R., Rock type connectivity estimation using percolation theory. *Mathematical Geosciences*, 45: 321–340, 2013a.

Sadeghnejad, S., Masihi, M. and King, P.R., Dependency of percolation critical exponents on the exponent of power law size distribution. *Physica A*, 392(24): 6189–6197, 2013b.

Sahimi, M., *Applications of Percolation Theory, Second Edition*. Taylor and Francis, London, 1994.

Saidi, A.M., *Reservoir Engineering of Fractured Reservoir: Fundamental and Practical Aspects*. Total Edition Presse, Paris, 1987.

Savik, P.N., Berre, I., Jakobsen, M. and Lien, M., A 3D computational study of effective medium methods applied to fractured media. *Transport in Porous Media*, 100: 115–142, 2013.

Stauffer, D. and Aharony, A., *Introduction to Percolation Theory*. Taylor and Francis, London, pp. 181, 1992.

Stell, G., Continuum theory of percolation. *Journal of Physics: Condensed Matter*, 8, A1–A17, doi: 10.1088/0953-8984/8/25A/002, 1996.

Taylor, W.L., Pollard, D. and Aydin, A., Fluid flow in discrete joint sets: field observations and numerical simulations. *Journal of Geophysical Research*, 104(B12): 28983–29006, 1999.

Torquato, S., *Random Heterogeneous Materials: Microstructure and Macroscopic Properties*. Interdisciplinary Applied Mathematics, vol. 16. Springer, New York, 2002.

Tóth, T.M. and Vass, I., Relationship between the geometric parameters of rock fractures, the size of percolation clusters and REV. *Mathematical Geosciences*, 43: 75–97, doi: 10.1007/s11004-010-9315-4, 2011.

Wen, R., Computer modelling of internal architecture in deep water reservoirs: a quantitative method to estimate connectivity and performance. In: AAPG Annual Convention, Houston, TX, USA, 2006.

Wilke, S., Guyon, E. and de Marsily, G., Water penetration through fractured rocks: test of a tri-dimensional percolation description. *Journal of the International Association for Mathematical Geology*, 17: 17–27, doi: 10.1007/BF01030364 1103/PhysRevA.38.2650, 1985.

Xia, W. and Thrope, M.F, Percolation properties of random ellipses. *Physical Review A*, 38: 2650–2656, 1988.

Chapter 4

The connectivity of overlapping sandbodies

The basic sandbody model and its features as developed in the framework of continuum percolation (Meester and Roy, 1996; Sahimi, 1994a) have been described in Chapter 3. It consisted of constant-size squares in 2D. Of course, in reality, the sandbodies may have different sizes, orientations and shapes. Also they are not spatially independent; they have spatial correlation. These changes to the simple model can alter the percolation behaviour and must be taken into account. This chapter explains how to incorporate these effects and what their impact is on percolation quantities.

4.1 Effects of sandbody size variation

We have so far assumed that the system is occupied by bodies of the same size. In reality, the sandbodies have a distribution of sizes which is related to the sedimentological environment in which they were deposited. Figure 4.1 shows an overlapping sandbody model at occupation fraction $p = 0.64$ made up by distributing squares with uniform size distribution in the range of 5–15 in a medium of size $L = 100$.

It should be recalled that for constant-size sands, there are only two dimensionless lengths involved in the scaling analysis, namely, the system size, L, and the correlation length, ξ. In the case of

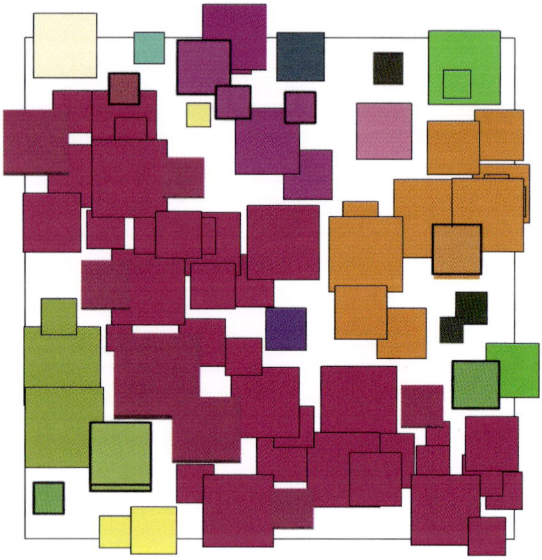

Figure 4.1. A realisation of the isotropic sandbody model of size $L = 100$ at occupation fraction $p = 0.64$ made by squares with uniform size distribution in the range of 5–15.

variable sandbody sizes, this changes. We shall consider two kinds of size distribution, continuous (e.g., uniform, Gaussian, lognormal or exponential) and discontinuous (where the bodies have two, or more, different sizes).

If the distribution of sizes is continuous, then it was suggested that a single effective size could be used (Roach, 1968). The hypothesis is that the connectivity of sandbodies of variable size is identical to the connectivity of constant-size sandbodies with this effective size. To investigate this, consider squares of variable size $a \times a$, sampled from a uniform distribution in the range $a \in [u, v]$ placed in an area A. The probability that a point chosen at random does *not* lie inside a body of area a_i^2 is

$$1 - a_i^2/A. \tag{4.1}$$

Assume that the sandbody positions follow a Poisson distribution; then the probability, $1 - p$, that a point does *not* lie within N such sandbodies becomes

$$1 - p = \prod_{i=1}^{N} \left(1 - \frac{a_i^2}{A}\right). \tag{4.2}$$

Taking logarithms and using the approximation $\ln(1 - x) \approx -x$ for $a_i^2 \ll A$ gives

$$\ln(1 - p) = \sum_{i=1}^{N} \ln\left(1 - \frac{a_i^2}{A}\right) = -\frac{1}{A}\sum_{i=1}^{N} a_i^2. \tag{4.3}$$

So the occupied fraction, or the probability that a point chosen at random *does* lie within at least one sandbody, is

$$p = 1 - e^{\frac{-N\langle a^2\rangle}{A}}, \tag{4.4}$$

where N is the total number of bodies and $\langle a^2\rangle$ is the average area of each body (angular brackets denote the mean) and $A = X \times X$ is the area of the region in which the bodies are located. It is the mean area of the sandbodies that is important. Hence, comparing this with the occupied fraction of a system made of constant-size sands gives an effective size for an equivalent sandbody:

$$l_{\text{eff}} = \sqrt{\langle a^2\rangle}. \tag{4.5}$$

We make the hypothesis that the connectivity of square sandbodies of variable size is identical to the connectivity of square bodies of the same size whose linear dimension is given by the square root of the average area of the variable size bodies. This modifies the system dimensionless size in the finite-size scaling (as used, e.g., in Eq. (2.10)) as

$$L = X/l_{\text{eff}}. \tag{4.6}$$

To show the applicability of this hypothesis, we should generate various realisations of overlapping sandbodies. To generate these, it is necessary to generate random variables with a given distribution. To do this, one can use the Marsaglia polar method, Mersenne Twister, the inversion method or more rigorous algorithms (Jansson, 1966;

Table 4.1. The definition and effective length (Eq. (4.5)) of different length distributions used to analyse the effects of sandbody size variation on percolation properties.

Distribution type	Distribution function $P(l)$	Parameters	Effective length, l_{eff}	Eq.
Uniform	$\begin{cases} \frac{1}{v-u}, & u \le l \le v \\ 0, & \text{otherwise.} \end{cases}$	$\bar{l}, f$	$\sqrt{\frac{u^2+uv+v^2}{3}}$	(4.7)
Gaussian	$\frac{1}{\sqrt{2\pi}\sigma} e^{\frac{-1}{2}\left(\frac{l-\bar{l}}{\sigma}\right)^2}$	$\bar{l}, \sigma$	$\sqrt{\bar{l}^2+\sigma^2}$	(4.8)
Lognormal	$\frac{1}{\sqrt{2\pi}\sigma_{\ln l}} e^{\frac{-1}{2}\left(\frac{\ln l-\ln\bar{l}}{\sigma_{\ln l}}\right)^2}$	$\bar{l}, \sigma$	$\sqrt{l^2 e^{2\sigma^2}}$	(4.9)
Exponential	$\frac{1}{\bar{l}} e^{-l/\bar{l}}$	$\bar{l}$	$\sqrt{2\bar{l}^2}$	(4.10)
Power law	$cl^{-\alpha}$ for $l_{\min} \le 1 \le l_{\max}$	$l_{\min}, l_{\max}, c$	$\sqrt{\frac{c}{3-\alpha}(l_{\max}^{3-\alpha} - l_{\min}^{3-\alpha})}$	(4.11)

Fisz, 1963; Matsumoto and Nishimura, 1998; Jensen *et al.*, 2000; Bassett *et al.*, 2000). Moreover, the effective length depends on the type of distribution and can be determined by using Eq. (4.5). Table 4.1 summarises the definition and the calculated effective length of uniform, Gaussian, lognormal, exponential and power law continuous-length distributions.

As an example, we demonstrate this hypothesis for the case of a uniform distribution (squares have sides size $a \in [u, v]$). Then the effective length for the sandbodies is given by Eq. (4.7) (Table 4.1). Figure 4.2 compares two realisations of squares distributed randomly in a square region of side size 100. One has squares with a uniform length distribution in the range of 5–15; the other has a constant effective length $l_{eff} = 10.4$. In both cases, the net-to-gross ratio is $p = 0.56$.

Using the hypothesis that the effective sandbody size is controlling the connectivity allows us to use the finite-size-scaling laws (Eqs. (2.11) and (2.14)) to estimate the connectivity. For example, the mean and standard deviation of connectivity for a system with uniform size distribution in the range 0.5–1.5 with the effective length

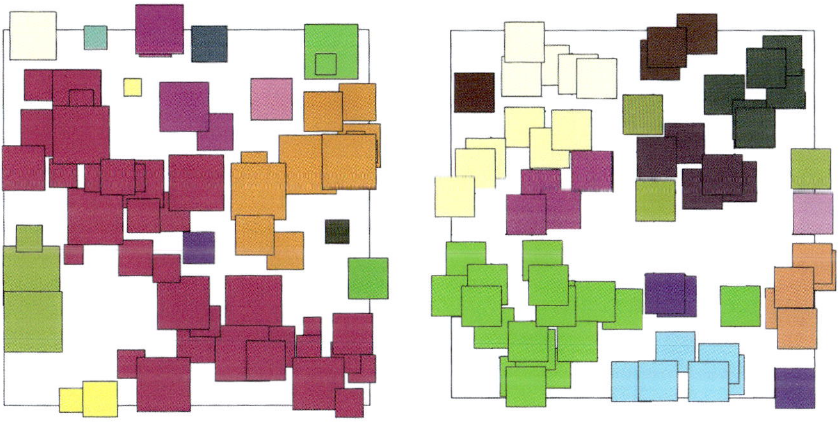

(a) (b)

Figure 4.2. Two realisations of the isotropic sandbody model at net-to-gross ratio of $p = 0.56$: (a) one made by variable size sandbodies and (b) the other made by sandbodies with constant effective length $l_{\text{eff}} = 10.4$.

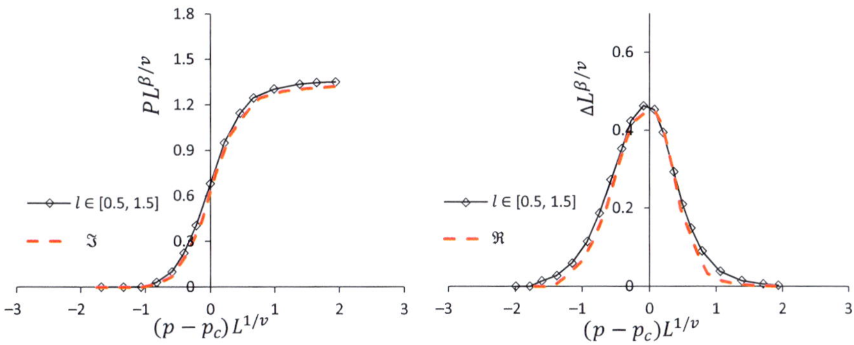

(a) (b)

Figure 4.3. Illustration of finite-size scaling of (a) the mean and (b) the standard deviation in the connected sand fraction results for variable sizes (solid line) and sand size l_{eff} (dotted line) in a sandbody model with uniform size distribution of sands in the range 0.5–1.5.

$l_{\text{eff}} = 1.04$ with a dimensionless size $L_x = 24/1.04 = 23.08$ are shown in Figure 4.3.

This is a consistent with the hypothesis that the percolation behaviour for systems of variable sand size can be modelled as though they had a single effective length. This enables us to use the same

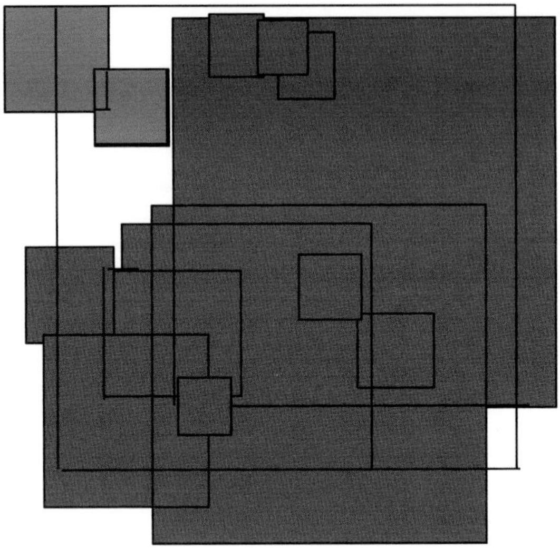

Figure 4.4. A realisation of sandbody model with a power law size distribution for square objects $\alpha = 2.3, p = 0.75, P = 0.89$.

isotropic master curves for connectivity (Figures 3.11 and 3.12) for variable size sands.

This analysis can be applied to different continuous distributions; clearly the effective length depends on the distribution as summarised in Table 4.1.

One problem with this approach is the power law distribution ($P(l) \sim l^{\alpha}$) where a few very large bodies completely dominate the connectivity of the system. This alters the nature of the finite-size-scaling law, and subsequently universality cannot be relied on. This is illustrated in Figure 4.4 where the power law exponent $\alpha = 2.3$. the largest square length, $l_{\max}$, is similar to the system size, L.

We generate a number of realisations of systems with a power law length distribution and measure the occupancy probability and connectivity. To check the dependency of the threshold on the power law exponent, we plot $\tilde{p}_c(L)$ as a function of system size, L, as shown in Figure 4.5. It can be seen that the change is not very significant. The error may be due to truncation of the size of the objects by the boundaries. By extrapolating these plots, it is found that the infinite threshold is unchanged.

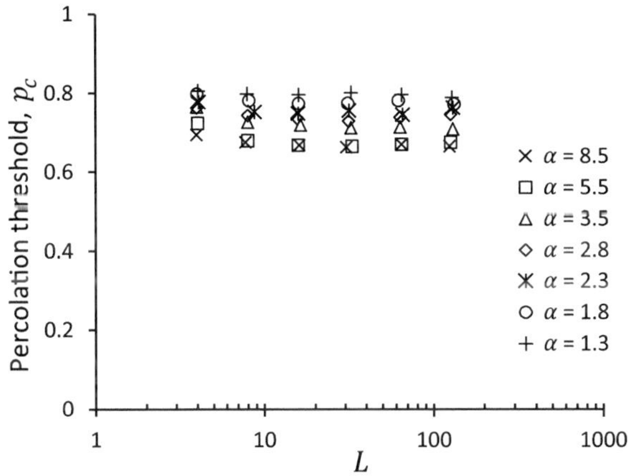

Figure 4.5. Variation of the percolation threshold, $\tilde{p}_c$, as a function of system size, L, for various length exponents, α, for overlapping squares.

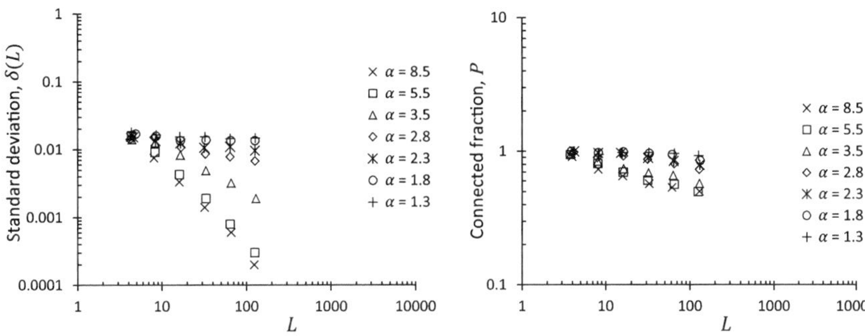

Figure 4.6. Variation of the (a) $\delta(L)$ and (b) P as a function of system size, L, for various length exponents of square objects.

We use the scaling (Eq. (2.28)) of the spread $\delta(L)$ of $\tilde{p}_c(L)$ to determine the critical exponent ν at each value of the power law exponent α. The scaling (Eq. (2.31)) at the threshold for the connectivity results can also be used to estimate the critical exponents (β/ν) (Figure 4.6).

As can be seen in Figure 4.7, the critical exponents do not have their standard values; instead they depend on the length exponent, α.

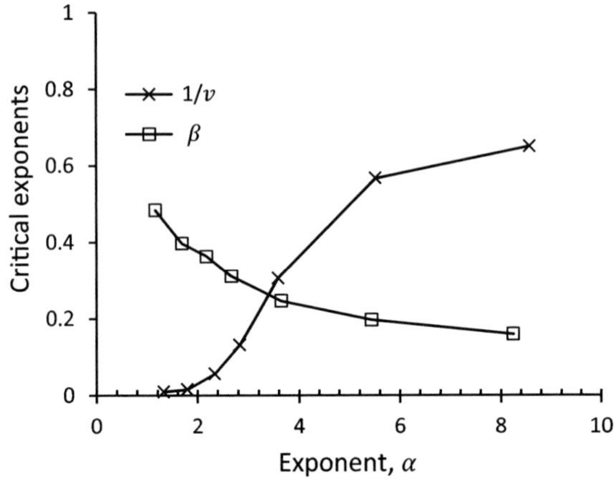

Figure 4.7. Dependency of the critical exponents, through the percolation analysis of the system made by squares with a power law distribution of size, on the exponent α (Sadeghnejad *et al.*, 2013b).

We can then use the usual finite-size transformations with these exponents and the effective length defined in Eqs. (4.5) and (4.6). We compare the results of these transformations on a system made of squares with a power law length exponent $\alpha = 2.3$ by using, first, the universal exponents (Table 2.2) and, second, the modified exponents depending on the power law exponent (Figure 4.7). The results of these two cases are shown in Figure 4.8.

Power law length distributions are commonly used to describe fracture lengths, so we postpone further discussion to Chapter 5.

Finally, the distribution of the sandbody sizes can be discontinuous. For example, there are two different sizes of bodies, u and v, described by

$$P(\ell) = \rho\delta(\ell - u) + (1 - \rho)\delta(\ell - v), \tag{4.12a}$$

where δ is the Dirac delta function and ρ is the fraction of sandbodies of size u. In this case, the effective length, l_{eff}, as defined previously, is

$$l_{\text{eff}} = \sqrt{\rho u^2 + (1 - \rho)v^2}. \tag{4.12b}$$

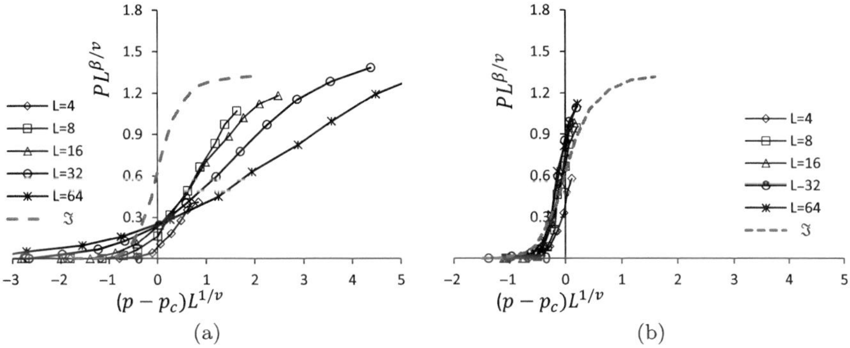

Figure 4.8. Finite-size scaling for squares with a power law distribution of lengths ($\alpha = 2.3$) using (a) the literature values ($\beta = 0.139$, $\nu = 1.33$) for the critical exponents (Table 2.2) and (b) modified values ($\beta = 0.35$, $\nu = 14.3$) from Figure 4.7.

However, in this case, this length is not found in either population of sandbodies, so it may not be a good representative value to use. There are basically two systems: one made up of the small bodies of size u, and other made up of the large bodies of size v. If the difference between u and v is large then most of the sandbody area is taken up by the large bodies and the small bodies are almost irrelevant, except that they slightly enhance the effective size of the large bodies by one correlation length of the smaller bodies as shown in Figure 4.9 (Phani and Dhar, 1984).

This leads to an effective shift in the threshold because the system behaves as though it were made up just of bodies of the size $v + \xi_u$ where ξ_u is the correlation length of the small bodies. If the small bodies are very much smaller than the large ones, the correlation length of small bodies, ξ_u, will only be significant if the area fraction of the small bodies is comparable to the percolation threshold, p_c. Hence, we expect that the connectivity of this mixed size system to be the same as the connectivity of the large bodies at the corrected area fraction defined by

$$p_\nu = 1 - e^{-(1-\rho)N\nu^2/A}. \tag{4.13}$$

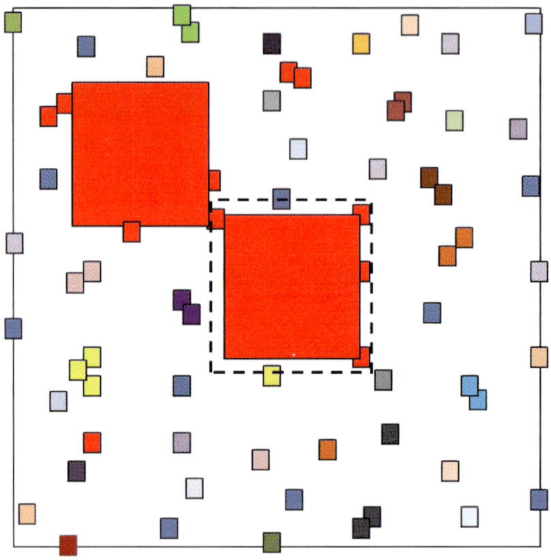

Figure 4.9. Illustration of sandbodies with two very different sizes in which large sandbody size is enhanced by one correlation length of the smaller ones.

A comparison of the mean connectivity determined for a mixed system with sand sizes 2 and 20 in the ratio 9:1 and the connectivity of sands of size 20 but with the corrected area fraction (Eq. (4.13)) is shown in Figure 4.10.

4.2 Effects of anisotropy

Realistic sandbodies are neither simple squares in 2D (or cubes in 3D) nor aligned. Consequently, their percolation behaviour is rarely isotropic. As described in Chapter 2, anisotropy leads to "easy" connectivity in the short direction and a "difficult" connectivity along the long axis of the system (Monetti and Albano, 1991; Marrink and Knackstedt, 1999; Masihi *et al.*, 2006). We use the description of finite-size effects in anisotropic lattice percolation from that chapter as the basis of the analysis here.

Instead of squares, we now use rectangles to represent the sandbodies. We set the size of the external region (reservoir) $X \times Y$. In this, we place a number of rectangular bodies of size $a \times b$ independently and uniformly. We then define the two dimensionless

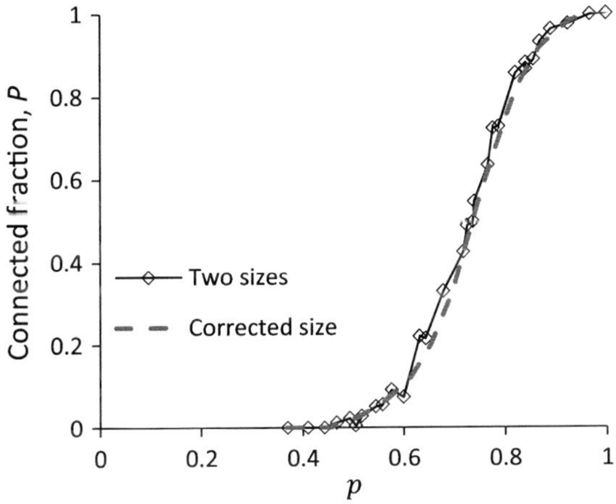

Figure 4.10. Predicted connectivity of overlapping sandbodies consists of sands with two sizes 2 and 20 in the ratio 9:1 by using the effective area fraction (Eq. (4.13)).

lengths:

$$L_x = X/a, \tag{4.14}$$

$$L_y = Y/b. \tag{4.15}$$

We define the aspect ratio, ω, representing the anisotropy as

$$\omega = L_x/L_y. \tag{4.16}$$

It is important to remember that it is the ratio of the *dimensionless* lengths that defines the aspect ratio. In reality, sandbodies can be hundreds of metres long and tens of metres thick giving an apparent anisotropy of 10 or more. However, the reservoir itself has a similar anisotropy, so in terms of the connectivity and percolation properties, the system is much more isotropic.

The other details of this model are the same as those described for the basic sandbody model in Chapter 3 except that the grid used to generate the Poisson process is now of size $a \times b$. Since the system is no longer isotropic, the connectivity in the x direction, P_x, is the area fraction of sandbodies belonging to clusters that connect the system in the x direction. Similarly, the connectivity in the y

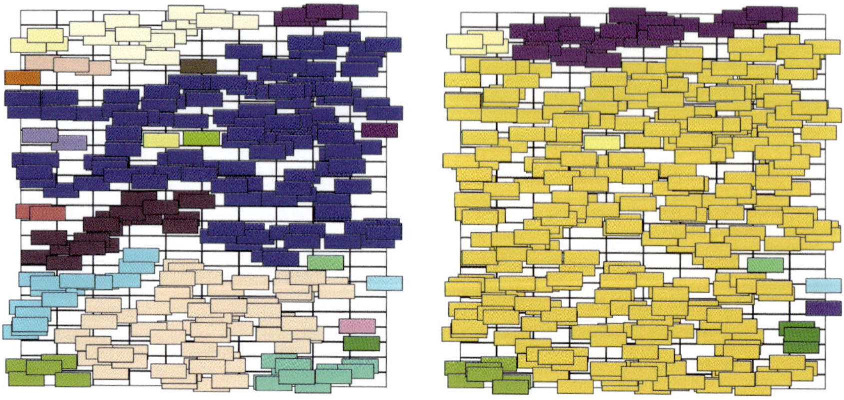

(a) (b)

Figure 4.11. Two realisations of anisotropic sandbodies model in which sands are represented by rectangles (a) below the threshold ($N = 280$, $p = 0.659$, $P_x = 0.449$, $P_y = 0$) and (b) just above the threshold for horizontal connectivity ($N = 370$, $p = 0.749$, $P_x = 0.868$, $P_y = 0$).

direction, P_y, is the area fraction of sandbodies belonging to clusters that connect the system in the y direction. We usually think of x as being the horizontal direction and y is vertical. Figure 4.11 shows two realisations of anisotropic sandbody models.

As in Chapter 2, we conjecture that there is a finite-size shift in the apparent threshold of the form $\tilde{p}_c^x - p_c^\infty = \Lambda(\omega)L_x^{-1/\nu}$ (Eq. (2.35)). We plot the apparent threshold against $L_x^{-1/\nu}$ in Figure 4.12, which demonstrates that this is the case.

Using the same arguments as before, we assume that the constant of proportionality, Λ, has the functional form $\Lambda(\omega) = c(\omega^{1/\nu} - 1)$ (Eq. (2.38)). In Figure 4.13, we plot Λ against $\omega^{1/\nu} - 1$ which gives $c \approx 0.5$.

We carry out the same data collapse for the mean and standard deviation in connectivity as before and find two curves, one for the x and one for the y connectivity (Figure 4.14).

If instead of the infinite threshold in these scalings we use the value corrected for finite size and anisotropy, we get collapse onto a single master curve as shown in Figure 4.15.

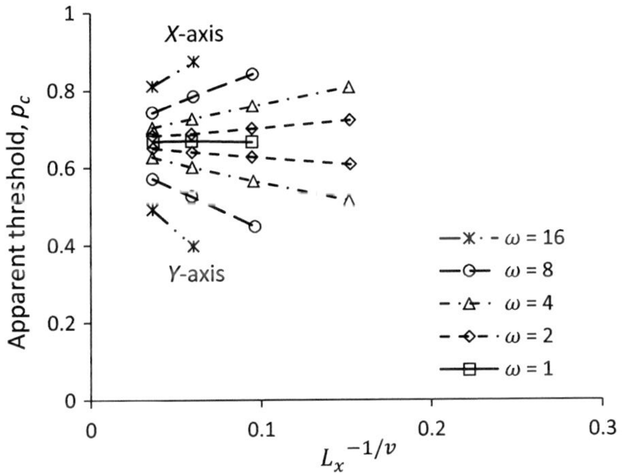

Figure 4.12. The plot of the apparent threshold against system size $L^{-1/\nu}$ shows the symmetry property (Eq. (2.35)) for the constant of proportionality Λ.

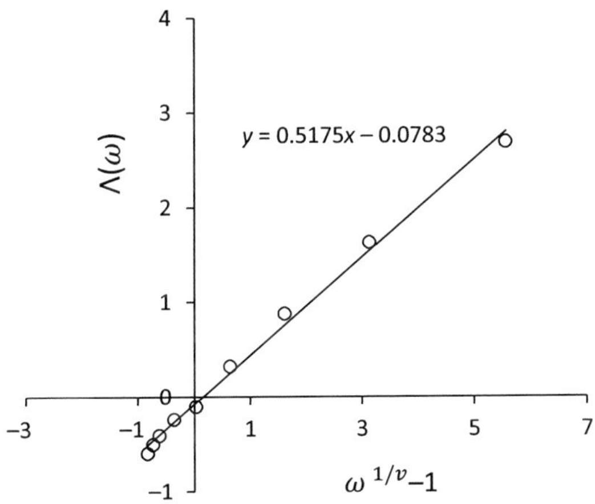

Figure 4.13. Plot of the constant of proportionality Λ against aspect ratio $\omega^{1/\nu} - 1$ indicating the functional form given in Eq. (2.38).

As a practical example of how to use these scaling relations, we consider a 2D cross section of a reservoir with thickness 10 m. The sandbodies have length 50 m and thickness 1 m. We consider a range of well spacings from 500 to 6,000 m. At a well spacing of 500 m, both dimensionless lengths are equal to 10 ($L_x = 500/50$ and $L_y = 10/1$).

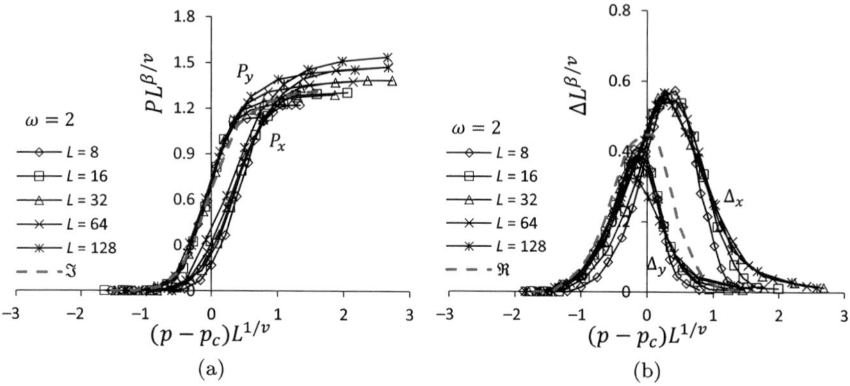

Figure 4.14. Illustration of the usual finite-size scaling (at a fixed aspect ratio $\omega = 2$) by using Eqs. (2.11) and (2.14) which shows two sets of data collapse in two directions. (a) Mean and (b) standard deviation.

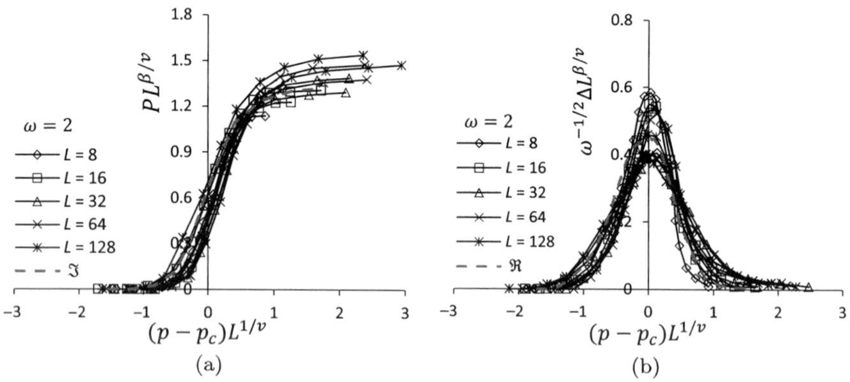

Figure 4.15. Illustration of the corrected finite-size scaling (at a fixed aspect ratio $\omega = 2$) by using Eqs. (2.39) and (2.40) which shows data collapse. (a) Mean and (b) standard deviation.

So the system is isotropic from a percolation point of view. For a 1,000 m well spacing, the dimensionless lengths are $L_x = 20$ and $L_y = 10$, which gives an aspect ratio of $\omega = 2$.

We can use the connectivity master curves finite-size scaling to plot the connected sand fraction as a function of well spacing at

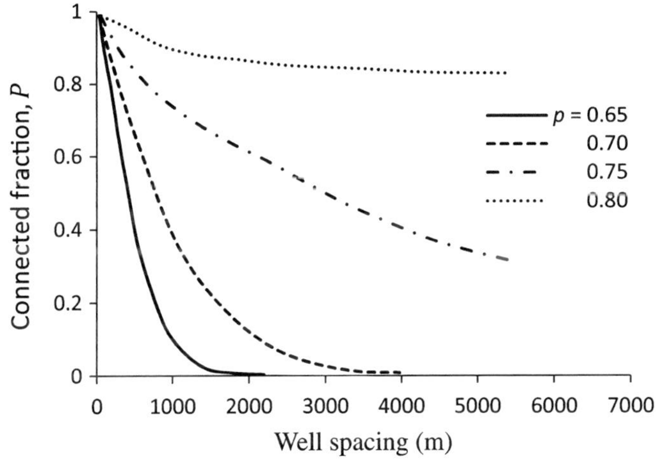

Figure 4.16. An example for the variation of the mean connected sand fraction as a function of well spacing at four different net-to-gross ratios.

different net-to-gross ratios such as $p = 0.65, 0.7, 0.75$ and 0.8. The result is shown in Figure 4.16.

4.3 Effects of orientation disorder

So far, we have assumed that all the sandbodies are aligned. In reality, their orientation is related to the depositional process and the sedimentary environment. The effect of orientation disorder in sandbodies is more relevant to 3D (we will describe this later) and to areal models in 2D. This is only important for continuum percolation, as in lattices, the orientation is fixed by the grid.

To account for this disorder, we assume that each sandbody has an orientation θ_i between θ_1 and θ_2 where $i \in \{1, 2, \ldots, n\}$ and n is the total number of sandbodies in the system. Figure 4.17 shows a realisation of square bodies with a uniform distribution of orientation in a system of size $L = 20$ near the threshold.

This orientational disorder may enhance the connectivity, particularly for the systems with long, thin sandbodies. In percolation terminology, this means that the percolation threshold has been reduced. Moreover, the orientational disorder makes the bodies

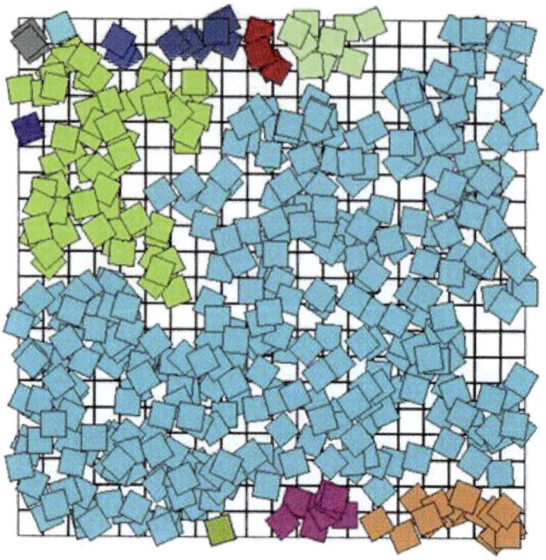

Figure 4.17. A realisation showing uniform sandbody orientation distribution between $-\pi/2$ and $+\pi/2$ where $L = 20$, $N = 500$, $p = 0.683$ and $P = 0.744$.

appear to be larger (as can be seen in Figure 4.17), and this alters the finite-size effects.

To consider the effect on the threshold, we use a conjecture about the universality of percolation thresholds (Scher and Zallen, 1970; Balberg *et al.*, 1983, 1984; Balberg, 1985; Sreenivasan *et al.*, 2003). The conjecture says that at the percolation threshold, there is a universal value of the excluded area, A_{ex}, defined as the area in which the centre of another body must not lie to avoid intersection (Figure 4.18).

We look at how the excluded area varies with the orientation of bodies. From Figure 4.18, the excluded area for simple convex *aligned bodies* like rectangles of size a and b is

$$A_{\text{ex}} = 4ab, \tag{4.17}$$

while for *non-aligned* rectangles with relative angle θ, the excluded area becomes

$$A_{\text{ex}} = 2ab(1 + \cos\theta) + (a^2 + b^2)\sin\theta. \tag{4.18}$$

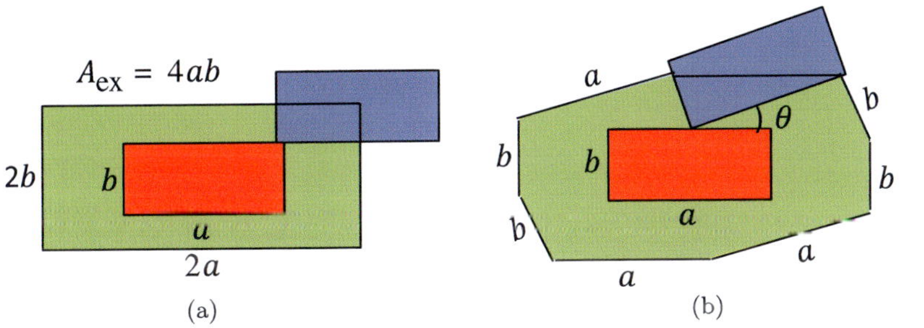

Figure 4.18. The excluded area for (a) aligned and (b) non-aligned rectangles is the green area surrounding the red rectangle.

When the bodies have different sizes or orientations, we need the average excluded area. For example, the average excluded area for rectangular bodies of size $a \times b$ with a uniform orientation distribution in the range $\theta \in [-\theta_0, +\theta_0]$ is

$$\langle A_{\text{ex}} \rangle = 2ab\left(1 + \frac{\sin^2 \theta_0}{\theta_0^2}\right) + \left(\frac{a^2 + b^2}{2\theta_0^2}\right)(2\theta_0 - \sin 2\theta_0). \qquad (4.19)$$

The excluded area fraction, ρ_{ex}, can be defined as the fraction of excluded area to the total area of the system, A,

$$\rho_{\text{ex}} = \frac{A_{\text{ex}}}{A}, \qquad (4.20)$$

which for sandbodies placed randomly in space gives an equivalent occupancy probability (from Eq. (4.4)) of

$$p = 1 - e^{-\rho_{\text{ex}}/4}. \qquad (4.21)$$

The conjecture about the universality of percolation thresholds states that the excluded area fraction, ρ_{ex}, takes a universal value, $\rho_{\text{ex}} = 4.41$, for all shapes at the percolation threshold. Equation (4.21) can be then rewritten as a function of the actual area fraction, ρ; for example, the critical net-to-gross ratio becomes

$$p_c = 1 - e^{-\rho_c}, \quad \text{where} \quad \frac{\rho_c}{ab} \approx \frac{4.41}{\langle A_{\text{ex}} \rangle}. \qquad (4.22)$$

This gives the critical net-to-gross ratio for square sandbodies equal to $p = 1 - e^{-\rho_{\text{ex}}}/4 \approx 0.668$, which is about right. Now with this conjecture, we can determine the shift in percolation threshold by using Eq. (4.22) as a function of the orientation disorder. For example, the threshold for rectangles with an aspect ratio ω and a uniform angular spread, from Eq. (4.19), becomes

$$p_c = 1 - \exp\left\{-\frac{4.41}{2\left(1+\frac{\sin^2\theta_0}{\theta_0^2}\right)+\frac{1+\omega^2}{2\omega\theta_0^2}(2\theta_0 - \sin 2\theta_0)}\right\}. \quad (4.23)$$

From this, we can find the modified infinite threshold for different distributions of the orientation. For an isotropic system made by randomly oriented squares in the range -90 to 90 ($\omega = 1$), the corrected threshold is $p_c = 0.661$.

Figure 4.19 shows the shift in threshold based on Eq. (4.23) as a function of the aspect ratio for different amounts of orientational disorder.

To test the corrected infinite threshold from Eq. (4.23), we take an overlapping sandbody model consisting of rectangles with a fixed aspect ratio $\omega = b/a = 2$ and angular dispersion $\theta_0 = \pi/8$ (Figure 4.20).

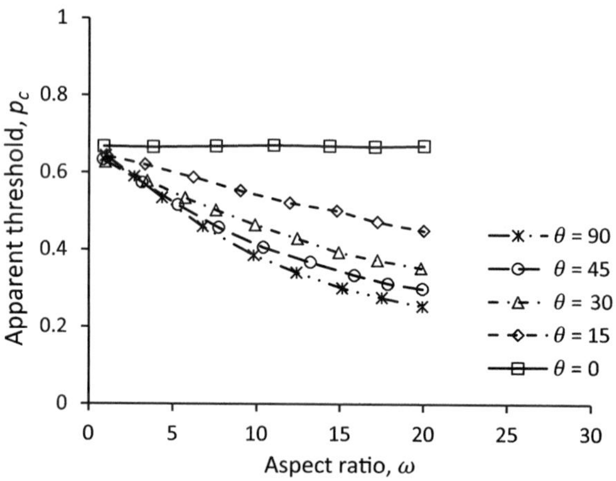

Figure 4.19. The shift in the threshold due to the orientational disorder of rectangular sandbodies with an aspect ratio ω and a uniform distribution in angular dispersion.

Figure 4.20. A realisation of an overlapping sandbody model of size $L_x = 20$, $\omega = 2$, $N = 210$, $p = 0.619$ and $P_y = 0.316$ where sandbodies have uniform orientation distribution between $-\pi/8$ and $+\pi/8$.

We then estimate the threshold value numerically for various system sizes. Then we use the finite-size scale dependency of the threshold (Eq. (2.27)) to extrapolate to the infinite system to estimate the infinite threshold. From the numerical results, the extrapolated value from the finite-size results gives $p_c = 0.619 \pm 0.004$. The corrected infinite threshold calculated from Eq. (4.23) for $\omega = 2$ and $\theta_0 = \pi/8$ is $p_c = 0.622$ which is in good agreement with the numerical results.

There are also finite-size effects for the orientation disorder. We use the concept of an effective length as before. Consider a rectangular sandbody that has angle θ as shown in Figure 4.21.

The maximum dimensions of the sandbody l_x and l_y are

$$\ell_x = a\cos\theta + b\sin|\theta|, \quad \ell_y = a\sin|\theta| + b\cos\theta. \tag{4.24}$$

The effective body size in each direction is obtained by taking averages of all of these maximal lengths. For example, for a uniform orientation distribution in the range $\theta \in [-\theta_0, \theta_0]$, the effective body

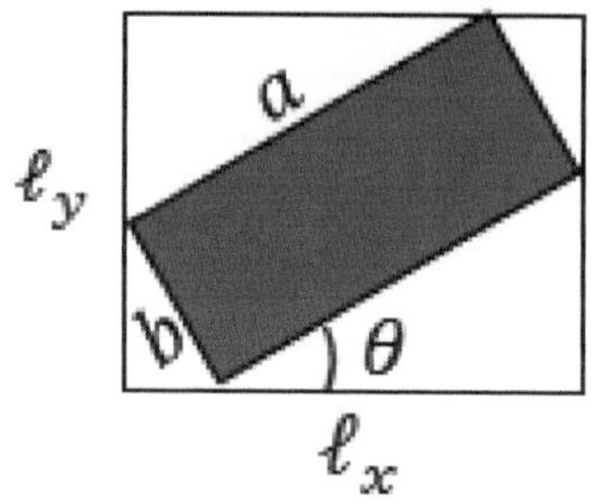

Figure 4.21. Illustration of the effective size of a sandbody of angle θ in two directions.

size becomes

$$\langle \ell_x \rangle = \frac{a \sin\theta_0 + b - b\cos\theta_0}{\theta_0}, \quad \langle \ell_y \rangle = \frac{a - a\cos\theta_0 + b\sin\theta_0}{\theta_0}. \quad (4.25)$$

The effective size of the system is now defined by two dimensionless lengths $L_x = X/\langle \ell_x \rangle$ and $L_y = Y/\langle \ell_y \rangle$ in the x and y directions. The ratio of these two effective system sizes due to the orientational disorder results in an anisotropy that is described by the aspect ratio, ω, in Eq. (4.16).

Now this aspect ratio, the corrected threshold and the effective system size can be used with the finite-size-scaling transformations described in Chapter 2. A comparison of the results of the mean horizontal connectivity for squares of size 1×1 in a region of size 16×16 for two different uniform angular distributions, one of $\pm 15°$ and the other of $\pm 30°$, is shown in Figure 4.22. For such a system, the average lengths along the x and y directions are $L = 1.118$ (for $\pm 15°$) and 1.21 (for $\pm 30°$). The corrected infinite threshold is $p_c = 0.641$ (and 0.626 for $\pm 30°$). As the aspect ratio is equal to 1 ($\omega = 1$), there is no effective finite-size shift in this case. These results are consistent with these hypotheses, and so we can use the isotropic connectivity master curves as before.

4.4 Other effects in 2D

In addition to the sand size and orientation distributions, there are other issues that need to be consider for making predictions in practical problems. These include sand shapes, sands correlations or well patterns. We shall discuss some of these effects in this section.

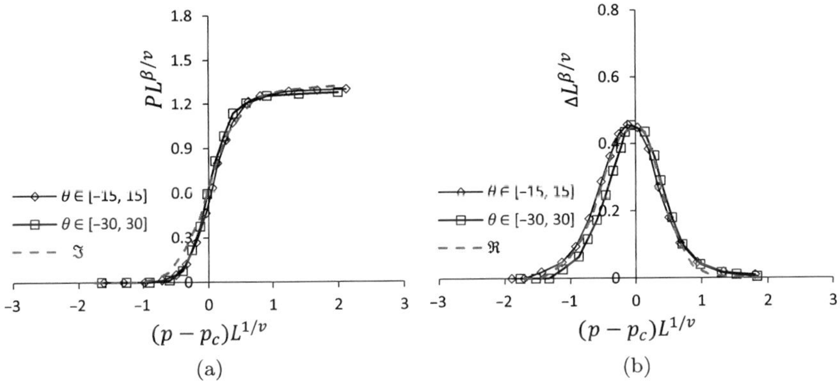

Figure 4.22. Finite-size scaling of the connectivity results for sandbodies with uniform orientation between $-15°$ and $15°$ and between $-30°$ and $30°$ compared with the universal connectivity curves in 2D. (a) Mean and (b) standard deviation.

4.4.1 *Sandbody shapes*

The sandbodies can be represented by any shape such as squares, circles or ellipses in 2D and cubes or ellipsoids in 3D (Balberg and Binenbaum, 1983; Charlaix, 1986; Xia and Thrope, 1988; King, 1990; Lee and Torquato, 1990; Quintanilla *et al.*, 2000; Garboczi *et al.*, 1995; Baker *et al.*, 2002; Watanabe *et al.*, 2004; Nurafza *et al.*, 2006; Sadeghnejad *et al.*, 2012, 2013a). The main effect of sandbody shape is to change the percolation threshold and in some cases to make the system anisotropic. However, the critical exponents do not change, and the universal values can be used. The numerical values for the thresholds of various sandbody shapes in 2D and 3D were given in Table 3.1. In general, for simple convex shapes, the threshold values only depend on the dimension of the system.

Finite-size scaling is unaffected by the shape of the objects. Also, fortuitously, the master curve is almost identical regardless of the shape of the bodies, Figure 4.23.

4.4.2 *Spatial correlation*

So far, we have assumed that there is no spatial correlation in the locations of the objects. However, for many natural systems,

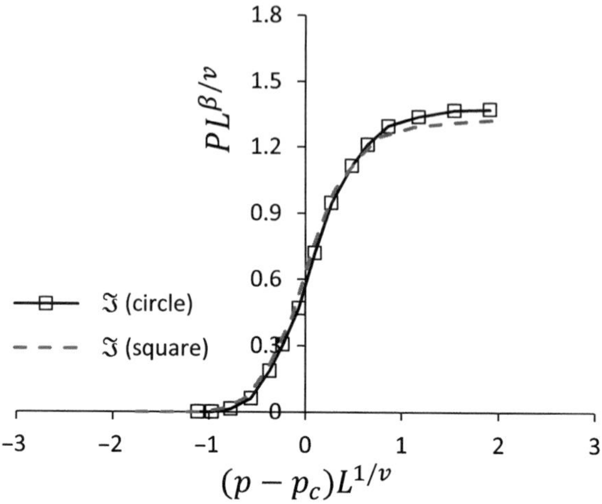

Figure 4.23. A comparison of the isotropic connectivity master curves for the bodies represented by circles and squares.

there is a correlation between the positions of the objects. For example, lateral stacking of channels can lead to spatial correlation (Mayall *et al.*, 2006). Figure 4.24 shows stacked channel sands and a representation by spatially correlated rectangles. Earlier in Chapter 1 (Figure 1.3), we showed the strong correlation observed in fluvial point bar deposits (Kuuskraa *et al.*, 1997).

The correlations observed in geological systems (Koltermann and Gorelick, 1996; Kuuskraa *et al.*, 1997; de Marsily *et al.*, 2005; Ronayne and Gorelick, 2006) may be studied using correlated percolation (Lee and Torquato, 1990; Prakash *et al.*, 1992; Ioannidis *et al.*, 1993; Sahimi and Mukhopadhyay, 1996; Masihi and King, 2007).

This correlation can be either of finite or very long range. Harter (2005) showed that the percolation threshold in Markov chain random fields with short-range correlation decreases as the correlation scale increases but the critical exponents are unaffected. However, for systems with long-range correlations, the percolation parameters can be changed significantly (Prakash *et al.*, 1992; Schmittbuhl *et al.*, 1993; Sahimi, 1994b, 1995; Makse *et al.*, 1996; Bour and Davy, 1997; Moukarzel, 2006; Masihi and King, 2007; Flukiger *et al.*, 2008). Lee

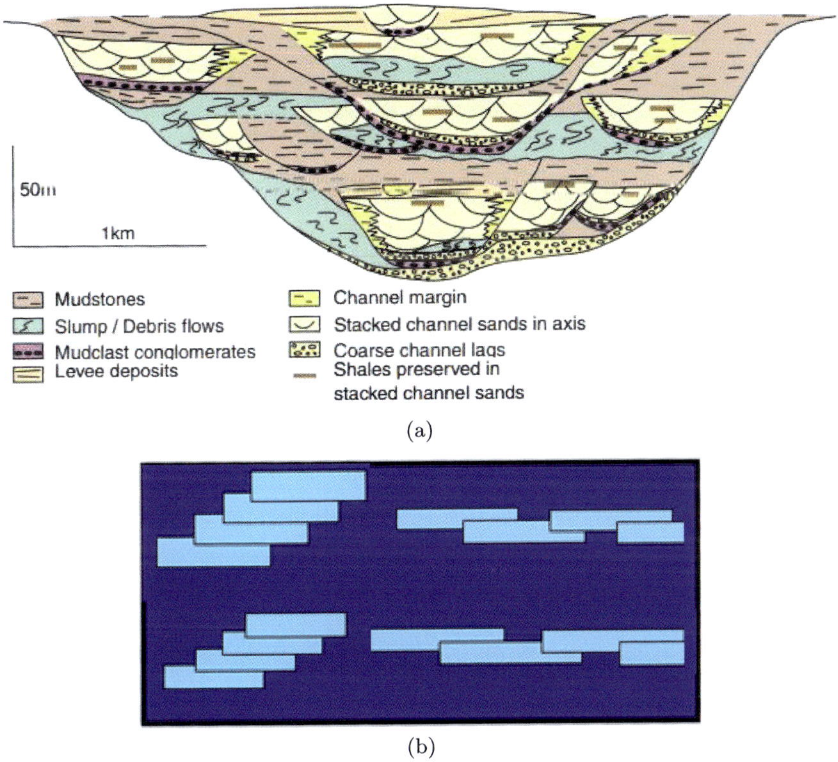

Figure 4.24. The sandbodies stacking patterns: (a) stacked channel sands in axis (Mayall *et al.*, 2006) and (b) a schematic of stacking patterns of sandbodies using rectangles.

and Torquato (1990) worked on correlated continuum percolation and found the universal curves and percolation thresholds. The main challenge is that the right model for long-range correlation is very dependent on the system under consideration and so has to be treated in a case-by-case basis.

4.4.3 *Well patterns*

So far, we have considered the connectivity between two lines located at the opposite sides of the system in 2D. These lines can be thought of wells at either side of a vertical cross section. In real reservoirs, wells are lines in 3D. They can take a variety of trajectories depending

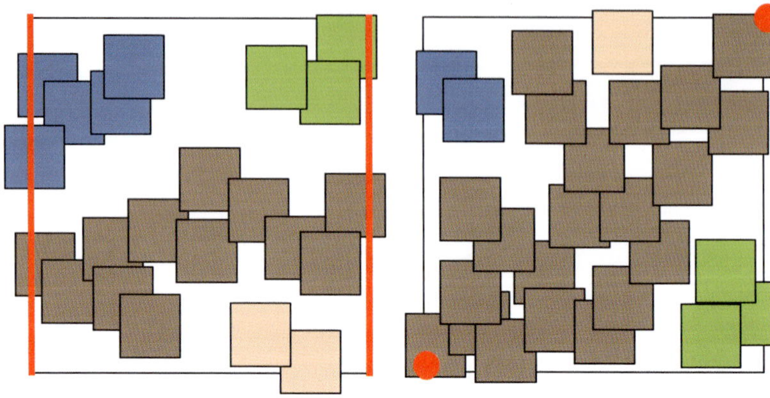

(a) (b)

Figure 4.25. The geometrical differences in the (a) L2L and (b) P2P connectivities. The connected cluster is in brown.

on the well pattern used. In a 2D areal map, wells could be considered to be points. Moreover, there may be more than two wells in the reservoir. We need to investigate how to modify the previous results for these more general well configurations.

First we establish a 2D continuum percolation model in which the connectivity is achieved through point-to-point (P2P) connection instead of the conventional line-to-line (L2L) connection. Figure 4.25 illustrates the geometrical difference between L2L and P2P connectivities. It is clear that the connectivity is more difficult for P2P connection than in the L2L case. So we might anticipate that the threshold is changed.

The same algorithm is used to generate ensembles of realisations and to calculate the percolation quantities. The only difference is that the spanning cluster is checked between two points at the two corners rather than the two opposite sides of the system. As in other percolation problems, we need to determine the percolation threshold, the critical exponents and appropriate scaling laws. Sadeghnejad and Masihi (2016) suggested that the threshold be defined as the lowest occupation probability over all realisations for which the system is connected. They observed that the previous definition (where the occupancy probability at which half of the realisations

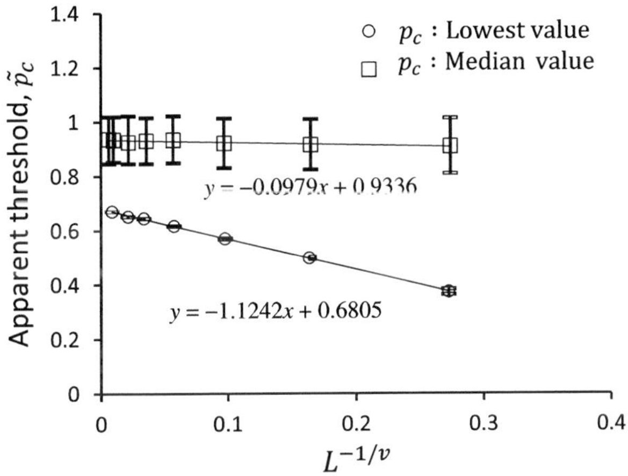

Figure 4.26. Variation of the threshold as a function of system size. The error bars show the variation observed over all realisations at each system size.

percolate) is not appropriate in this model. The reason is that the probability distribution for threshold over all realisations is no longer Gaussian as in the previous L2L problem. Figure 4.26 compares the percolation threshold variations against systems size for these two definitions.

From the finite-size dependency of the apparent threshold, $\tilde{p}_c(L)$ (Eq. (2.27)), the slope gives the percolation threshold $p_c = 0.680$ which is in agreement with the threshold value for square sandbodies found before (Table 3.1). However, the critical exponents may change; for example, in the case of the correlation length exponent, the typical size of the largest cluster levels off before the threshold because it is constrained by the boundary of the system (Figure 4.27). As the correlation length above the threshold represents the size of holes in the largest cluster, we expect a similar correlation length exponent in both the P2P and the L2L cases. However, the other critical exponents such as connectivity exponent may not be the same (Tavagh-Mohammadi *et al.*, 2016).

Hence, the appropriate exponents can be estimated from finite-size effects as before (Eqs. (2.1) and (2.31)) and then used to predict the connectivity in these systems (see Sadeghnejad and Masihi (2016) for more details).

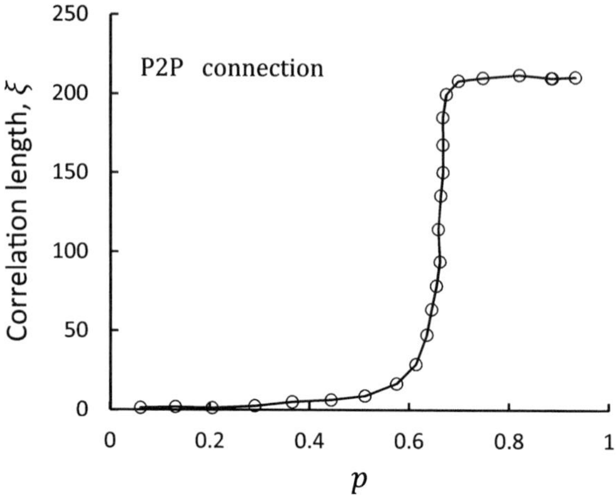

Figure 4.27. Variation of the correlation length by measuring the radius of gyration (Eq. (2.30)) of the largest cluster against occupancy.

4.5 Analysis of connectivity in 3D

So far, we have only considered 2D systems. We expect many of the rules to be directly transferable to 3D and so a similar approach can be used. From a practical point of view, the computational time in 3D is much higher than in 2D (Lin and Hu, 1998; Tavagh-Mohammadi *et al.*, 2016; Sadeghnejad and Masihi, 2017). All we need are the new exponents, the new threshold value and the new universal scaling curves for the connectivity and its associated uncertainty. One complication is the configuration of the wells. In 2D, we assumed connectivity between two lines located at the left- and right-hand boundaries, whereas in 3D, the connection is assumed to be between the two opposing planes.

4.5.1 *Idealised overlapping sandbody model*

As in 2D, we shall start with the simplest model and discuss more realistic features later. We consider a simple isotropic sandbody model in 3D where sandbodies are represented by cubes distributed randomly and independently in space. Again sandbodies are assumed to be entirely permeable with perfect hydraulic contact between them.

Take an arbitrary cubic region of size $X \times X \times X$ within which cubes of length a are placed independently. As before, we define the dimensionless size of the system as $L = X/a$. The centre point of each body is assigned randomly, and they are allowed to overlap. Connectivity between the opposing planes is used to define the connection in each direction. In this simple isotropic model, we expect the connection to be the same in all three directions. Two realisations of cubic sandbodies below and above the threshold are shown in Figure 4.28.

The rest of the model is identical to the 2D case as are the computational algorithms to get the connectivity. In this case, we need to check whether two cubes intersect. For aligned cuboids, this is easily done from the distance between the two centres and the lengths of the sides. The intersection criterion for more complicated sandbody shapes such as two ellipsoids is far more difficult and usually has to be determined numerically (Choi *et al.*, 2003). An analytical method based on the contact function defined by Perram *et al.* (1996) can be used to check the intersection of two ellipsoids. Detailed algorithms and numerical results are described by Sadeghnejad *et al.* (2013a). For non-aligned cubes or parallelepipeds

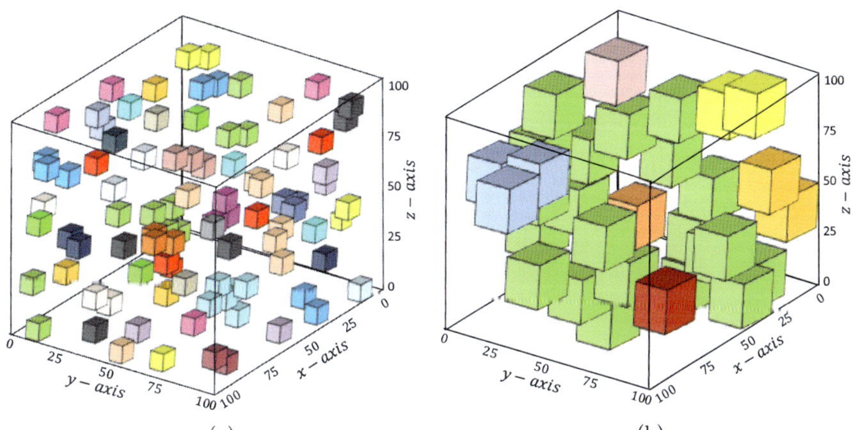

Figure 4.28. Two realisations of simple 3D cubic sandbodies model (a) below the threshold with $N = 100$, $L = 10$ and $p = 0.13$ and (b) above the threshold with $N = 40$, $p = 0.4071$ and $P = 0.7624$.

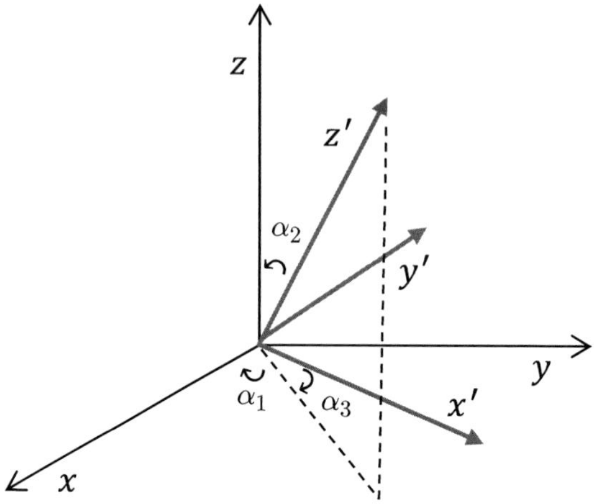

Figure 4.29. Illustration of three Euler angles $(\alpha_1, \alpha_2, \alpha_3)$ to show the rotation in 3D.

of any orientation, the intersection criterion can be achieved by checking the intersection of two rectangles in 3D space. It should be noticed that three Euler angles (azimuth, elevation and tilt denoted by $[\alpha_1, \alpha_2, \alpha_3]$ with $0 < \alpha_1 < 360°$, $0 < \alpha_2 < 180°$ and $0 < \alpha_3 < 360°$) are used to represent the orientation of non-aligned sandbodies or to evaluate the rotational quantity associated with a body in 3D (Figure 4.29).

We expect the same framework for finite-size scaling to apply in 3D as it did in 2D but with new critical exponents given in Table 2.2. We plot the connected sand fractions, P, as a function of net-to-gross ratio, p, for a variety of system sizes in Figure 4.30. We see very similar behaviour to that in 2D.

We first have to estimate the percolation threshold using the finite-size dependency Eq. (2.27) and find $p_c^\infty = 0.275 \pm 0.004$ (Figure 4.31).

The numerical results for various definitions of the threshold in 3D indicate that there is an appreciable finite-size shift in the threshold compared to the negligible shift in 2D:

$$\tilde{p}_c - p_c \approx \Lambda L^{-1/\nu}. \tag{4.26}$$

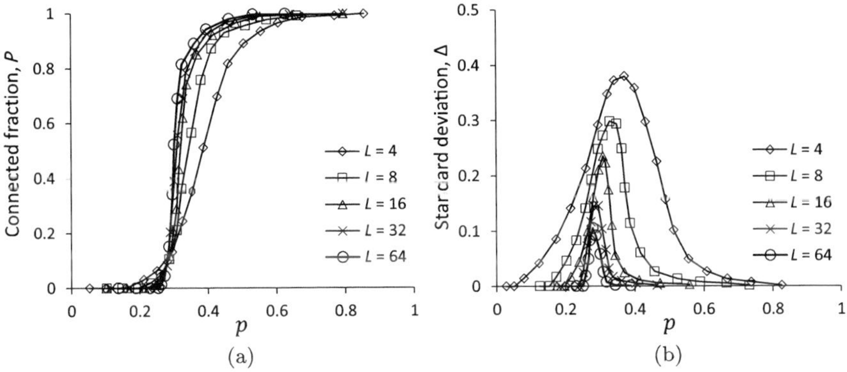

Figure 4.30. The connected sand fraction against p at various system sizes in 3D. (a) Mean and (b) standard deviation.

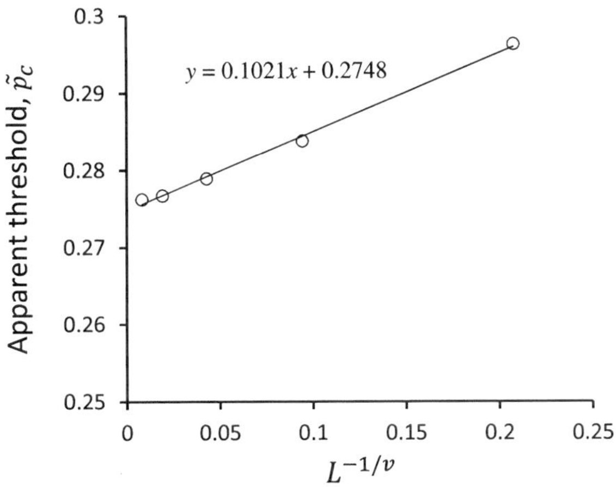

Figure 4.31. Determination of the threshold by extrapolation from finite-size systems by using the literature value for the critical exponent $\nu = 0.88$.

We find $\Lambda \approx 0.1$ (Sadeghnejad *et al.*, 2012). We use the same methods as in 2D to find the critical exponents. A log–log plot of the error obtained for the apparent threshold against system size, L, gives an estimate for the exponent ν (Figure 4.32). The value of 0.87 obtained this way is in agreement with literature value in Table 2.2. Also, a log–log plot of the connected sand fraction against $(p - p_c)$ for a

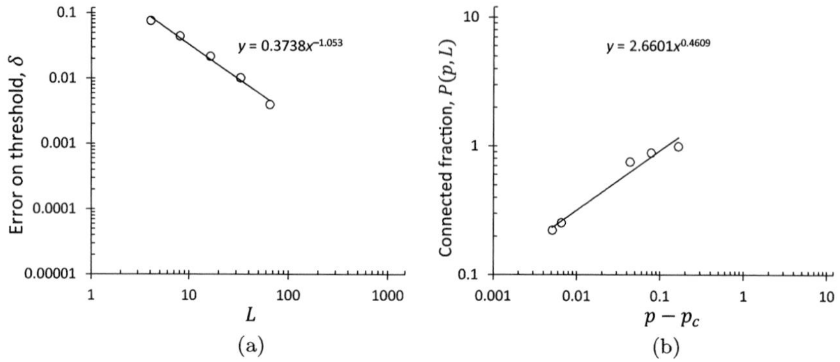

Figure 4.32. Illustration of using the scaling laws (Eqs. (2.1) and (2.28)) to estimate the critical exponents (a) ν and (b) β for the cubic sandbodies model in 3D.

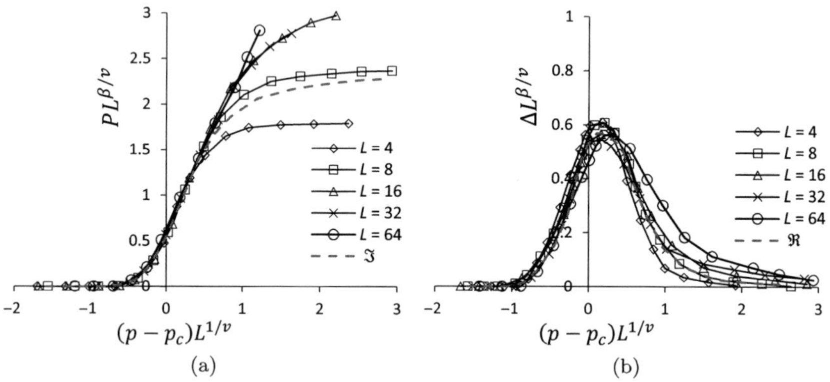

Figure 4.33. Rescaling the connectivity using the finite-size-scaling law (Eqs. (2.11) and (2.14)) in 3D. (a) Mean and (b) standard deviation.

relatively large system size $L = 20$ gives an estimate for exponent $\beta = 0.41$ which again is close to the literature values in Table 2.2.

Now by using the percolation threshold and critical exponents along with the finite-size scaling (Eqs. (2.11) and (2.14)), we can rescale the curves to collapse the results onto isotropic universal connectivity curves ($\Im$ and $\Re$) in 3D.

From these connectivity master curves, we can reconstruct the connected sand fraction against the net-to-gross ratio for an arbitrary system size in 3D, without actually performing any further realisations.

4.5.2 *Anisotropy effects in 3D*

To account for anisotropy, we extend the basic 3D sandbody model. An arbitrary external cuboidal region with size $X \times Y \times Z$ is considered, and a number of cuboid bodies of size $(a \times b \times c)$ are placed independently and uniformly (a realisation is shown in Figure 4.34).

The effective system size is defined by three dimensionless lengths in the X, Y and Z directions as

$$L_x = X/a, \quad L_y = Y/b, \quad L_z = Z/c. \tag{4.27}$$

The procedure to account for the effects of anisotropy is similar to that explained in 2D (Section 4.2) except that there is an extra degree of freedom available. There are two aspect ratios (ω_1, ω_2) that can

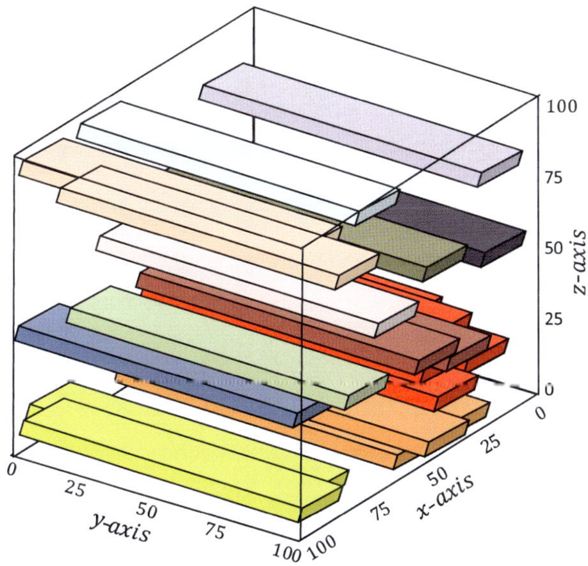

Figure 4.34. A realisation of the anisotropic overlapping sandbody model in 3D.

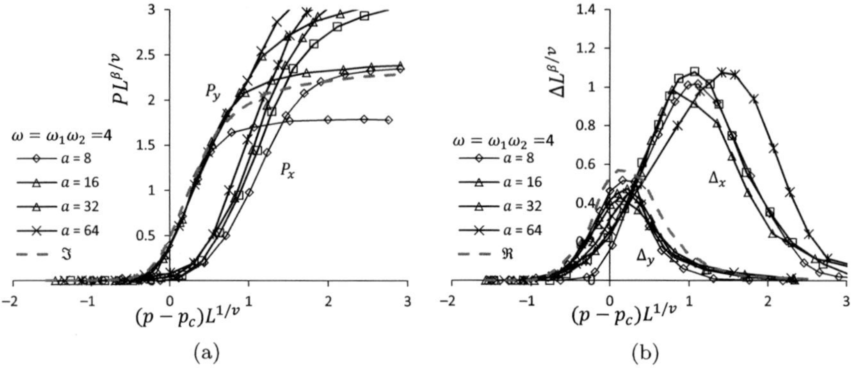

Figure 4.35. Rescaled horizontal P_x and vertical P_y mean connectivity curves using infinite threshold for the aspect ratio of 4 with its associated uncertainty, where the dotted curves are the isotropic mean connectivity curve and its associated uncertainty curve respectively. (a) Mean and (b) standard deviation.

be defined as

$$\omega_1 = L_x/L_y, \tag{4.28}$$

$$\omega_2 = L_x/L_z. \tag{4.29}$$

We shall consider a particular case where the two aspect ratios are the same (i.e., $b = c$). This enables us to compare the x direction connectivity to the y direction connectivity and the finite-size scaling. If we rescale the mean connectivity results $P(p, L)$ with L_x as before, we see that again there is a data collapse in each direction. Figure 4.35 shows the mean connected sand fraction and error associated with the mean connectivity for a fixed value of the aspect ratios of $\omega_1 = \omega_2 = 2$.

We would expect that the amount of the displacement about the isotropic universal connectivity curves ($\Im$ and $\Re$) will be a function of the two aspect ratios ω_1 and ω_2. As in 2D, we conjecture that the same isotropic universal curve may be used but with an apparent threshold, which we would expect to be dependent on both the system size and the two aspect ratios,

$$\tilde{p}_c^i(\omega_1, \omega_2, L) - p_c^\infty = \Lambda_i(\omega_1, \omega_2)L_x^{-1/\nu}, \tag{4.30}$$

where i denotes the x, y or z direction and p_c^∞ is the infinite threshold. We expect similar symmetries as found in 2D. For example, keeping L_z fixed and interchanging two lengths L_x and L_y, ω_1 goes to $1/\omega_1$ and ω_2 goes to ω_2/ω_1. This simply means that we have swapped the labelling of the axes. Therefore the x direction connectivity before this interchange should be equivalent to the connectivity in the y direction after the interchange,

$$\tilde{p}_c^x(\omega_1, \omega_2) = \tilde{p}_c^y(1/\omega_1, \omega_2/\omega_1). \tag{4.31}$$

However, this interchange does not alter the connectivity in the z direction,

$$\tilde{p}_c^z(\omega_1, \omega_2) = \tilde{p}_c^z(1/\omega_1, \omega_2/\omega_1). \tag{4.32}$$

Using the scaling relationship for the apparent threshold (Eq. (4.30)), these two properties (Eqs. (4.31) and (4.32)) give

$$\Lambda_x(\omega_1, \omega_2) = \omega_1^{1/\nu} \Lambda_y \left(\frac{1}{\omega_1}, \frac{\omega_2}{\omega_1} \right), \tag{4.33}$$

$$\Lambda_z(\omega_1, \omega_2) = \omega_1^{1/\nu} \Lambda_z \left(\frac{1}{\omega_1}, \frac{\omega_2}{\omega_1} \right). \tag{4.34}$$

Using the other possible rotations of L_x, L_y and L_z would give the following symmetry relations:

$$\Lambda_x(\omega_1, \omega_2) = \omega_2^{1/\nu} \Lambda_z \left(\frac{\omega_1}{\omega_2}, \frac{1}{\omega_2} \right), \tag{4.35}$$

$$\Lambda_y(\omega_1, \omega_2) = \omega_2^{1/\nu} \Lambda_y \left(\frac{\omega_1}{\omega_2}, \frac{1}{\omega_2} \right), \tag{4.36}$$

$$\Lambda_x(\omega_1, \omega_2) = \Lambda_x(\omega_2, \omega_1), \tag{4.37}$$

$$\Lambda_y(\omega_1, \omega_2) = \Lambda_z(\omega_2, \omega_1). \tag{4.38}$$

In 2D, the numerical results showed that the apparent thresholds were symmetrically placed about the isotropic case. However, this is not the case in 3D (Figure 4.36). In 2D, the finite-size shift in the isotropic threshold was negligible in comparison to the effect

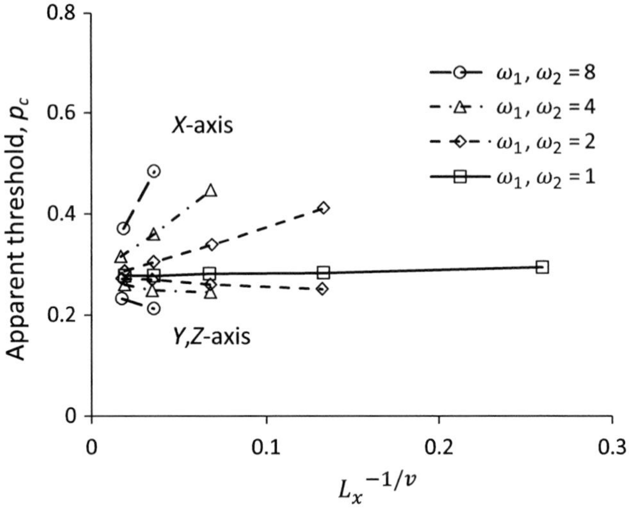

Figure 4.36. Illustration of the finite-size dependency of the apparent threshold (Eq. (4.30)) in the x and the y (or z) directions as a function of $L_x^{-1/\nu}$.

of anisotropy. The numerical results shown in Figure 4.31 indicate that this is not the case in 3D, where the shift was $\Lambda \approx 0.1$ (Eq. (4.26)). Hence, it is not possible to define a single function for Λ as in 2D. In 3D, we have three functions Λ which are interrelated through the symmetries mentioned above. By determining one of these Λ functions, the others can be obtained from these relations (Eqs. (4.35)–(4.38)).

We start by determining the function Λ_x. From the above equations, we expect that Λ_x is composed of the product and (or) sum of the two aspect ratios ω_1 and ω_2. However, it is unlikely for the sum to be important as the aspect ratios are basically defined by the ratios of the lengths. Plotting Λ_x against ω (defined as the product of ω_1 and ω_2) gives a single curve as shown in Figure 4.37 from which we deduce that

$$\Lambda_x(\omega_1, \omega_2) = a\left[(\omega_1\omega_2)^b - 1\right] + c, \tag{4.39}$$

$$a = 0.25, \quad b = 0.77, \quad c = 0.1. \tag{4.40}$$

The above analysis shows how the apparent threshold varies as a function of system size and the aspect ratios in 3D. With this

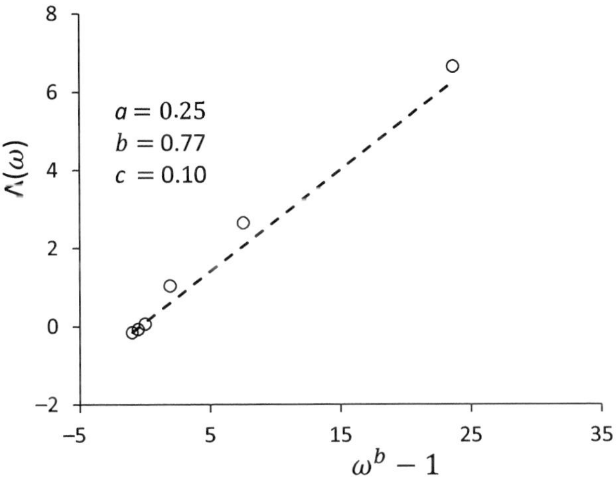

Figure 4.37. Variation of Λ_x against $\omega^b - 1$ (ω is the product of ω_1 and ω_2).

corrected apparent threshold, the finite-size-scaling rules to estimate the mean connected sand fraction in 3D becomes

$$P(p, L_x, \omega_1, \omega_2) = L_x^{-\beta/\nu} \Im[(p - \tilde{p}_c)L_x^{1/\nu}]. \tag{4.41}$$

Figure 4.38 shows the data collapse in the critical region for a variety of system sizes. The standard deviation of connectivity obeys a similar finite-size scaling as in 2D. We use the finite-size apparent threshold (corrected by the amount of shift from Eq. (4.39)) as well as a change in magnitude, which can be accounted for by rescaling with the geometric mean length, $(L_x L_y)^{0.5}$. The resulting finite-size scaling for the standard deviation of connectivity in 3D becomes

$$\Delta(p, L_x, \omega_1, \omega_2) = \omega^{1/2} L_x^{-\beta/\nu} \Im[(p - \tilde{p}_c)L_x^{1/\nu}]. \tag{4.42}$$

The data collapse shown in Figure 4.38 is a good enough approximation for engineering purposes which enable us to use the same 3D isotropic universal connectivity curves ($\Im$ and $\Re$ in Figure 4.33)

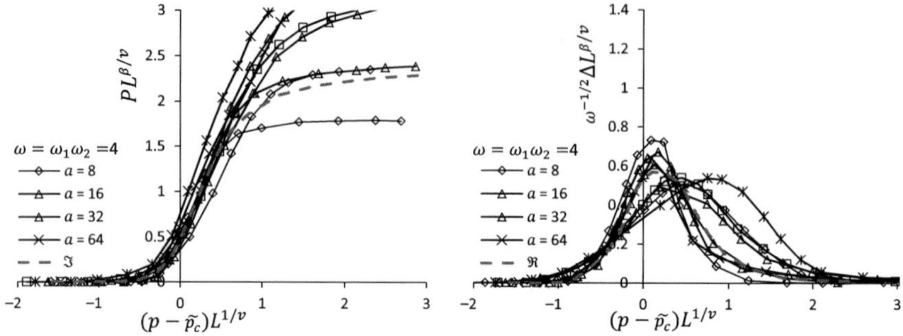

(a) (b)

Figure 4.38. Data collapse of (a) the mean and (b) the standard deviation in connectivity using the corrected finite-size scaling for the aspect ratio of 4. The isotropic the connectivity curve is shown in dotted line (Sadeghnejad *et al.*, 2010).

for predicting the connectivity of anisotropic overlapping sandbodies in 3D.

4.5.3 *Sandbody size variation*

As in 2D, we now relax some of the assumptions made so far. First we take into account the effect of sand size variation. As sandbodies are placed randomly in the system, the same arguments used to derive Eq. (4.4) can be used to determine the volume fraction of sands in 3D,

$$p = 1 - e^{\frac{-N\langle a^3\rangle}{V}}, \tag{4.43}$$

where N is the number of sandbodies, $\langle a^3\rangle$ is the average volume of each body and V is the volume of the region in which the bodies are placed. This leads to the following definition for the effective sand lengths (Balberg *et al.*, 1984; Bour and Davy, 1998; Mourzenko *et al.*, 2005),

$$l_{\text{eff}} = \sqrt[3]{\langle a^3\rangle}. \tag{4.44}$$

We can then use the hypothesis that this effective sandbody size leads to the correct dimensionless length to use in the finite-size-scaling

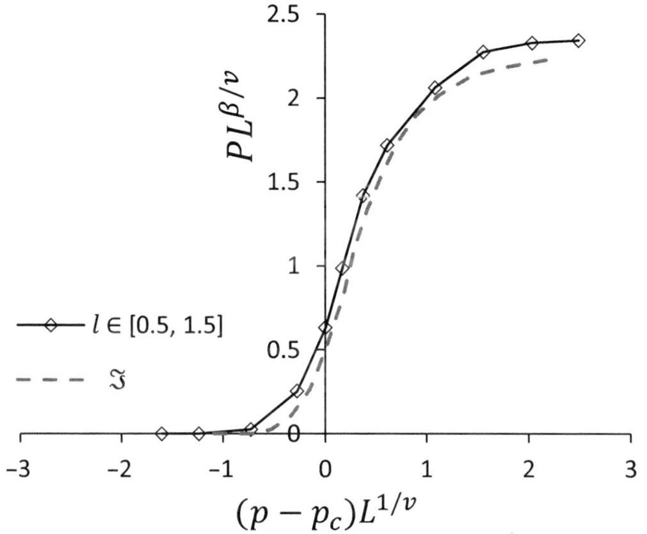

Figure 4.39. Finite-size scaling of the mean connected sand fraction results for a 3D sandbody model with uniform size distribution of sands in the range 0.5–1.5.

laws (Eqs. (2.11) and (2.14)). For example, the mean connectivity for a sandbody model of size $L = 24$ with uniform size distribution of sands in the range 0.5–1.5 with the effective length from Eq. (4.44) $\left(l_{\text{eff}} = \sqrt[3]{\frac{u^4-v^4}{4(u-v)}} = 1.07\right)$ in a system of dimensionless size $L_x = 24/1.07 = 22.5$ is shown in Figure 4.39.

4.5.4 *Effects of orientational disorder*

To account for the orientational disorder of sandbodies in 3D, a similar analysis to that used in 2D is made. Consider a simple case where the orientational disorder is along one axis. For example, a uniform orientation distribution in the range $\theta \in [\theta_1, \theta_2]$ around the x axis (Figure 4.40). The algorithm to identify if two bodies intersect is based the criterion for two non-aligned rectangles.

In this case, the effective size of the system along the X, Y and Z directions can be calculated by $L_x = \frac{X}{\langle l_x \rangle}, L_y = \frac{Y}{\langle l_y \rangle}, L_z = \frac{Z}{\langle l_z \rangle}$. As usual, we define the aspect ratios, ω_1 and ω_2, for the anisotropy in Eqs. (4.28) and (4.29). Then the effects of these effective lengths

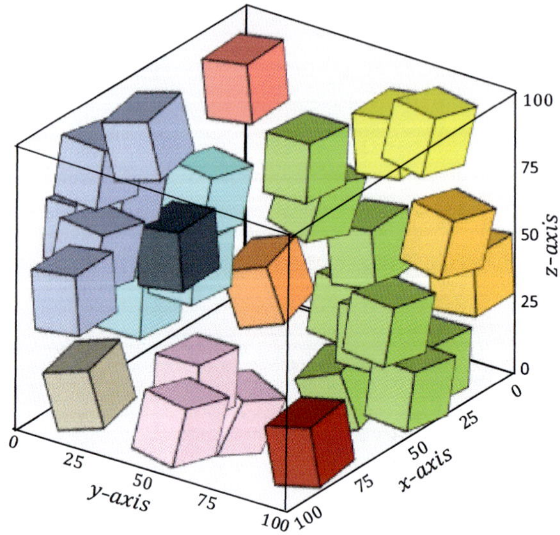

Figure 4.40. A realisation of overlapping sandbodies made by cubes with orientation disorder along the x axis.

can be considered through the finite-size-scaling laws of anisotropic systems (Eqs. (4.41) and (4.42)).

In addition to the finite-size effects, this orientation disorder can affect the infinite percolation threshold. This was discussed in 2D. The same arguments hold in 3D except we now use excluded volume. For example, the excluded volume of *aligned* cubic bodies of size $a \times b \times c$ is $V_{\mathrm{ex}} = 8abc$. Then the excluded volume fraction, ρ_{ex}, can be defined as the fraction of excluded volume to the total volume of the system, V,

$$\rho_{\mathrm{ex}} = V_{\mathrm{ex}}/V, \tag{4.45}$$

which leads to the relation with the occupied sand fraction (as in Eq. (4.43):

$$p = 1 - e^{\rho_{\mathrm{ex}}/8}. \tag{4.46}$$

In 3D, the conjecture for the percolation thresholds states that the excluded volume fraction, ρ_{ex}, takes a universal value for all shapes of $\rho_{\mathrm{ex}} \approx 2.58$. Assuming this conjecture, the critical net-to-gross

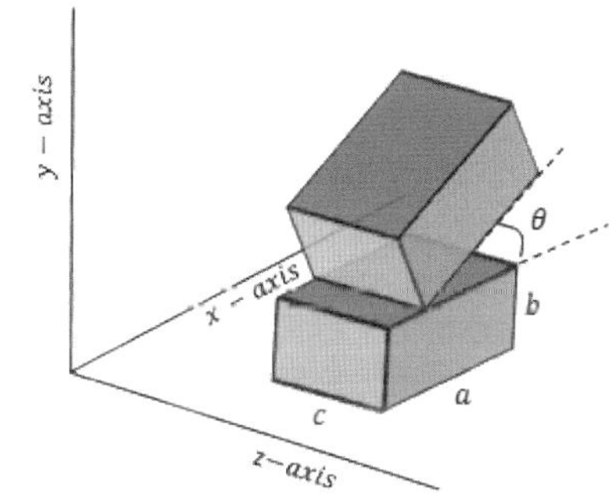

Figure 4.41. The effective size of a cubic bodies with rectangular facets of size $a \times b$ at an angle θ to each other in the XY plane in 3D.

ratio becomes

$$p_c = 1 - e^{\rho_c}, \quad \text{where } \frac{\rho_c}{abc} = \frac{2.58}{\langle V_{\text{ex}} \rangle}. \tag{4.47}$$

With this conjecture, we can determine the shift in percolation threshold from Eq. (4.47) as a function of orientation disorder.

We examine this hypothesis for a simple model of cubic bodies with rectangular facets of size $a \times b$ at an angle θ to each other in the XY plane (Figure 4.41). Assume a uniform orientation distribution in the range $\theta \in [-\theta_0, \theta_0]$; then the average excluded volume becomes

$$\langle V_{\text{ex}} \rangle = \left[2ab \left(1 + \frac{\sin^2 \theta_0}{\theta_0^2} \right) + \left(\frac{a^2 + b^2}{2\theta_0^2} \right) (2\theta_0 - \sin 2\theta_0) \right] c. \tag{4.48}$$

Now, from Eqs. (4.47) and (4.48), we determine the corrected threshold arising from the orientational disorder as a function of the aspect ratio. A comparison of the computed results of the mean horizontal connectivity for a set of cubes of size $1 \times 1 \times 1$ with two different uniform angular distributions (one between -15° and 15° and the other between -30° and 30°) around the x direction in a region of size $16 \times 16 \times 16$ with the predictions from finite-size-scaling rules, which uses the effective lengths and the corrected threshold (Eqs. (4.47) and (4.48)), is shown in Figure 4.42.

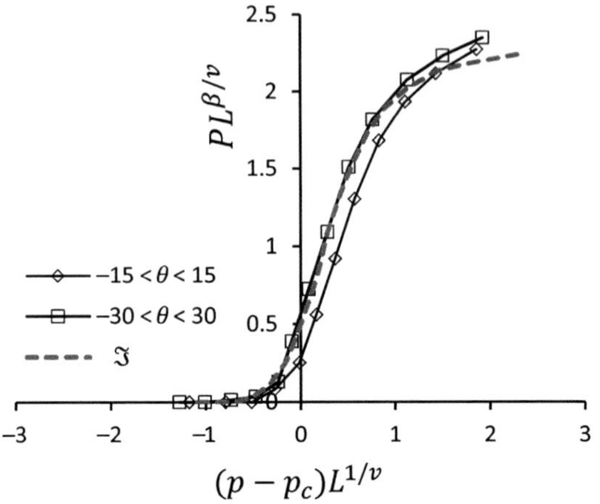

Figure 4.42. Comparison of the computed connectivity results for a simplified model of sandbodies with two different uniform orientation distribution (one between $-15°$ and $15°$ and the other between $-30°$ and $30°$) with the corrected scaling rules.

4.5.5 *Other effects in 3D*

In addition to the sand size and orientation variations, there may be other issues such as shape, spatial correlations and well patterns to be considered. In this section, we shall only consider the impact of the sandbody shape. Further research is still required for a complete investigation of all of the other effects.

Technically, we can represent sandbodies with any shape that is geologically appropriate. For example, we could use ellipsoids distributed randomly as shown in Figure 4.43.

We then follow the same computational framework to generate realisations and analyse the statistics of percolation quantities. However, the only new complexity is to define and implement the intersection criterion between two new sand shapes in an efficient way. As discussed before, checking whether two ellipsoids intersects leads to a constrained minimisation technique (Perram *et al.*, 1996; Choi *et al.*, 2003).

The percolation threshold for ellipsoidal sandbodies was found to be about 0.286 (Sadeghnejad *et al.*, 2013a) which should be compared with the value for cubes of 0.276. The simulation results

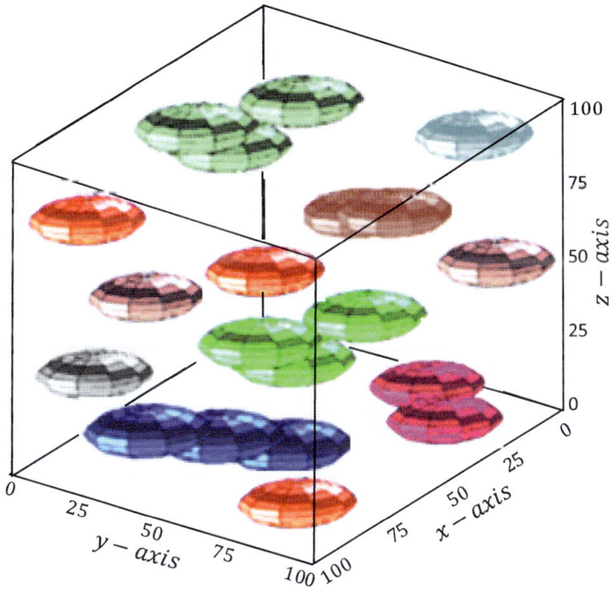

Figure 4.43. A realisation of a sandbody model using ellipsoids below the threshold.

that the critical exponents were $\beta = 0.42$ and $\nu = 0.86$ which are in good agreement with the literature values (Table 2.2). Figure 4.44 compares the master curve of mean connectivity for the ellipsoidal sandbodies with that for the cubic sandbodies. There is a negligible difference, so we conjecture that the percolation analysis is largely the same for any convex shape.

Another issue is the effect of the well pattern. So far, we have considered the connectivity between the two faces (face-to-face (F2F) connectivity) on the opposite sides of the system. In 3D geological formations, the production and injection wells are represented by two lines and well trajectories can be far more complicated. Hence, L2L connectivity is needed (see Figure 4.45). In such cases, the percolation properties such as spanning cluster, backbone or shortest path may be changed compared to F2F models (Soares *et al.*, 2004; Sadeghnejad and Masihi, 2017). However, it is yet unclear how to transform F2F connectivity results to L2L connectivity in order to

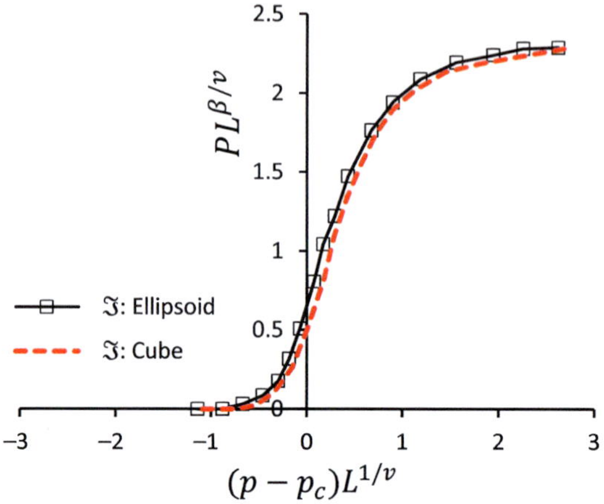

Figure 4.44. Comparison of connectivity master curve of a system composed of ellipsoidal sandbodies with the connectivity of cubic sandbodies.

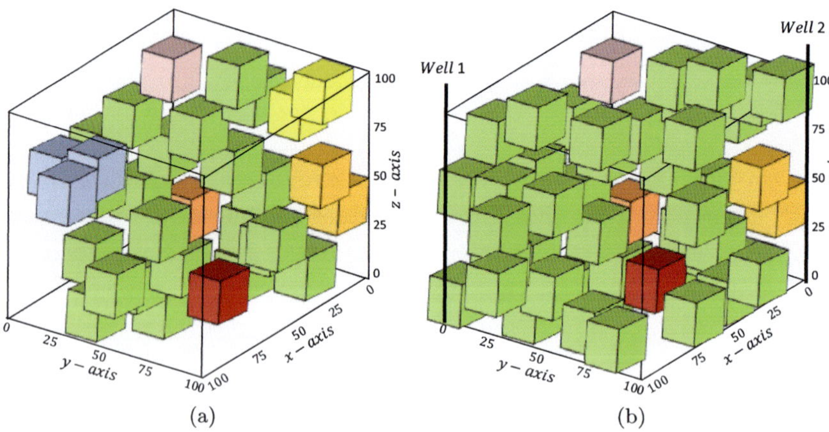

Figure 4.45. Illustration of (a) F2F connection along x axis and (b) L2L connection in 3D.

use the same master connectivity curves (Figure 4.33). There may also be a completion interval for a well or there may be more than two wells in the reservoir through specific well patterns among which we need the connected sand fraction. This is an ongoing research problem.

4.6 Case studies

There have been several attempts to use the universal curves of connectivity in 2D and 3D on real field-scale datasets (Nurafza *et al.*, 2006; Sadeghnejad *et al.*, 2012, 2013a). Three field-scale applications are described in this section.

4.6.1 *Case study 1: A carbonate gas condensate reservoir*

Nurafza *et al.* (2006) applied the scaling on a fine-grid model of the central simulation area ($48 \times 97 \times 108$ grids in x, y and z directions) of a carbonate gas condensate reservoir at an average depth of 3,000 m with tilted fault blocks that consists of three reservoirs: R1 superior reservoir called R1S (a 60 m thick package of highly layered carbonate deposits), R1 inferior reservoir called R1I (a 95 m thick homogenous succession of carbonates) and a lower Bireno reservoir. The central reservoir zone considered in this case study is $9{,}000 \times 18{,}500 \times 100\,\text{m}^3$ with three types of facies: biostrome (of typical size $500 \times 2{,}000 \times 1.9\,\text{m}^3$), miliolid limestone ($1{,}000 \times 3{,}000 \times 1.7\,\text{m}^3$) and foreshore ($1{,}000 \times 4{,}900 \times 1.6\,\text{m}^3$). However, both reservoir geometry and sandbodies were not exact cuboids in 3D (or rectangles in 2D) which causes some errors in the estimation of the connectivity.

The direct numerical simulation used the fine-grid permeability map of the reservoir from which various 2D permeability models were extracted. For each of these models, a critical value of the permeability based on the minimum value of $K = 0.1 \pm 0.05$ md that allows the flow in gas condensate reservoirs was used. This converts the permeability maps with a continuous distribution to a binary one. So $K = 0$ for $K < K_{\text{critical}}$ and $K = 1$ for $K > K_{\text{critical}}$. This idea of using a threshold permeability value will be further explored in Chapter 7.

The occupancy faction p is then defined as the fraction of cells above the critical value. The connected fractions P_x and P_y are defined in the usual way. Each facies type is given a uniform orientation distribution in the range of $330° \pm 30°$. The connectivity

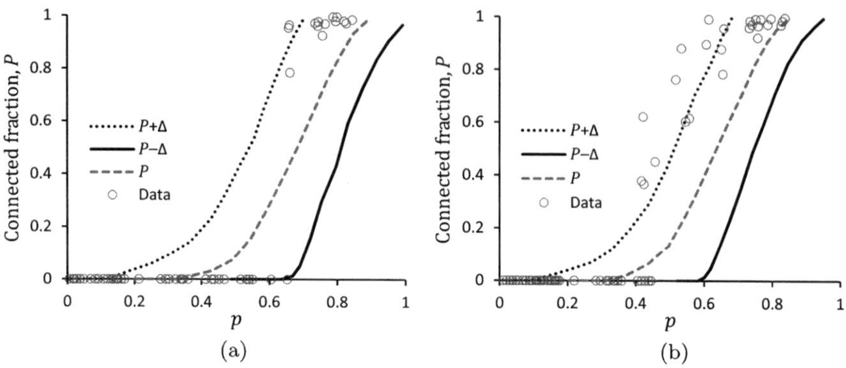

Figure 4.46. Comparison of the computed connected sand fraction with prediction from percolation theory for (a) facies body of size 1,000 m in the x direction and (b) facies body of size 4,900 m in the y direction within a system of size 9,000 × 18,500 m^2 (where L_x = 9,000/1,000 and L_y = 18,500/4,900).

results for miliolid limestone with linear size of 1,000 in the x direction and for Foreshore with linear size of 4,900 in the y direction within the system size of 9,000 × 18,500 m^2 (for which $L_x = 9{,}000/1{,}000$ and $L_y = 18{,}500/4{,}900$) are shown in Figure 4.46.

4.6.2 *Case study 2: An offshore oil reservoir*

Sadeghnejad *et al.* (2011a) studied an offshore oil reservoir (called B) located at the south of Iran at a depth of 2,337 m TVDSS. This was formed from a mixed carbonate–clastic shelf system which extended over the whole Arabian Plate during the Early and Middle Cretaceous. It is characterised by a low ramp gently dipping to the north. The paleocoast was generally oriented from WNW–ESE to N–S and the source of the clastics sediments was the Arabian Shield to the southwest.

Core and palynological data indicate that the oil reservoir consists of a series of incision-fill sequences occurring in an estuarine/coastal plain/deltaic environment which results in a thick stack of excellent quality sands incising into each other with few remaining shalier sediments locally separating these sequences as schematically shown in Figure 4.47 (Huerlimann, 2004).

Figure 4.47. Schematic illustration of various time scales of sedimentation of the offshore oil reservoir B where channels shown in red formed during clastic influx. Clastic progradations started on marls/limestone and then influxes were bounded by carbonate platform transgression at the top.

The sands of reservoir B consist mainly of valley-fill deposition sediments, where fresh water fluvial, coarse-grained sediment accumulated in the upper and middle reaches of the incised valley system, while tidally influenced sediments accumulated in the marine-influenced middle and lower reaches. As a result of this aggradation, less fluvial-derived sediment reached the lower reaches of the valley system, which resulted in clean, well-sorted shoreline material being reworked and deposited in tidal channels in this area. Using various phases of an incision-fill sequence and conventional core analysis five rock types were identified: loose sands (ϕ: 27–31%, K: 10–13 md), massive clean sands (ϕ: 27–31%, K: 1.8–2.2 md), deded sand and clay (ϕ: 27–31% K: 1.3–1.5 md), laminated sand and clay (ϕ: 12–20% K: 0.05–0.1 md) and silty shale (negligible ϕ and K).

Data from different sources (such as geology, geophysics and petrophysics) can be used through a deterministic or stochastic framework to build a fine-grid model. This fine-grid model can then be used to evaluate the connectivity. To evaluate the permeability of the field, various 2D (370 areal maps, 410 cross-sectional maps) and 3D (20 sector maps) were used. For each 2D or 3D model, an appropriate critical value of the permeability distribution has been selected to change the permeability map to a binary system. The permeabilities lower than this threshold are set to 0 and the values greater than this are set to 1. It should be emphasised that the choice of cut-off would affect the connectivity predictions. We choose the mean of the permeability probability distribution (K_c = 3,000 md)

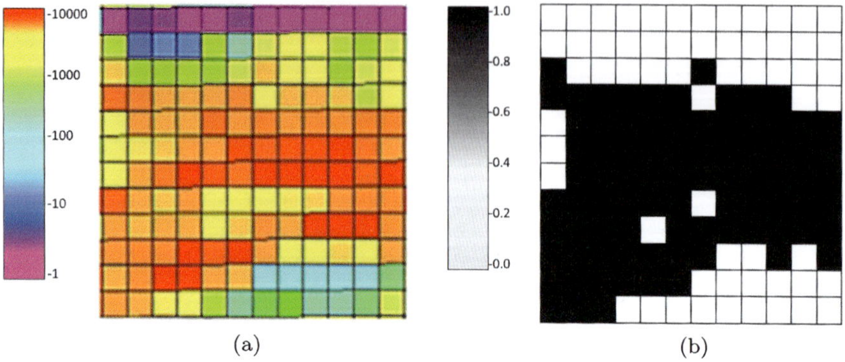

Figure 4.48. (a) An arbitrary 12×12 vertical permeability cross section of reservoir B and (b) its binary map after applying the threshold value of 3,000 md.

as the critical value. A sensitivity analysis can be carried out to find the impact of this threshold value on the percolation predictions (Sadeghnejad *et al.*, 2013b). A realisation of a permeability cross section and its equivalent binary map by applying a permeability cut-off of $K = 3{,}000$ md is shown in Figure 4.48.

By computing the connected fraction of cells above the threshold of each permeability map, a scatter graph of connectivity versus the net-to-gross ratio, p, is obtained. Each point on this graph represents a specific cross section of the map.

To compare these results with those from percolation theory, we need to have the sandbody shapes, orientations and sizes. These can be inferred by looking at the facies map in the reservoir. In this analysis, high permeability and porosity facies are selected as the main reservoir. Figure 4.49 illustrates how to estimate the sandbody sizes in each direction either manually or by using commercial reservoir modelling software.

The typical size of sandbodies estimated from different cross sections is $1{,}200 \times 1{,}300 \times 15\,\text{m}^3$. This is consistent with previous geological studies (Huerlimann, 2004). The average sizes of sandbodies in the vertical cross sections and the areal models are $1{,}200 \times 15$ and $1{,}200 \times 1{,}300\,\text{m}^3$ respectively. Now we can perform finite-size-scaling transformations. For example, Figure 4.50 compares the computed connectivity values (points) for 2D areal and cross-sectional maps

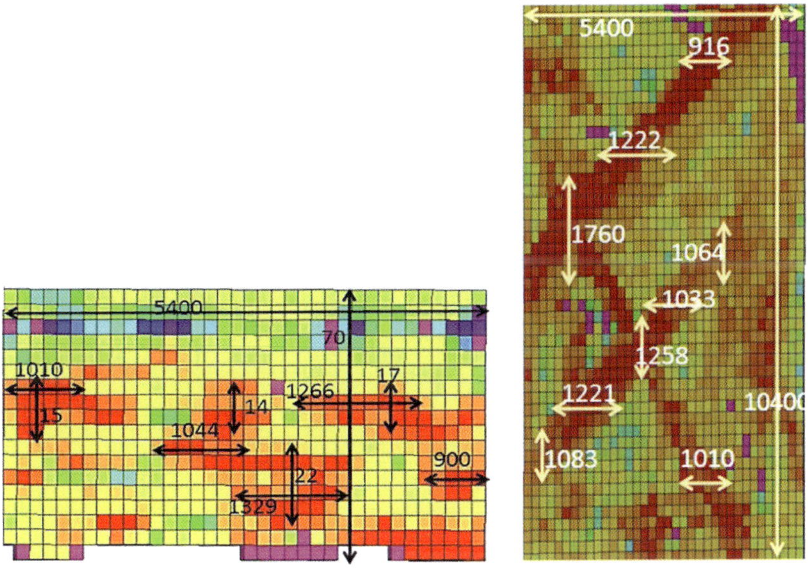

(a) (b)

Figure 4.49. Illustration of sandbody size estimation in various directions. The numbers are in metres. (a) cross section model, (b) areal model.

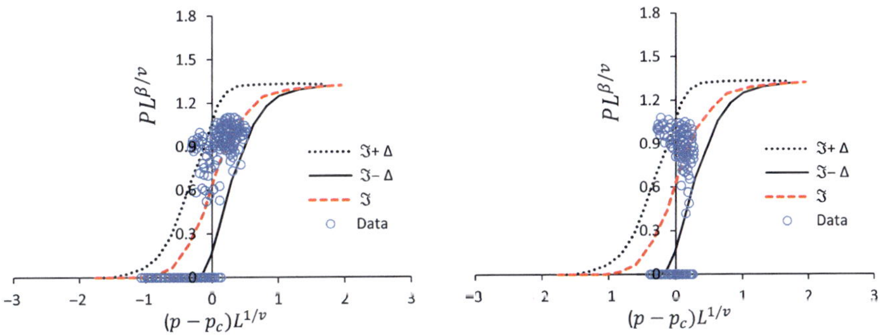

(a) (b)

Figure 4.50. A comparison of the computed connectivity values (points) for 2D areal (a) and cross-sectional (b) maps with the connectivity master curves in 2D, when the mean permeability value is used as the cut-off ($K = 3{,}000$ md).

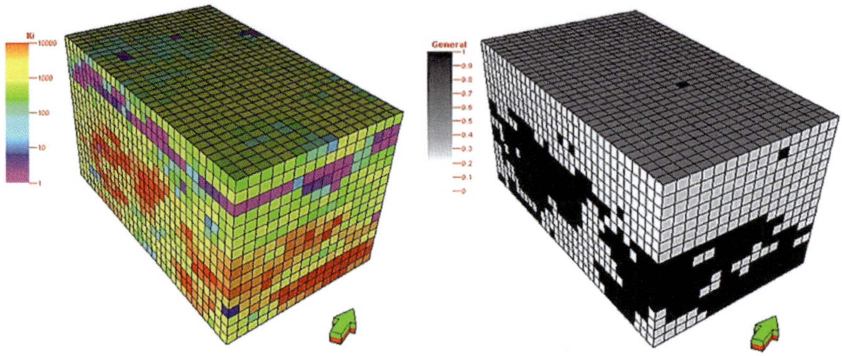

(a) (b)

Figure 4.51. (a) An arbitrary $32 \times 16 \times 16$ permeability sector of oil reservoir B and (b) its binary map after applying the threshold value of 1,800 md.

with the mean connectivity master curves (Figures 3.11 and 3.12) and one standard deviation either side (Sadeghnejad *et al.*, 2011a).

Similarly, the connectivity predictions from percolation theory approach have been validated against the field data in 3D. A realisation of the permeability map of 3D sector of oil reservoir B of size $32 \times 16 \times 16\,\mathrm{m}^3$ and its equivalent binary map (using permeability cut-off of $K = 1{,}800$ md) is shown in Figure 4.51. The estimated size of sandbodies for the finite-size-scaling calculations is $1{,}200 \times 1{,}300 \times 15\,\mathrm{m}^3$.

Again we carry out finite-size-scaling transformations for this 3D sector model. Figure 4.52 compares the computed connectivity values (points) for 20 3D maps with the mean connectivity master curves plus and minus one standard deviation (Sadeghnejad *et al.*, 2012).

This analysis can be extended to check the sensitivity of the results to the value of the cut-off used or to consider the effects of sandbody size variation and orientation disorder. Similar analyses have been performed on using different rock type definitions (Sadeghnejad *et al.*, 2010).

Finally, we examine the connected volume of hydrocarbons (CVH) between two wells. This is basically the product of the connected sand fraction and the volume of hydrocarbon initially in

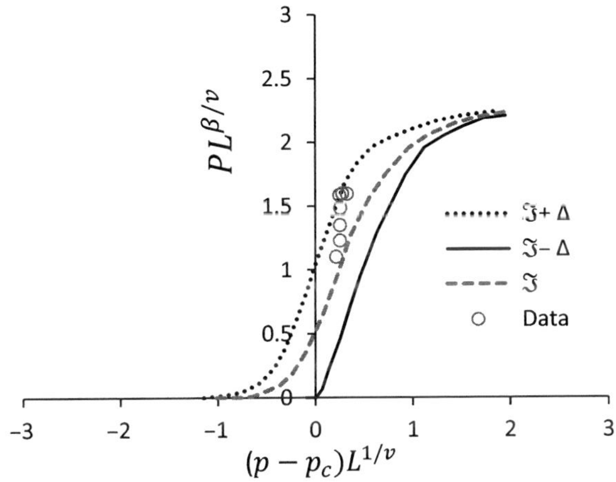

Figure 4.52. A comparison of the computed connectivity values (points) for 20 3D maps with the connectivity master curves. The average size of sandbodies is $1{,}200 \times 1{,}300 \times 15\,\text{m}^3$ and the permeability cut-off value is 1,800 md.

place (VHIIP): CVH = VHIIP × P. To estimate the CVH against well spacing at various net-to-gross ratios, we use the connected sand fraction from the finite-size-scaling laws (Sadeghnejad *et al.*, 2014). Figure 4.53 shows connected sand fraction and the CVH against well spacing at various net-to-gross ratios for a 3D sector of size $2{,}265 \times 2{,}850 \times 15\,\text{m}^3$. This prediction is consistent with the calculation of the CVH in a conventional approach for reservoir B with an average net-to-gross value of 0.41 and well spacing of 2,265 m. The value obtained conventionally was in the range of 7,801,131–9,905,997 bbl. The advantage of the percolation method is that it could be done on a spreadsheet without expensive computations required by the usual approach.

4.6.3 *Case study 3: A gas reservoir C*

We take a gas reservoir C in the south of Iran Sadeghnejad *et al.*, 2011b) as our final example. The reservoir, discovered in 1960s, covers an area of about $21 \times 17\,\text{km}^2$ with relatively low dips (less than 20°)

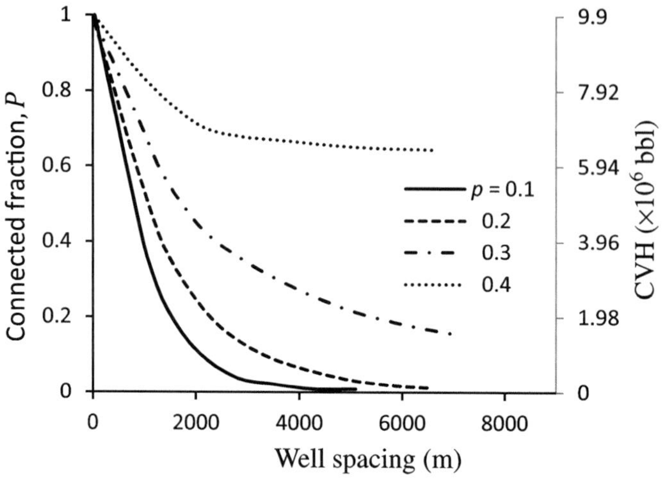

Figure 4.53. The connected sand fraction and the CVH versus well spacing of a 3D cross section of reservoir B at different net-to-gross ratios.

and consists of mainly limestone, dolomite, dolomitic limestone and anhydrite.

2D cross sections (of size 1,600–12,800 m) of the reservoir model in different locations were used to evaluate the connectivity results. As before, a cut-off on the permeability histogram is chosen to change the permeability maps to a simple permeable/impermeable map. Having collected the statistics and the percolation parameters, the results are compared with predictions from finite-size scaling. The size of sandbodies is 4,500 $\times$ 800 m^2 in the X and Y directions. The results for the connectivity are shown in Figure 4.54 which indicates a reasonable match. The fit may not be perfect, but the CPU time required for the percolation method is a fraction of a second compared with computationally intensive conventional approach.

4.7 Problems

4.1. Consider a simple 2D isotropic sandbody model with sands of unit size just above the threshold with $p = 0.6$ and $P = 0.52$.

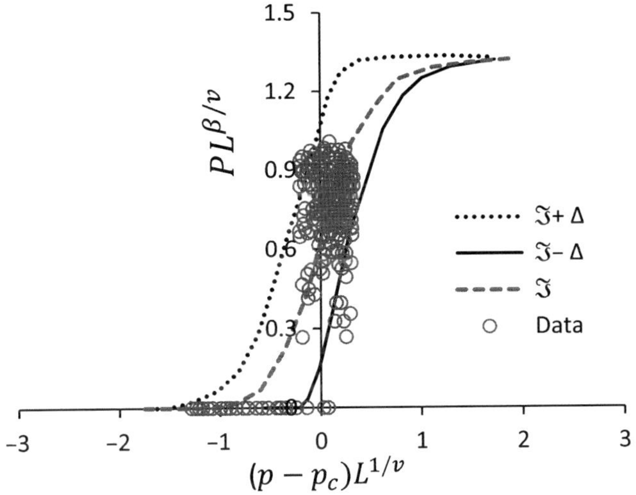

Figure 4.54. Reservoir connectivity results from cross-sectional maps of reservoir C compared with percolation predictions.

Now let us make anisotropic models by assuming that the sands have length $l = 2, 3, 4, \ldots$ Use finite-size scaling to determine the connected sand fraction of these anisotropic models, and plot the connected sand fraction against the aspect ratio ω.

4.2. Plot the variation of the connected sand fraction against well spacing at net-to-gross values $p = 0.65, 0.7, 0.75$ and 0.80 for a cross-sectional reservoir model of 10 m thickness which contains sandbodies of 50 m length and 1 m thickness. You may calculate the connected sand fraction at well spacing of 500 m, 1,000 m and 2,500 m.

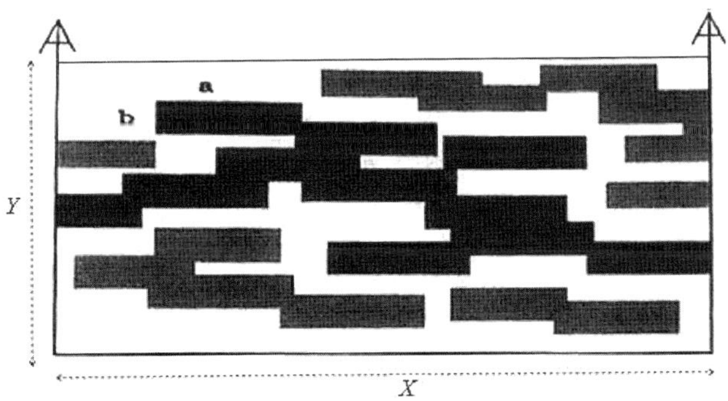

4.3. A 2D overlapping sandbody model is defined by squares of the same size but with a uniform orientation distribution in the range of θ_1 and θ_2 with $0 \leq \theta_1 \leq \theta_2 \leq \pi/2$ distributed randomly in an infinite medium. Use the excluded area argument with Eq. (4.18) to find the effect of orientation disorder on the percolation threshold.

4.4. Consider a 3D overlapping sandbody model that is made by distributing the cubes of size $10 \times 50 \times 100\,\text{m}^3$ randomly in the cuboid region of size $200 \times 500 \times 2{,}000\,\text{m}^3$. Estimate the percolation threshold in the x, y and z directions.

4.5. Consider a 3D overlapping sandbody model that is made by distributing the sands of size $10 \times 10 \times 10\,\text{m}^3$ in a cuboid region of the same thickness and width of 100 m. Plot the variation of the connected sand fraction against well spacing at net-to-gross ratio values $p = 0.4, 0.5, 0.6, 0.7$ and 0.80 along the x axis. You may calculate the connected sand fraction at well spacing 100 m, 200 m and 500 m. Compare your results with problem 4.1.

4.6. Consider a 2D overlapping sandbody model made by constant-size sands with dimensionless size of 10 and occupancy probability of $p = 0.6$. To investigate the effect of sand size variation, compare the connected sand fraction of this simple case with the connected sand fractions estimated in the following two models:

- sands having a Gaussian size distribution with mean size of 10 and variance of 25
- sands having an exponential size distribution with mean size of 10.

4.7. An engineer tried to apply the connectivity master curve for a real 2D reservoir map with a lognormal permeability distribution with the statistics shown in figure below. He examines the mode and the geometric mean as the cut-off on the permeability distribution to make the medium binary consistent with percolation approach. What is the uncertainty in the estimated connected sand fraction due to fact that he does not know which cut-off is correct?

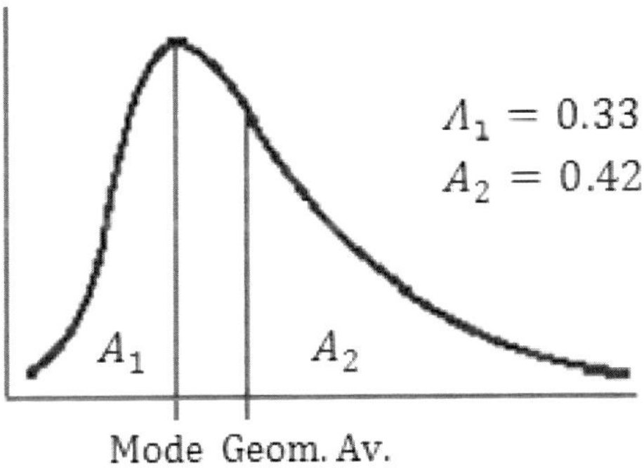

References

Baker, D.R., Paul, G., Sreenivasan, S. *et al.*, Continuum percolation threshold for interpenetrating squares and cubes. *Physical Review E*, 66: 0546136, doi: 10.1103/PhysRevE.66.046136, 2002.

Balberg, I., Universal percolation threshold limits in the continuum. *Physical Review B*, 31(6): 4053–4055, 1985.

Balberg, I. and Binenbaum, N., Computer study of the percolation threshold in a two dimensional anisotropic system of conducting sticks. *Physical Review B*, 28(7): 3799–3811, 1983.

Balberg, I., Binenbaum, N. and Anderson, C.H., Critical behaviour of the two dimensional sticks system. *Physical Review Letters B*, 51(18): 1605–1608, 1983.

Balberg, I., Anderson, C.H., Alexander, S. and Wagner, N., Excluded volume and its relation to the onset of percolation. *Physical Review B*, 30(7): 3933–3943, 1984.

Bassett, E.E., Bermner, J.M., Morgan, B.J.T., Jolliffe, I.T. *et al.*, *Statistics: Problems and Solutions, Second Edition.* World Scientific Publishing Co., Singapore, 2000.

Bour, O. and Davy, P., Connectivity of random fault networks following a power law fault length distribution. *Water Resources Research*, 33: 1567–1583, 1997.

Bour, O. and Davy, P., On the connectivity of three dimensional fault networks. *Water Resources Research*, 34: 2611–2622, 1998.

Charlaix, E., Percolation threshold of random array of discs: a numerical simulation. *Journal of Physics A: Mathematical and General*, 19(9): L533–L536, 1986.

Choi, Y.K., Wang, W., Kong, H. *et al.*, Exact collision detection of two moving ellipsoids under rational motions. International Conference on Robotics & Automation Taipei, Taiwan, September 14–19, 2003.

de Marsily, G., Delay, F., Gonsalves, J. *et al.*, Dealing with spatial heterogeneity. *Hydrogeology Journal*, 13: 161–183, doi: 10.1007/s10040-004-0432-3, 2005.

Fisz, M., *Probability Theory and Mathematical Statistics.* John Wiley & Sons, New York, 1963.

Flukiger, F., Plouraboué, F. and Prat, M., Nonuniversal conductivity exponents in continuum percolating Gaussian fractures. *Physical Review E*, 77: 047101, 2008.

Garboczi, E.J., Snyder, K.A., Douglas, J.F. and Thorpe, M.F., Geometrical percolation threshold of overlapping ellipsoids. *Physical Review E*, 52(1): 819–827, 1995.

Harter, T., Finite-size scaling analysis of percolation in three-dimensional correlated binary Markov chain random fields. *Physical Review E*, 72: 026120.1–026120.7, 2005.

Huerlimann, A., Master development plan report of Shell Company, Appendix seven of offshore reservoir rock properties, Tehran, Iran, 2004.

Ioannidis, M.A. and Chatzis, I., The effect of spatial correlations on the accessibility characteristics of three-dimensional cubic networks as related to drainage displacements in porous media. *Water Resources Research*, 27: 1777, 1993.

Jansson, B., *Random Number Generators.* Almqvist and Wiksell, Stockholm, 1966.

Jensen, J.L., Lake, L.W., Corbett, P.W.M. and Goggin, D.J., *Statistics for Petroleum Engineers and Geoscientists.* Elsevier, Amsterdam, 2000.

Lee, S.B. and Torquato, S., Monte Carlo study of correlated continuum percolation: universality and percolation thresholds. *Physical Review A*, 41: 5338–5344, 1990.

Lin, C.Y. and Hu, C.K., Universal finite size scaling functions for percolation on three dimensional lattices. *Physical Review E*, 58: 1521–1527, 1998.

King, P.R., The connectivity and conductivity of overlapping sandbodies. In: Buller, A.T. *et al.* (eds.), *North Sea Oil and Gas Reservoirs — II.* Graham & Trotman, London, pp. 353–362, 1990.

Koltermann, C.E. and Gorelick, S.M., Heterogeneity in sedimentary deposits: a review of structure-imitating, process-imitating, and descriptive approaches. *Water Resources Reserach*, 32: 2617–2658, doi: 10.1029/96WR00025, 1996.

Kuuskraa, V.A., Barrett, T., Mueller, R. and Hansen, J., Reservoir characterization for development of Mesaverde Group sandstones of the Piceance Basin, Colorado. In: Coalson, E.B., Osmond, J.C. and Williams, E.T. (eds.), *Innovative Applications of Petroleum Technology in the Rocky Mountain Area.* Rocky Mountain Association of Geologists, Denver, CO, pp. 61–72, 1997.

Makse, H., Davies, G., Havlin, S., Ivanov, P., King, P. and Stanley, H., Long-range correlations in permeability fluctuations in porous rock. *Physical Review E*, 54: 3129, 1996.

Marrink, S.J. and Knackstedt, M.A., Percolation thresholds on elongated lattices. *Journal of Physics A: Mathematical and General*, 32: L461–L466, 1999.

Masihi, M., King, P.R. and Nurafza, P., The effect of anisotropy on finite size scaling in percolation theory. *Physical Review E*, 74: 042102, 2006.

Masihi, M. and King, P.R., A correlated fracture network: modeling and percolation properties. *Water Resources Research Journal*, 43, doi: 10.1029/2006WR005331, 2007.

Matsumoto, M. and Nishimura, T., Mersenne twister: a 623-dimensionally equidistributed uniform pseudo-random number generator. *ACM Transactions on Modeling and Computer Simulation*, 8(1): 3–30, doi: 10.1145/272991.272995, 1998.

Mayall, M., Jones, E. and Casey, M., Turbidite channel reservoirs — key elements in facies prediction and effective development. *Marine and Petroleum Geology*, 23(8): 821–841, 2006.

Meester, R. and Roy, R., *Continuum Percolation*. Cambridge University Press, London, 1996.

Monetti, R.A. and Albano, E.V., Critical behaviour of the site percolation model on the square lattice in a $L \times M$ geometry. *Zeitschrift für Physik B: Condensed Matter*, 82: 129–134, 1991.

Moukarzel, C.F., Percolation in networks with long-range connections. *Physica A: Statistical Mechanics and its Applications*, 372: 340–345, 2006.

Mourzenko, V.V., Thovert, J.F. and Adler, P.M., Percolation of three dimensional fracture networks with power law size distribution. *Physical Review E*, 72: 036103.1–036103.14, 2005.

Nurafza, P., King, P.R. and Masihi, M., Facies connectivity modelling: analysis and field study. Paper SPE 100333, Proceedings of the SPE Europec, Vienna, Austria, 12–15 June, 2006.

Perram, J.W., Rasmussen, J., Præstgaard, E. *et al.*, Ellipsoid contact potential: theory and relation to overlap potentials. *Physical Review E*, 54: 6565–6572, 1996.

Phani, M.K. and Dhar, D., Continuum percolation with discs having a distribution of radii. *Journal of Physics A: Mathematics and General*, 17: L645–L649, 1984.

Prakash, S., Havlin, S., Schwartz, M. and Stanley, H.E., Structural and dynamical properties of long-range correlated percolation. *Physical Review A*, 46: R1724, 1992.

Quintanilla, J., Torquato, S. and Ziff, R.M., Efficient measurement of the percolation threshold for fully penetrable discs. *Journal of Physics A*, 33: L399–L407, doi: 10.1088/0305-4470/33/42/104, 2000.

Ronayne, M.J. and Gorelick, S.M., Effective permeability of porous media containing branching channelnetworks. *Physical Review E*, 73: 026305, doi: 10.1103/PhysRevE.73.026305, 2006.

Sadeghnejad, S., Masihi, M., King, P.R. *et al.*, Effect of anisotropy on the scaling of connectivity and conductivity in continuum percolation theory. *Physical Review E*, 81: 061119, doi: 10.1103/PhysRevE.81.061119, 2010.

Sadeghnejad, S., Masihi, M., King, P.R. *et al.*, Reservoir conductivity evaluation using percolation theory. *Petroleum Science and Technology*, 29: 1041–1053, doi: 10.1080/10916460903502506, 2011a.

Sadeghnejad, S., Masihi, M., Pishvaei, M., Shojaei, A. and King, P.R., Utilization of percolation approach to evaluate the reservoir connectivity and effective permeability: a case study on North Pars Gas Field. *Scientia Iranica C*, 18(6): 1391–1396, 2011b.

Sadeghnejad, S., Masihi, M., Shojaei, A. *et al.*, Field scale characterization of geological formations using percolation theory. *Transport in Porous Media*, 92: 357–372, doi: 10.1007/s11242-011-9907-6, 2012.

Sadeghnejad, S., Masihi, M., Pishvaie, M. and King, P.R., Rock type connectivity estimation using percolation theory. *Mathematical Geosciences*, 45: 321–340, 2013a.

Sadeghnejad, S., Masihi, M. and King, P.R., Dependency of percolation critical exponents on the exponent of power law size distribution. *Physica A: Statistical Mechanics and its Applications*, 392(24): 6189–6197, 2013b.

Sadeghnejad, S., Masihi, M., Pishvaei, M., Shojaei, A. and King, P.R., Estimating connected volume of hydrocarbon during early reservoir life by percolation theory. *Energy Sources, Part A: Recovery, Utilization, and Environmental Effects*, 36(3): 301–308, 2014.

Sadeghnejad, S. and Masihi, M., Point to point continuum percolation in two dimensions. *Journal of Statistical Mechanics: Theory and Experiment*, 10: 103210, doi: 10.1088/1742-5468/2016/10/103210, 2016.

Sadeghnejad, S. and Masihi, M., Analysis of a more realistic well representation during secondary recovery in 3D continuum models. *Computational Geosciences*, 21(5–6): 1035–1048, doi: 10.1007/s10596-017-9640-5, 2017.

Sahimi, M., *Applications of Percolation Theory.* Taylor and Francis, London, pp. 258, 1994a.

Sahimi, M., Long-range correlated percolation and flow and transport in heterogeneous porous media. *Journal of Physics I*, 4: 1263–1268, 1994b.

Sahimi, M., Effect of long-range correlations on transport phenomena in disordered media. *AICHE Journal*, 41: 229, 1995.

Sahimi, M. and Mukhopadhyay, S., Scaling properties of a percolation model with long range correlations. *Physical Review E*, 54: 3870, 1996.

Scher, H. and Zallen, R., Critical density in percolation processes. *Journal of Chemical Physics*, 53: 3759–3761, 1970.

Schmittbuhl, J., Vilotte, J.P. and Roux, S., Percolation through self-affine surfaces. *Journal of Physics A: Mathematical and General*, 26, 6115, 1993.

Soares, R.F., Corso, G., Lucena, L.S., Freitas, J.E., da Silva, L.R., Paul, G. and Stanley, H.E., Distribution of shortest paths at percolation threshold: application to oil recovery with multiple wells. *Physica A: Statistical Mechanics and its Applications*, 343: 739–747, 2004.

Sreenivasan, S., Bakera, D.R., Paula, G. and Stanley, H.E., The approximate invariance of the average number of connections for the continuum percolation of squares at criticality. *Physica A: Statistical Mechanics and its Applications*, 320: 34–40, 2003.

Tavagh-Mohammadi, B., Masihi, M. and Ganjeh-Ghazvini, M., Point-to-point connectivity prediction in porous media using percolation theory. *Physica A: Statistical Mechanics and its Applications*, 460: 304–313, 2016.

Watanabe, H., Yukawa, S., Ito, N. and Hu, C.-K., Superscaling of percolation on rectangular domains. *Physical Review Letters*, 93(19): 190601.1–190601.4, 2004.

Xia, W. and Thrope, M.F., Percolation properties of random ellipses. *Physical Review A*, 38: 2650–2656, 1988.

Chapter 5

Percolation and fracture systems

Flow through fractured rocks is important in many disciplines such as hydrology, reservoir engineering, nuclear waste disposal, CO_2 sequestration and geothermal reservoirs. The presence of fractures in the medium may affect all aspects of the flow behaviour or be critical for the suitability of a rock mass for storage.

We described a very basic fracture network model in Chapter 3 in a simple continuum percolation framework. The model discussed there used constant-size line segments distributed randomly in a 2D system. In reality, fractures have different sizes, apertures, orientations, spacings, shapes, matrix/fracture permeability contrast. Most crucially, they have a strong spatial correlation. Previous investigations of intersecting line segments include geometrical studies (Huseby *et al.*, 1997), permeability studies (Koudina *et al.*, 1998; Leung and Zimmerman 2012), effects of fracture size (Mourzenko *et al.*, 2004 2005), anisotropic effects (Khamforoush *et al.*, 2008; Mourzenko *et al.*, 2011), the contribution of the matrix permeability (Bogdanov *et al.*, 2003, 2007) and fractures with different aperture or permeability variation (Hamzehpour *et al.*, 2009; Leung and Zimmerman 2012). It is important to know how these factors alter the percolation behaviour.

Consider the orientations of fractures. Typically fractures form in two or more families of approximately parallel fractures called fracture sets. Different fracture sets often result from successive tectonic events (Bear *et al.*, 1993; Sahimi, 1995; Saidi, 1987; Adler and Thovert, 1999) which can cause anisotropy in the connectivity or conductivity.

The connectivity of fracture networks also depends on the fracture spacing, defined as the perpendicular distance between two neighbouring fractures of the same set. This is often related to both the degree of rock deformation and the layer thickness (Rives *et al.*, 1992; Wu and Pollard, 1995). As pointed out by Rives *et al.* (1992), at the early stages of fracture set development, fractures have a negative exponential distribution for the spacing due to the random positioning. By increasing the degree of rock deformation, the fractures become more organised and the spacing evolves to a log normal and then a normal. Field observations and numerical analysis have also shown the existence of different forms of distribution from normal to exponential and power law as well as a variation in the dependence of spacing on the bed thickness (Priest and Hudson, 1976; Hudson and Priest, 1979; Sen and Kazi, 1984; Narr and Suppe, 1991). It is believed that the incorporation of unrefined data sets of joints from several joint sets leads to the negative exponential distribution for the fracture spacing, whereas data for a single joint set provide a lognormal distribution (Gillespie *et al.*, 1993). The statistics of fracture spacing can also be used to describe the clustering of the fracture system (Odling *et al.*, 1999) and also influences the fracture network connectivity.

Another important feature which controls the connectivity is the distribution of fracture lengths. Several distributions have been found in nature for fracture lengths. These include the power law (Segall and Pollard, 1983; Heffer and Bevan, 1990; Odling, 1997; van Dijk *et al.*, 2000), lognormal (Rouleau and Gale, 1985; Odling, 1997) or exponential (Priest and Hudson, 1981; Rouleau and Gale, 1985). It should be noticed that truncation of the short lengths (due to resolution) and the censoring of the long fractures (due to finite sample size) alter the appearance of the power law distribution. This is why field data sometimes appear to fit better with something other than a power law distribution.

For hydraulic conductivity, the fracture aperture (defined as the perpendicular distance between the two opposite walls of the fracture) is important. It is positively correlated with the fracture size (Vermilye and Scholz, 1995; Olson, 2003). Both lognormal (Rouleau and Gale, 1985; Bear *et al.*, 1993) and power law distributions (Bonnet *et al.*, 2001) have been used to fit the apertures. The conductivity inside a single fracture may also be non-uniform due to the distribution of roughness along the fracture wall.

The spatial distribution of fractures may not be uniform which would affect the connectivity. There may also be correlations between the geometrical properties of the fracture. For example, there may be a correlation between fracture positions and their orientation or size (Darcel *et al.*, 2003a). This can be supported by geomechanical arguments dealing with fracture interactions and growth (Renshaw and Pollard, 1994). For instance, Bour and Davy (1999) analysed the fracture network and found a cross-correlation between the position of a facture and its length so that large fractures have their neighbours located at a greater distance than small fractures.

We wish to investigate how these new features of fracture networks affect the percolation properties. Field data obtained from large-scale investigations show that the fracture network structure is close to the threshold and displays strong channelling patterns which can be explained by percolation theory (Cacas *et al.*, 1990). It was found that in some natural fracture networks, only a small percentage of fractures contributed to the permeability of the system (Sahimi, 1995). The large uncertainty associated with sparse data and the lack of distinction between faults and joints makes it debatable whether or not natural fractures are well above the percolation threshold (highly interconnected) or near the threshold (poorly connected) (Berkowitz *et al.*, 2000). As pointed out by Berkowitz (2002), in some circumstances, fracture networks seem to be highly connected, whereas in many other cases, where the fractures are created as a results of stress, the network is poorly connected which indicates that it is near the percolation threshold. These studies indicate that the application of percolation theory in many realistic fracture networks is reasonable. In this chapter, we will extend the basic fracture model to incorporate the various effects described above.

5.1 Effects of fracture size variation

In reality, fractures have a distribution of lengths. Often a power law is found to be a good fit for the fracture length distribution (Heffer and Bevan, 1990; Renshaw, 1999; van Dijk *et al.*, 2000; Bonnet *et al.*, 2001; Bogdanov *et al.*, 2007). The effect of the fracture length distribution on the fracture connectivity and the percolation threshold of fracture networks has been extensively studied (Robinson, 1983; Balberg *et al.*, 1984; Charlaix *et al.*, 1984; Balberg, 1985; Berkowitz, 1995; Berkowitz and Balberg, 1993; Bour and Davy, 1997, 1998; Renshaw, 1999; Rossen *et al.*, 2000; Sisavath *et al.*, 2004; Mourzenko *et al.*, 2005; Leung and Zimmerman 2012). When the fractures have a length distribution with no characteristic length (such as a power law), the size of the boundaries introduces the only length scale (Rossen *et al.*, 2000).

In 2D, the excluded area as shown in Figure 3.14 was used to account for the effects of fracture orientation distribution (Balberg *et al.*, 1984; Adler and Thovert, 1999). The average excluded area for constant-length fractures with a random orientation distribution is $l^2(2/\pi)$. The general form of the average excluded area for randomly oriented fractures can be defined as (Robinson, 1983; Balberg *et al.*, 1984; Adler and Thovert, 1999; Masihi *et al.*, 2006a)

$$\langle a_{\mathrm{ex}}\rangle = \langle l^2\rangle(2/\pi). \tag{5.1}$$

In Chapter 4, we used the hypothesis that there is an effective length l_{eff} that controls the connectivity when the fractures have different lengths. So, comparing Eq. (5.1) with the excluded area for constant-length fractures suggests an effective length of

$$l_{\mathrm{eff}} = \sqrt{\langle l^2\rangle}. \tag{5.2}$$

We expect that the average excluded area based on the second moment given by Eq. (5.1) is more appropriate than one based on the first moment, e.g., $\langle l\rangle^2(2/\pi)$, when the length distribution has a large standard deviation. This is because for many distributions (such as the uniform or Gaussian), the first moment is independent of the standard deviation of the distribution. However, observations indicate that very long fractures have an undue influence of larger fractures (Balberg *et al.*, 1984). For very broad length distributions, there is reasonable probability that one very long fracture exists that connects the whole system.

In fracture networks of varying length, the dimensionless effective system size is defined as

$$L_x = \frac{X}{\langle l_x \rangle} = \frac{X}{(2/\pi) l_{\text{eff}}}, \tag{5.3}$$

where X is the system size along the x axis.

We also need to define the equivalent of the percolation probability P. It is clear that the fracture length distribution in the spanning cluster is different from the fracture length distribution of the whole network. This means that we cannot use the fracture number ratio (number density) for the connected fraction of fractures P. Using the excluded area arguments, the connected fraction P is defined as

$$P = \frac{n_s \langle l^2 \rangle_s}{n_w \langle l^2 \rangle_w} = \frac{\sum_s l_i^2}{\sum_w l_i^2}, \tag{5.4}$$

where the subscripts s and w denote respectively the spanning cluster of fractures and the whole fracture network. For isotropic constant-length fractures, the average in Eq. (5.4) is irrelevant, and so Eq. (5.4) reduces to Eq. (3.6). Clearly, the length distribution in both the spanning cluster and the whole network is the same.

For finite-length-scale distributions, we examine the hypothesis that connectivity is controlled by the effective length given by Eq. (5.2). The effective lengths based on the second moment for uniform, Gaussian, lognormal and exponential length is given in Table 4.1. The same computational algorithm as described in Section 3.2.2 can be used to generate fracture realisations and calculate the percolation quantities.

To investigate the connectivity of random length fractures, we distribute fractures randomly in a square region of size 100×100. We consider two length distributions: one with a normal length

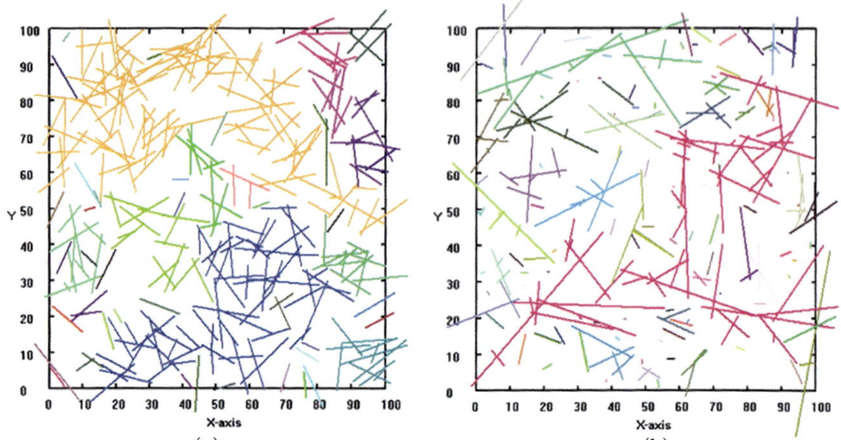

(a) (b)

Figure 5.1. Two realisations of randomly oriented fractures at the threshold having fracture lengths with (a) a normal distribution ($\bar{l} = 12$, $\sigma = 4$, $N = 298$, $P = 0.38$) and (b) exponential distribution ($\bar{l} = 10$, $N = 235$, $P = 0.45$).

distribution (mean length 12 and standard deviation 4) and the other with an exponential length distribution (mean length 10). Figure 5.1 shows two realisations of randomly oriented fractures with fracture occupancy density of $p = 0.53$ (using definition from Eq. (3.3)) which is close to the threshold made by fractures with these two distributions of lengths.

From Table 4.1, the effective lengths controlling the connectivity of these fracture systems are $l_{\text{eff}} = 12.65$ (normal) and $l_{\text{eff}} = 14.14$ (exponential). By using the size dependence of the threshold and the other methods described in Chapter 2, the estimated values of the exponents and threshold in the case of normal (exponential) distributions are $1/\nu = 0.64 \pm 0.08(1/\nu = 0.60 \pm 0.13)$, $\beta/\nu = 0.14 \pm 0.05(\beta/\nu = 0.13 \pm 0.03)$ and $p_c = 0.598 \pm 0.011$ ($p_c = 0.594 \pm 0.037$). These all confirm that the universality hypothesis holds for these systems. The finite-size scaling for each of these systems is shown in Figure 5.2. These demonstrate that we can use the standard isotropic master curves ($\Im$ and $\Re$) that were found for the constant-length fractures when suitably modified with the effective lengths.

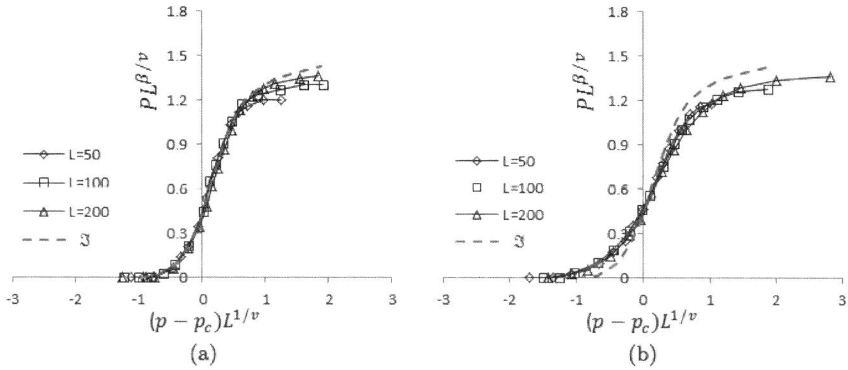

Figure 5.2. Illustration of the rescaled mean connected fraction in fracture networks with fracture length distributions: (a) normal distribution with effective length equals 12.65 and (b) exponential with effective length equals 14.14.

As stated previously, the use of a simple effective length may not be appropriate when there is a very broad distribution of lengths. In this case, a small number of very long fractures may dominate the connectivity. Indeed, if the length distribution has no characteristic scale, the only length scale of relevance is the system size. There is a probability that there will be at least one fracture that crosses the entire system and is solely responsible for the connectivity. This is most clearly demonstrated by using a power law length distribution (Renshaw 1999; Bonnet *et al.*, 2001; de Dreuzy *et al.*, 2004). This control over connectivity by one large fracture may change the nature of the finite-size-scaling law and the universality of percolation theory (Berkowitz, 1995; Adler and Thovert, 1999; Renshaw, 1999; Bonnet *et al.*, 2001; de Dreuzy *et al.*, 2004; Masihi *et al.*, 2007; Sadeghnejad *et al.*, 2013). The probability density function for the power law distribution is defined as

$$P(l) = cl^{-\alpha}, \tag{5.5}$$

where c is a normalisation coefficient and α is the (not necessarily integer) length exponent This exponent is typically found from outcrop studies to be in the range of 1.7–2.75. In nature, power laws cannot really hold over all length scales but must be limited by lower and upper bounds $l \in [l_{\min}, l_{\max}]$; other than these, there are no characteristic lengths. For a large value of the exponent α, almost all fractures will have the same (minimum) length, while for a small exponent α, the probability that most of the fractures are of the maximum length increases. To analyse the impact of the power law exponent on connectivity of fracture systems and see different regimes of connectivity, we use a wider range of power law exponents compared with those found in field observations. However, we do not consider the case $\alpha < 1$ because it generates more large fractures than small ones which would be inconsistent with geological observations (Bonnet *et al.*, 2001).

In this section, randomly oriented fractures with power law length distribution are distributed uniformly within a square. The lower cut-off for the length distribution is arbitrarily chosen to be $l_{\min} = 1$. The longest fracture included is much larger than the system size L. However, for such fractures, only that part inside the square ($\sim L$) affects the system connectivity. Three realisations of the fractures with a power law length distribution with exponent $\alpha = 5.5$, 2.8 and 1.8 in the system of size $L = 50$ are shown in Figure 5.3.

There are three regimes of connectivity depending on the exponent α. As can be seen in Figure 5.3, for $\alpha > 3$, almost all of the fractures generated are of the minimum length ($l_{\min} = 1$) and small fractures rule the connectivity In this case, the classic percolation theory, as described in previous chapters, applies. For $\alpha < 2$, the connectivity is controlled by the largest fractures in the system. For $2 < \alpha < 3$, both small and large fractures contribute to the connectivity. As can be seen, for $\alpha < 2$, the number of very long fractures increases and so the number required to connect opposite sides decreases significantly. Also, in this case, removing the small fractures does not alter the connectivity significantly. Comparing the length distribution in the whole network with those in the spanning cluster shows that for $\alpha > 3$, the length distribution in both networks are similar, but for $\alpha < 2$, the discrepancy between the statistics of these two classes increases as α decreases.

As before, we use the hypothesis that an effective length controls the connectivity. Hence, considering the cut-offs on the power law distribution, the second moment of fracture length distribution

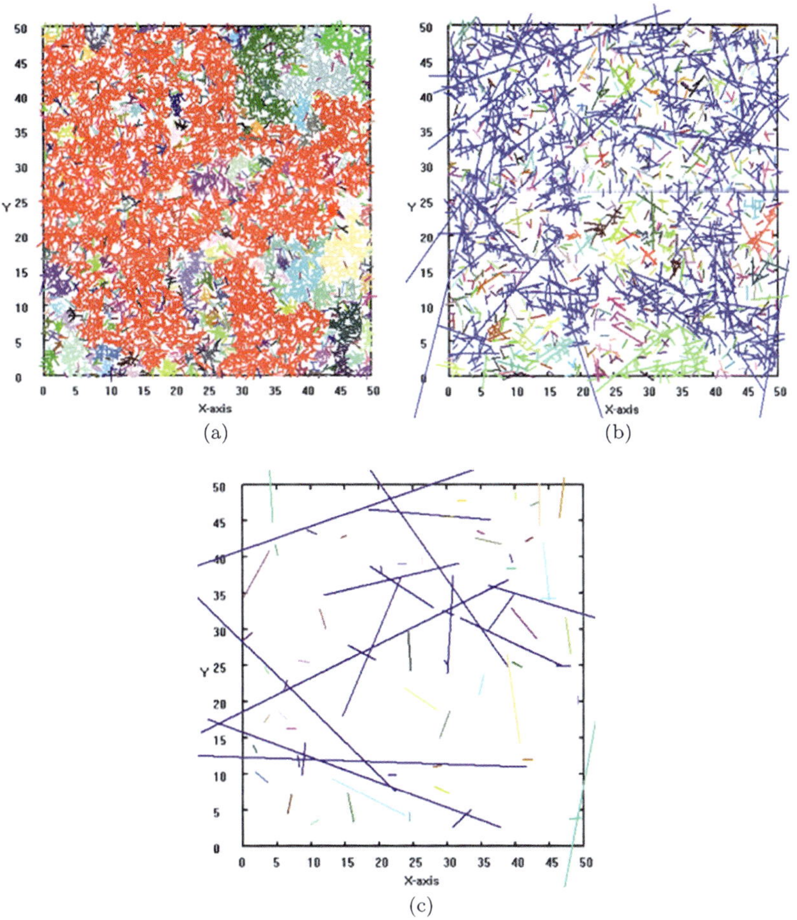

Figure 5.3. Three realisations of randomly oriented fractures at the threshold having a power law length distribution with (a) $\alpha = 5.5$, $N = 7{,}914$, $p = 0.596$, $P = 0.617$, (b) $\alpha = 2.8$, $N = 2{,}062$, $p = 0.815$, $P = 0.869$ and (c) $\alpha = 1.8$, $N = 72$, $p = 0.562$, $P = 0.83$.

(which is used to define the effective length) becomes

$$\begin{aligned} \langle l^2 \rangle &= \int_1^L l^2 P(l)dl + \int_L^{l_{\max}} L^2 P(l)dl \\ &= c(L^{3-\alpha} - 1)/(3-\alpha) + cL^2(l_{\max}^{1-\alpha} - L^{1-\alpha})/(1-\alpha) \quad (\alpha \neq 1, 3) \\ &= c \ln L - cL^2(l_{\max}^{-2} - L^{-2})/2 \quad (\alpha = 3) \\ &= c(L^2 - 1)/2 + cL^2 \ln(l_{\max}/L) \quad (\alpha = 1), \end{aligned} \tag{5.6a}$$

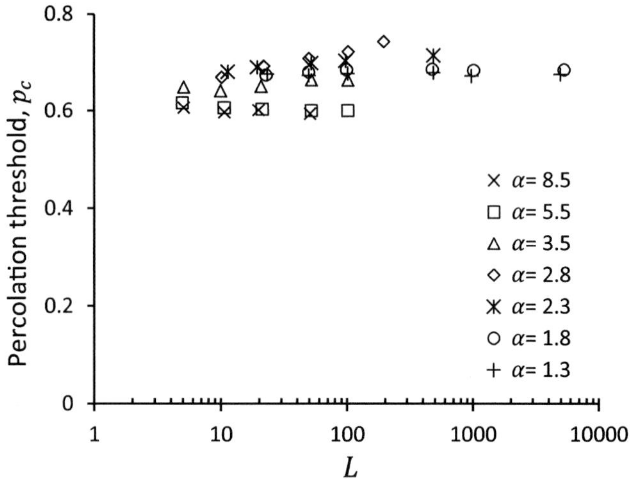

Figure 5.4. The threshold values in fracture networks with power law fracture length distribution as a function of system size for various values of the exponent α.

where the normalisation coefficient c is

$$c = \begin{cases} \dfrac{1-\alpha}{(l_{\max}^{1-\alpha} - 1)} & (\alpha \neq 1), \\ 1/\ln l_{\max} & (\alpha = 1). \end{cases} \tag{5.6b}$$

Given these, the value of the percolation threshold as a function of the system size is shown in Figure 5.4.

This indicates that the variation of the threshold with the system size is not very significant, especially given the errors in estimating the power from outcrop data. Also, the critical exponents (ν and β/ν) can be estimated through the scaling dependency of the error in the threshold (Eq. (2.28)) and the connected fraction at the threshold (Eq. (2.31)). The results of this scale dependency are shown in Figure 5.5.

The estimation of the critical exponents in Figure 5.5 and the estimation results given in Figure 4.6 are similar and as illustrated in Figure 5.6 show that the dependency of the critical exponents on the power law exponent α is independent of the object shape, i.e.,

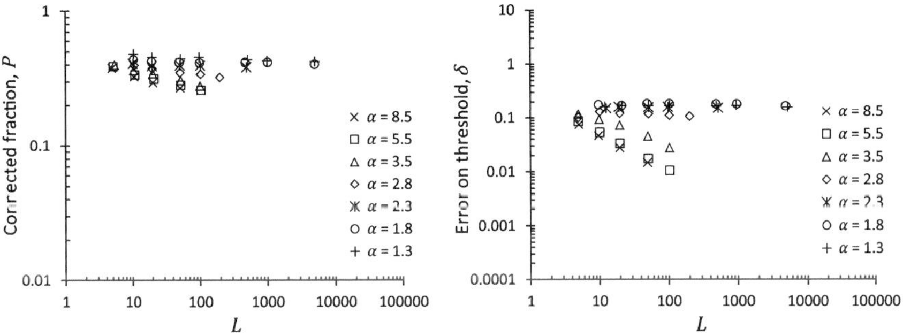

(a) (b)

Figure 5.5. (a) The connected fraction of fractures and (b) the variation of the error δ in the threshold against the system size for different values of the power law length exponent α.

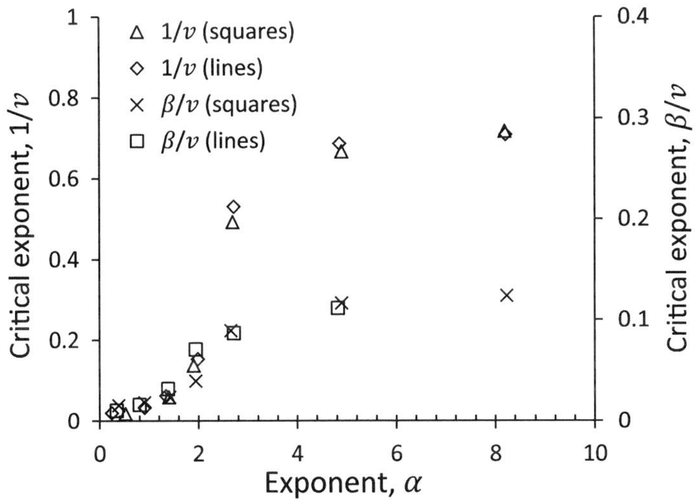

Figure 5.6. The dependency of the critical exponents on the power law exponent α in two different systems made by line segments and squares.

whether the objects are line segments or squares (Sadeghnejad *et al.*, 2013).

Applying the finite-size-scaling laws using the standard values for the critical exponents (Table 2.2) does not give a data collapse for the connectivity of fractures with a power law length

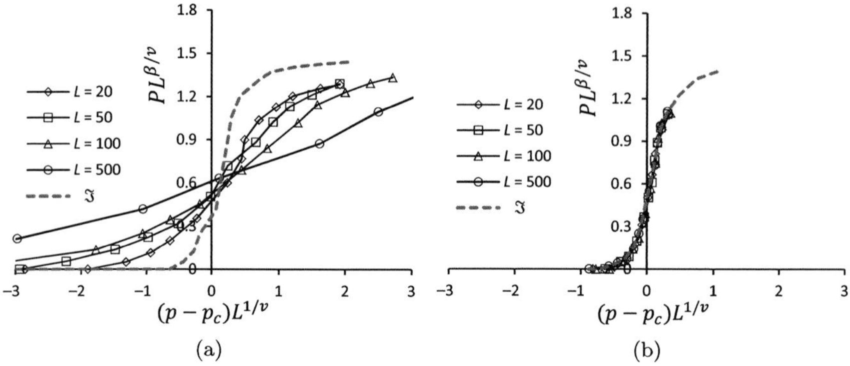

Figure 5.7. Illustration of the finite-size scaling for fracture systems with a power law distribution with exponent $\alpha = 2.3$ with (a) standard values for the critical exponents from Table 2.2 and (b) the corrected values for the critical exponents from Figure 5.6.

distribution (Figure 5.7). However, if the corrected critical exponents from Figure 5.6 are used ($1/\nu = 0.05$ and $\beta = 0.76$), a reasonable data collapse for the mean connectivity is found (Figure 5.7). This again shows that the hypothesis that the system can be controlled by the effective length is applicable to very wide length distributions provided that the corrected critical exponents are used.

5.2 Effects of fracture orientation variations

The simple fracture model described so far assumes that fractures have random orientations. In the field, fractures are often observed to lie in clear sets within which they have near parallel orientations. The orientation distribution of fractures affects the percolation threshold but not the critical exponents (Balberg and Binenbaum, 1983; Robinson, 1983; Balberg *et al.*, 1984; Masihi *et al.*, 2006b). Also, if there are any preferred directions around which the fractures' orientations are clustered, there will be a macroscopic anisotropy in the connectivity. This will be discussed later.

In this section, we concentrate on the simple case where there are just two orientations for which the orientation distribution is

$$P(\theta) = \rho\delta(\theta - \theta_1) + (1 - \rho)\delta(\theta - \theta_2), \tag{5.7}$$

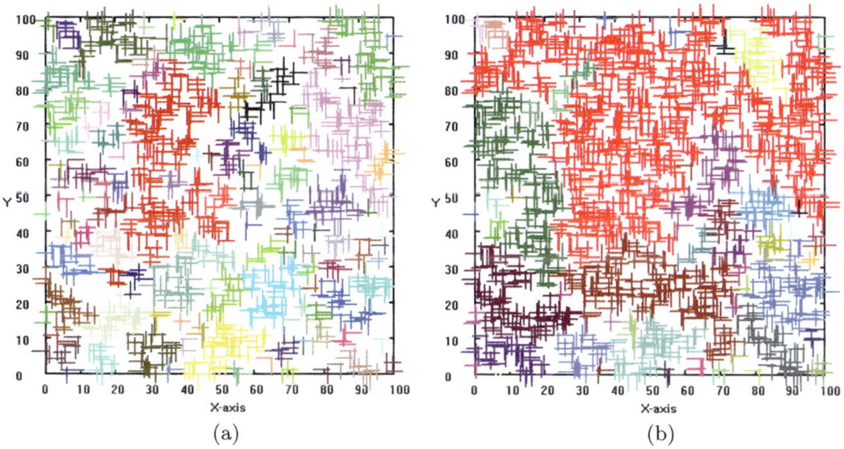

Figure 5.8. Two realisations of the horizontal–vertical fracture model with fracture of size $l = 5$ in the system of size $L = 100$ (a) before the threshold with $N = 1{,}915$ where different colours show different finite clusters and (b) at the threshold with $N = 2{,}443$ where connected cluster consists of 993 fractures.

where δ is the Dirac delta function and ρ is the fraction of fractures in the direction θ_1. To illustrate this (Figure 5.8), we use the following parameters $\rho = 0.5$, $\theta_1 = 0, \theta_2 = \pi/2$.

In these examples, the connectivity is isotropic, but if the fracture sets are not orthogonal or when the number density of each fracture set is different, then the resulting fracture model becomes anisotropic. The value of the percolation threshold for the constant-length orthogonal fractures is $I_c \approx 3.2$ (Robinson, 1984; Balberg, 1985; Adler and Thovert, 1999) and $p_c = 0.538 \pm 0.003$ which is lower than the threshold value of the fracture system with random orientation ($I_c \approx 3.6$ and $p_c = 0.590 \pm 0.003$). However, the critical exponents of this model are not changed) as shown in Figure 5.9.

Hence, the same connectivity scaling (Eqs. (2.11) and (2.14)) can be used to estimate the mean connectivity and standard deviation in connectivity. Figure 5.10 compares the mean connectivity and standard deviation in connectivity results for the randomly oriented fracture model and horizontal–vertical fracture model. The connectivity master curves of both models are very similar to those obtained

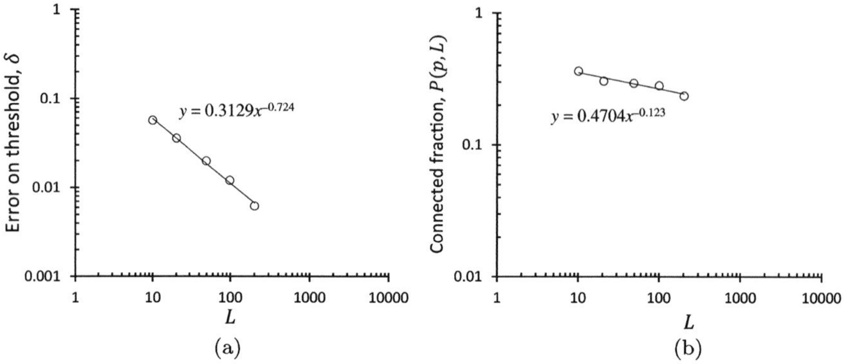

Figure 5.9. Estimation of critical exponents ν and β from (a) log–log plot of the error δ obtained on the threshold against L and (b) the scale dependency of the percolation probability P.

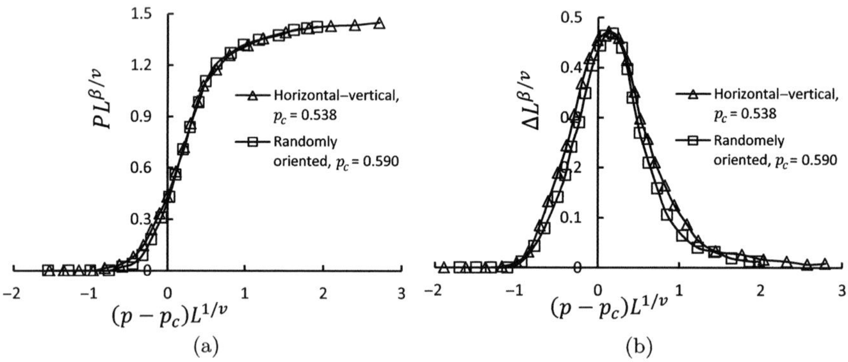

Figure 5.10. Comparison of master connectivity curves for the randomly oriented fracture model and horizontal–vertical fracture model. (a) Master curve for mean connectivity results and (b) master curve for the standard deviation in connectivity results.

before provided we use the correct percolation threshold for the model.

5.3 Effects of anisotropy in the system

In fractured media, fracture sets with particular orientations are typically formed as a result of the tectonic history (Harris, 1992; Bear *et al.*, 1993; Sahimi, 1995; Taylor *et al.*, 1999; Adler and Thovert,

1999; Zhang and Sanderson, 2002). It has been shown that the critical length required for the onset of percolation increases with the macroscopic anisotropy caused by a preferred orientation of line segments (Balberg and Binenbaum, 1983). Masihi *et al.* (2007) have shown how the existence of fracture sets with two different orientations may lead to anisotropy in the system by creation of an *easy* direction for the connected paths and a *difficult* direction. In this section, we discuss the effects of preferred orientation of fractures on connectivity.

If all the fractures were perfectly aligned, they would not intersect to form connected clusters as they are assumed to be infinitesimally thin. However, in reality, the fractures have some kind of orientational disorder. Assume that the orientation of the fractures has a uniform probability distribution in the range $[-\theta_0, +\theta_0]$ around the x axis, for various θ_0. The average projected length of the fractures in both x and y directions are

$$\langle l_x \rangle = \frac{l}{\theta_0} \int_0^{\theta_0} \cos\theta \, d\theta = l \frac{\sin\theta_0}{\theta_0}, \tag{5.8a}$$

$$\langle l_y \rangle = \frac{l}{\theta_0} \int_0^{\theta_0} \sin\theta \, d\theta = l \left\{ \frac{1 - \cos\theta_0}{\theta_0} \right\}. \tag{5.8b}$$

As with the other anisotropy models described in previous chapters, there are two effective length scales involved: $L_x = L/\langle l_x \rangle$ and $L_y = L/\langle l_y \rangle$. The aspect ratio representing the anisotropy of the system is $\omega = L_x/L_y$. This model, which is globally isotropic but locally anisotropic, becomes locally isotropic by shrinking the x direction by a factor of ω. This is equivalent to the globally anisotropic lattice model. The transformation does not alter the dimensionless lengths L_x and L_y.

The same definition as before can be used for the vertical connected fraction P_x and the horizontal connected fraction P_y. Two realisations of an anisotropic system with constant-length fractures ($l = 5$) with $1/\omega = 3.73$ ($\theta = 30°$) at the threshold p_c are shown in Figure 5.11. This shows that the "easy" direction is parallel to the x axis (total fractures $=$ 3,802 with 432 fractures in the

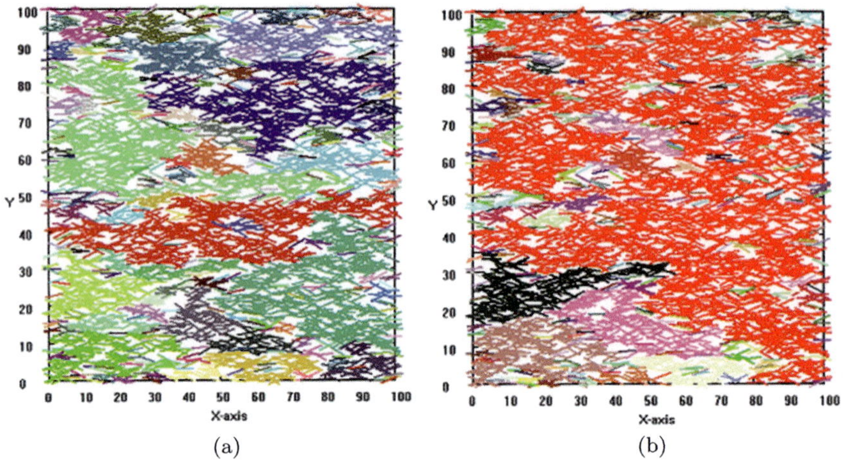

Figure 5.11. Two realisations of an anisotropic fracture system with the percolating cluster in red at the threshold showing (a) easy connectivity direction along the x axis and (b) difficult connectivity direction along the y axis.

percolating cluster) and a "difficult" direction is along the y axis (total fractures = 4,184 with 2,723 fractures in the percolating cluster).

Having collected the statistics of the horizontal and vertical connected fractions of fractures over all realisations, we carry out finite-size-scaling transformations (Eqs. (2.11) and (2.14)) by using the length scale L_x as before. The rescaled horizontal and vertical mean and standard deviation of connectivity results for anisotropic fractures with a uniform distribution of fracture orientation in range $[-\pi/6, \pi/6]$ around x axis is shown in Figure 5.12.

These connectivity results show a data collapse but in two sets as expected for anisotropic systems. As with the sandbody model, the connectivity curves are displaced from the universal connectivity curves (i.e., $\Im$ and $\Re$). This indicates that we should use two different apparent thresholds $\tilde{p}_c$ in each direction as before (Eq. (2.35)). The existence of two apparent percolation thresholds in the x and y directions can be seen by plotting the correlation length against the percolation parameter p; this is shown in Figure 5.13 for a system size $L = 20$ and $\theta_0 = \pi/6$.

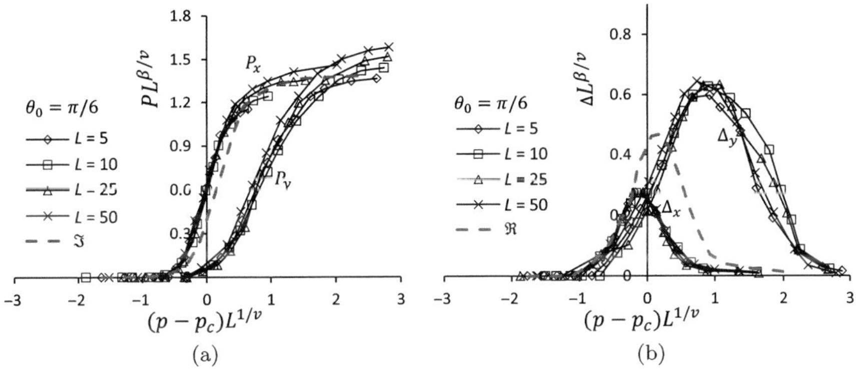

Figure 5.12. Finite-size scaling of the (a) mean and (b) standard deviation of connected fraction of fractures in the horizontal and vertical directions that shows displacement about the universal connectivity master curves.

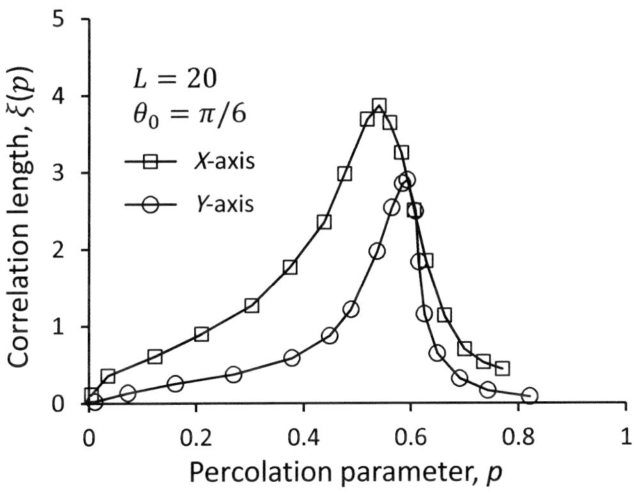

Figure 5.13. The correlation length in the horizontal and vertical directions against the percolation parameter in the fracture system showing two peaks corresponding to two thresholds.

The same correction for the apparent threshold of anisotropic systems (Eq. (2.38)) found previously is applicable here. It can be used to bring the curves back to the same isotropic universal mean connectivity curve. The numerical results obtained for the apparent thresholds in both x and y directions show the same symmetry as

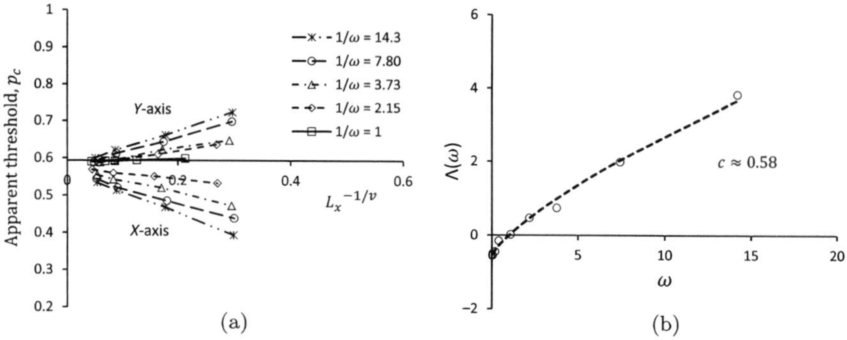

Figure 5.14. (a) The variation of the apparent thresholds of fracture systems in the x and y directions against $L_x^{-1/\nu}$ showing the certain symmetry and (b) the shift in the apparent threshold Λ as a function of aspect ratio (Eq. (2.38)) in fracture system.

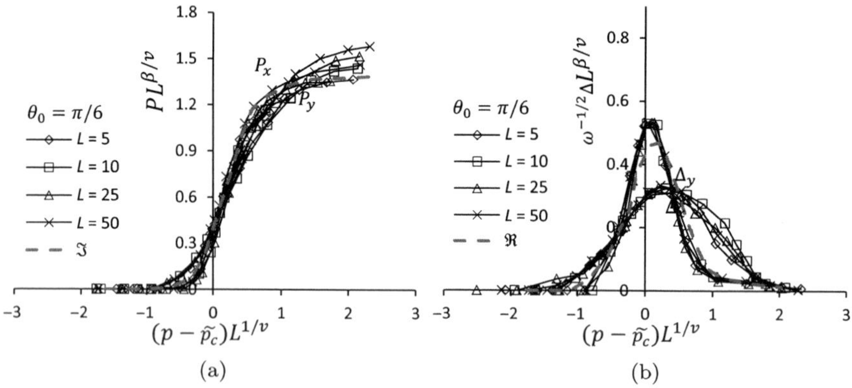

Figure 5.15. Illustration of correct finite-size scaling for the (a) mean and (b) standard deviation of connectivity results in fracture system.

Eq. (2.34) and also fit the suggested functional form (Eq. (2.38)) with coefficient $c \approx 0.58$ (Figure 5.14).

Using these corrected threshold values, the curves for the anisotropic fracture systems collapse onto the same universal curves (Figure 5.15). As before, we can use this finite-size scaling as a good approximation to predict the connectivity in anisotropic fracture systems.

5.4 Effects of spatial correlation on fracture connectivity

The fracture models considered so far all assume fractures to be spatially uncorrelated. They also neglected any relationship between model parameters such as the fracture position, length and orientation. These assumptions are not born out of observations of natural fractures over various scales (Bour and Davy, 1997).

There has been very little investigation on the effect of short- and long-range correlations on the percolation properties of fracture networks. Increasing the fracture clustering at small scale may improve the connectivity of the system, but at large scale, it produces zones with very few fractures. This may significantly decrease the connectivity. Fracture patterns have been described by using fractals (see Section 2.7). Fractal statistics can be used to describe the spatial distribution of fractures (Berkowitz, 1995; Bonnet *et al.*, 2001). As the fracture pattern is not regular, we can use the box counting method (Mandelbrot, 1982) as described in Section 2.7 to estimate the fractal dimension D of a fracture system (see Figure 5.16).

Field data shows that the distribution of fracture centres has long-range correlations with a fractal dimension in the range of $D \in [1.5, 2]$ (Bonnet *et al.*, 2001). The lower limit corresponds to the strong correlation (all the fractures cluster in a narrow band), whereas the upper limit ($D = 2$) shows the uncorrelated fractures.

The consequences of fractal correlation on the percolation quantities of fracture networks have been studied in different ways (Watanabe and Takahashi, 1995; Berkowitz *et al.*, 2000; Darcel *et al.*, 2003b; de Dreuzy *et al.*, 2004; Heffer and Bevan, 1990). Watanabe and Takahashi (1995) demonstrated that fewer fractures are required to connect two wells for their fracture network model compared with the uncorrelated fracture model. Darcel *et al.* (2003b) have shown that the regime of connectivity can be identified depending on the competition between the fractal dimension D and the exponent α of the power law length distribution. In the first regime ($\alpha < D + 1$), fractal correlation has a negligible contribution to the connectivity, whereas for $\alpha > D + 1$, the connectivity is determined by the contribution of small-scale fractures and significantly controlled by

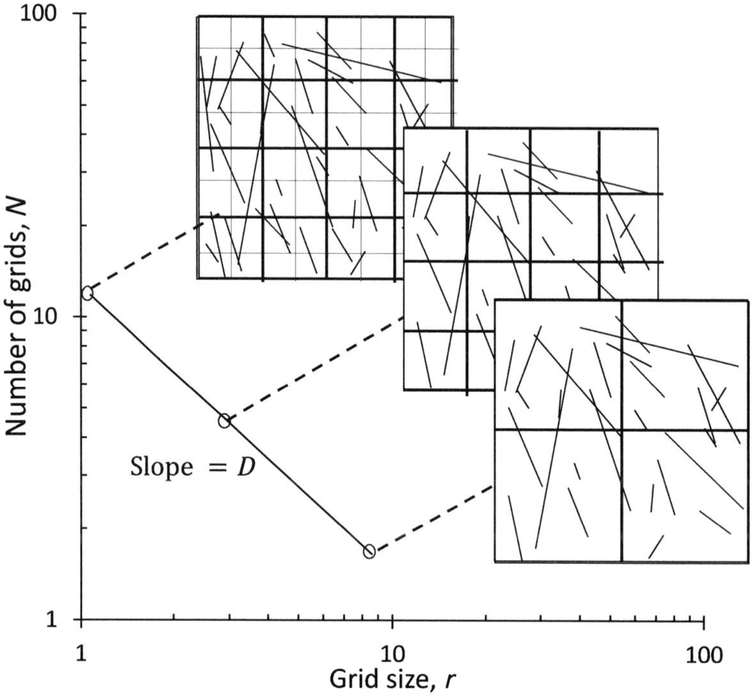

Figure 5.16. Illustration of box counting method to calculate the fractal dimension of a fracture system by covering the region with a grid of size r and counting the number of cells N inside which the fracture is presented and repeating it for different grid sizes. The slope of log–log plot of N versus r gives the fractal dimension D.

spatial correlation. Figure 5.17 shows three realisations of fracture systems with a power law length distribution of these regimes of connectivity at different fractal dimensions.

Moreover, as pointed out by Berkowitz *et al.* (2000), the connectivity of the fracture patterns in the first regime is scale dependent, whereas for the latter regime, the connectivity will be independent of the scale of measurement. On the large scale, the correlation and length distribution have opposite effects on connectivity (de Dreuzy *et al.*, 2004). Increasing the correlation strength decreases the connectivity, whereas increasing the width of the length distribution

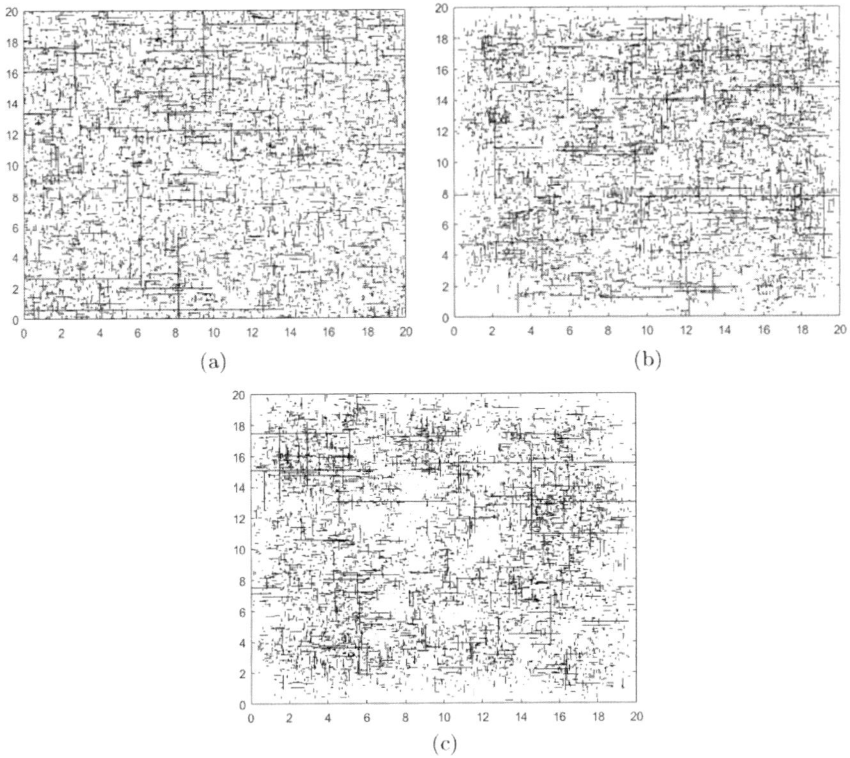

Figure 5.17. Illustration of the effect of the fractal dimension on the fracture network in a square of side 20 with a power law length distribution with an exponent $\alpha = 2.7$ and fractal dimension (a) $D = 2$, (b) $D = 1.8$ and (c) $D = 1.6$.

increases the connectivity. The numerical analysis of the percolation behaviour of systems with long-range correlation has shown that the critical exponents may change (Prakash *et al.*, 1992; Schmittbuhl *et al.*, 1993), whereas for systems with short-range correlation, the principle of universality is still valid (Harter, 2005).

The existence of potential correlation between geometrical properties of fractures (such as fracture local positions and their orientations or sizes) has been addressed by analysing fracture networks (Bour and Davy, 1999; Darcel *et al.*, 2003a; Masihi and King, 2007a,

2007b) or by means of geomechanical arguments which deal with the interactions between fractures (Renshaw and Pollard, 1994). Bour and Davy (1999), for instance, analysed data and found a correlation between the position of a fracture and its length so that large fractures have their neighbours located at greater distance than small fractures.

To account for the spatial distribution of fractures and to analyse its impact on the percolation properties, we first need an appropriate technique to produce an ensemble of realisations of correlated fracture systems. The fracture networks can be generated by using either a process-based method (by modelling the physics of the geomechanics) or a stochastic approach. These realisations are used within the framework of percolation theory to analyse connectivity and conductivity. In the presence of spatial correlations, the definition of the percolation parameter, equivalent to the occupancy probability p, is not straightforward. As pointed out by Darcel *et al.* (2003b), neither the mean fracture density $\rho = Nl^2/L^2$ proposed by Bour and Davy (1997) nor the mean fractal fracture density $\rho_D = Nl^D/L^D$ suggested by Berkowitz *et al.* (2000) is able to give the connectivity state of the constant-length fractal fracture networks. This needs further clarification.

Masihi and King (2007a) introduced a fracture model with long-range correlations. The fractures had a spatial correlation in the elastic displacement developed by Heffer and King (2006) by using simulated annealing. The algorithm produces fracture networks which follow a power law length distribution. As an example, the rescaled connectivity curves for four such fracture networks using literature values of the critical exponents ($\nu = 4/3, \beta = 5/36$ from Table 2.2) are shown in Figure 5.18. This shows that the calculated connectivity values are outside the master connectivity curves (the range $P \pm \Delta$). Masihi and King (2007b) showed that the critical exponents must be changed in order to bring the calculated connectivity curves along the principal axes of the system back to the master connectivity curves (Figure 5.18).

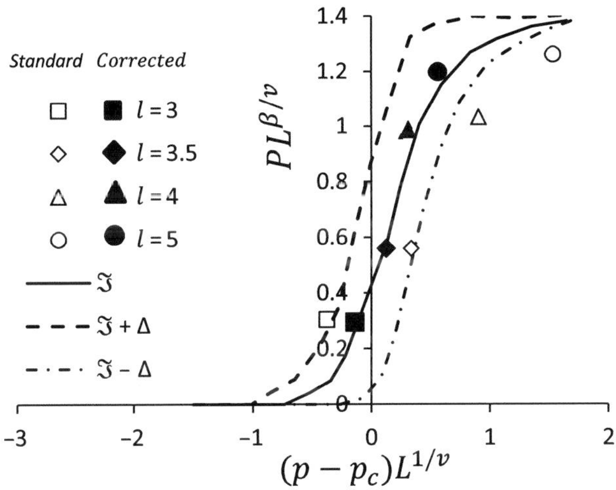

Figure 5.18. Comparison of the computed connectivity of the correlated fracture networks (with a power law length distribution at average fracture lengths 5, 4, 3.5 and 3) with the universal connectivity curves (solid and dotted curves) using standard values for the exponent (Table 2.2) and the corrected values $1/\nu = 0.4$ and $\beta/\nu = 0.09$.

5.5 Effective permeability of fracture systems

So far, we have discussed the connectivity of fracture systems. The percolation approach can also be used to find the dynamic behaviour of fracture networks, the simplest example of which is the effective (single-phase) permeability. Both fractures and the matrix can contribute to the permeability of a fractured porous medium. In some cases, the permeability contrast between the two media is not that large and so the permeability of the background matrix is also important. This has been studied previously (Zimmerman and Bodvarsson, 1996; Matthäi and Belayneh, 2004; Savik *et al.*, 2013). Moreover, the matrix may have a permeability distribution or the permeability of each fracture may not be the same (Bogdanov *et al.*, 2003; Mourzenko *et al.*, 2011; Savik *et al.*, 2013). We consider the case where flow only takes place in the fractures and the background matrix is completely impermeable. We will generalise this later.

The first estimate of the effective permeability for fractures was done by Snow (1969) who used smooth parallel plane walls of infinite extent to represent the fracture. Since then Snow's work has been extended to finite-length fractures. There have also been numerical simulation studies of fluid flow in 2D fracture networks to investigate the hydraulic conductivity dependency on the fracture interconnection, the average number of intersections per fracture (Long and Witherspoon, 1985; Hestir and Long, 1990; Leung and Zimmerman 2012). The effective medium approximation of Kirkpatrick in 1973 (Zimmerman and Bodvarsson, 1996; Adler and Berkowitz, 2000) and generalised power law average (Desbarats, 1992) have also been used to determine the effective permeability of fracture networks.

In this section, we assume steady-state single-phase laminar flow for which we can use Darcy's law. As described for classic percolation (Chapter 2), not all parts of the connected cluster contribute to the flow equally. The flow is concentrated in those fractures that make up the backbone of the connected cluster (Section 2.2.4). More details about the structure and scaling properties of the backbone will be described in Chapter 6. In order to reduce the computational effort, we first extract the backbone and then calculate its effective permeability. We use the "Burning method" (Herrmann and Stanley, 1984; Herrmann *et al.*, 1984) to find the backbone. Figure 5.19 compares the connected cluster and the backbone for a fracture model.

For simplicity, we first assume that all fractures have the same permeability equal to 1 and that there is prefect hydraulic connection at the intersection between fractures. Then the continuity equation is applied at each node in the fracture network as given by

$$\sum_j q_{ij} = 0, \tag{5.9}$$

where the Darcy's law gives the flowrate (q_i) in a fracture with length l_i as

$$q_{ij} = -\frac{k_f}{\mu}\frac{p_j - p_i}{l_{ij}}. \tag{5.10}$$

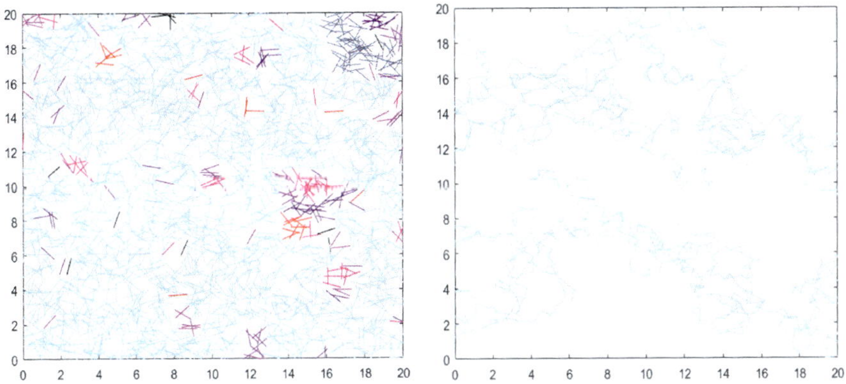

Figure 5.19. A realisation of a fracture system made by randomly oriented fracture of size $l = 1$ in a square region of size $L = 20$ at the threshold point showing (a) the connected cluster and the other finite clusters and (b) the backbone of the connected cluster.

The subscript j denotes all the neighbouring nodes connected to node i, k_f is the single fracture permeability, A is the fracture cross-sectional area, μ is the liquid viscosity and p_i and p_j are two neighbour node pressures. The single fracture permeability is given by (van Golf-Racht, 1982)

$$k_f = \frac{b^2}{12}, \tag{5.11}$$

where b is the fracture aperture. Again for simplicity, we assume the same aperture value for all fractures in the system. The boundary conditions are constant pressure for the left and right sides of the system and no flow for the upper and lower boundaries. The combination of Eqs. (5.9) and (5.10) gives the continuity equation $\sum_j \frac{p_j - p_i}{l_{ij}} = 0$.

The resulting system of linear equations can be solved by using an appropriate numerical method from which the pressure at each node and consequently flowrate along each fracture can be calculated. We then calculate the total flow rate across the system and equate the flux with those from an equivalent homogenous system from which the effective permeability k_{eff} can be determined. To compare

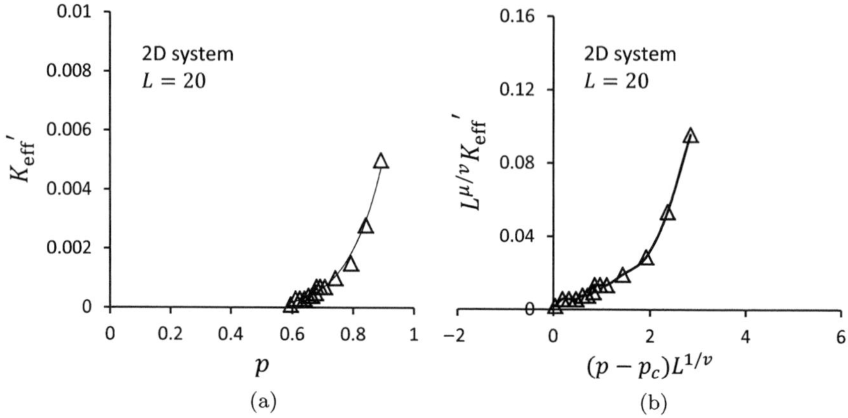

Figure 5.20. Illustration of the normalised effective permeability of fracture system of size $L = 20$. (a) The normalised permeability of fracture systems against percolation parameter and (b) the rescaled normalised permeability using finite-size-scaling law.

different fracture systems, we normalise this value by dividing it by the single fracture permeability (k_f) (Mourzenko *et al.*, 2011; Leung and Zimmerman 2012).

As expected, below the threshold, there cannot be any overall conductance as there is no connected flow pathway from inlet to outlet. For very large systems, above the threshold, the effective permeability is expected to rise monotonically following a power law (Eq. (2.7)) with conductivity exponent given in Table 2.2. The normalised effective permeability plotted against the fracture density is shown in Figure 5.20(a).

To estimate the conductivity exponent μ, we plot the normalised permeability versus $p - p_c$ for a relatively large system size. Figure 5.21 shows that the slope (for $L = 50$) is $\mu = 1.25$ which is in agreement with the reported literature values in Table 2.2.

Alternatively, the finite-size dependency of the effective permeability at the threshold can be used to estimate the critical exponents,

$$K_{\text{eff}} \sim L^{-\mu/\nu}, \tag{5.12}$$

using the literature value (Table 2.2) for the correlation length exponent ν. The finite-size scaling for the effective permeability

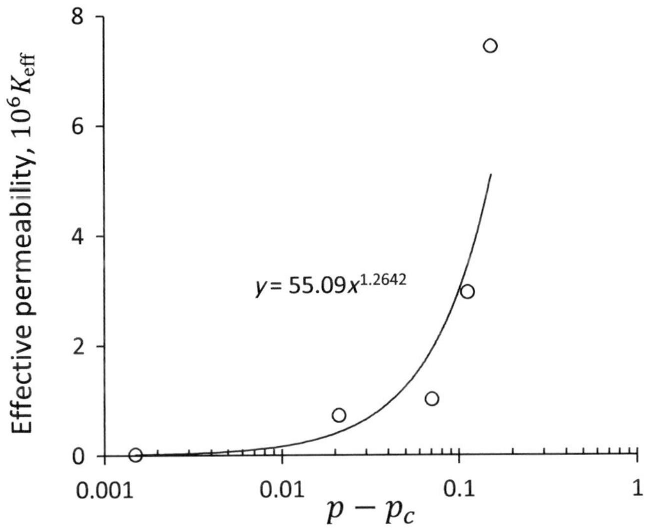

Figure 5.21. Variation of normalised permeability of fracture systems against $p - p_c$ (for $L = 50$).

shown in Figure 5.20(a) using the estimated critical exponents ($\mu = 1.25$, $\nu = 1.33$) is shown in Figure 5.20(b). The rescaled effective permeability curve can be used to estimate the permeability of fracture networks in other system sizes. However, it is strictly only valid for the region close to the threshold.

Away from the critical region, other techniques (such as effective medium theory) can be applied to estimate the effective permeability (Zimmerman and Bodvarsson, 1996). Examples include fracture networks where the fracture density is well above the threshold as well as cases of permeable media with a very low density of fractures.

5.6 Analysis in 3D

So far, we have only considered 2D systems. Clearly, the real world is 3D. Often, stereological methods are used to relate the 2D outcrop data to 3D (Berkowitz and Adler, 1998; Thovert and Adler, 2004). In 3D, we expect many of the rules from 2D to hold, so a similar approach can be used; however, the computational effort is much more time consuming.

5.6.1 *Idealised fracture network models*

A simple model of fractures in 3D is a cube of size $L \times L \times L$ in which squares of size $l \times l$ represent the fractures. They have random orientations and are spatially distributed with Poisson statistics. The random orientations for squares are assigned by choosing three Euler angles α_1, α_2 and α_3 as shown in Figure 4.29. Then the location of four corners and the centre of each fracture is stored in a data structure. To determine whether or not two squares intersect, we implement the algorithm used by Baker *et al.* (2002) that is based on the intersection of a line and a fourvertex facet.

A simple extension of 2D models is used to consider the connectivity between two opposing planes of a cube. The dimensionless system length is defined by $L_x = L/l$. In 3D, the equivalent of the occupancy probability has been defined in several ways (Bour and Davy, 1998; Adler and Thovert, 1999; Mourzenko *et al.*, 2005; Masihi *et al.*, 2008):

- the number of fractures per volume, $\rho = N/L^3$, which is not able to give a unique value for the percolation threshold as it is shape dependent,
- the dimensionless density, $\rho = Nl^3/8L^3$,
- the average number of intersections per fracture, I, with other fractures in the network,
- the probability density function that a point is in the effective volume of a fracture,

$$p = 1 - \exp\left(\frac{-N\langle V_{\text{ex}}\rangle}{8L^3}\right), \tag{5.13}$$

- the dimensionless density ρ' defined as the number of fractures per average excluded volume,

$$\rho' = \rho\langle V_{\text{ex}}\rangle, \tag{5.14}$$

where N is the number of fractures in the region and $\langle V_{\text{ex}}\rangle$ is the average excluded volume over the fracture population. The average excluded volume for the randomly oriented squares is $\langle V_{\text{ex}}\rangle = 2l^3$

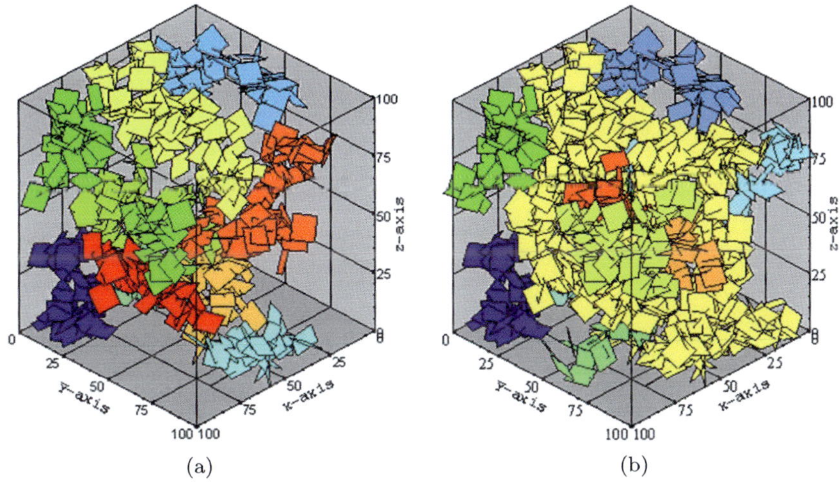

Figure 5.22. Two realisations of random fractures of length $l = 10$ in a cubic system of size $L = 100$ showing the 10 largest clusters (a) below the threshold with $N = 1{,}200$ and (b) at the threshold with $N = 1{,}348$ where the connected cluster consists of 488 fractures.

(Thovert and Adler, 2004). The excluded volume of other geometrical objects representing fractures such as discs, hexagons, squares, equilateral triangles and rectangles can be found in Mourzenko *et al.* (2011). The same computational algorithm as in Chapter 3 (Figure 3.3) is used to generate an ensemble of realisations and to determine the statistics of the percolation quantities. Two realisations of randomly oriented fractures of size $l = 10$ in a system of size $L = 100$ are shown in Figure 5.22.

As usual, we plot the percolation probability against the occupancy fraction p over all of the realisations at a particular system size and find a scatter of points. We take an ensemble average and standard deviation of the results. Then we plot the mean connectivity P against the occupancy parameter p for a variety of system sizes L and find the typical "smeared" percolation transition (Figure 5.23). Similarly, plotting the standard deviation in connectivity $\Delta(p, L)$ against occupancy p at different system sizes gives different *bell-shaped* curves.

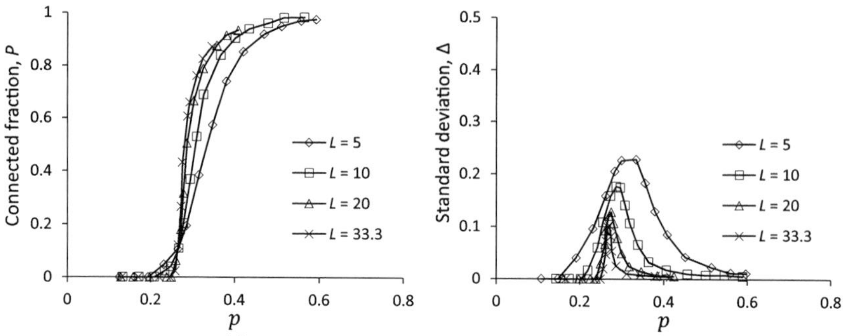

(a) (b)

Figure 5.23. The finite-size dependency of (a) the mean connected fraction and (b) the standard deviation in connectivity results in fracture systems against occupancy p.

The values of the percolation threshold in 3D for constant-length randomly oriented squares for the different occupancy parameter definitions is $\rho_c = 0.15 \pm 0.03$, $I_c = 2.40 \pm 0.04$, $p_c = 0.26 \pm 0.03$ (Robinson, 1984; Masihi *et al.*, 2007). The estimate of the threshold based on the definition in Eq. (5.13) for the connectivity results shown in Figure 5.23 can also be made by using the scale dependency of the percolation threshold (Eq. (2.27)) which gives $p_c = 0.260 \pm .003$.

By using the finite-size-scaling relations (Eqs. (2.28) and (2.31)), we found the numerical values of the exponents as $1/\nu = 1.11 \pm 0.04$ and $\beta/\nu = 0.56 \pm 0.11$ which confirms the universality principal since they agree with the literature values for the critical exponents in 3D (Table 2.2).

Using these values, we carry out the usual finite-size-scaling transformation (Figure 5.24) to show that all the mean connectivity curves $P(p, L)$ collapse on to the universal mean connectivity curve. Similarly, the standard deviation of connectivity curves $\Delta(p, L)$ collapses on to a single universal standard deviation of connectivity curve.

Before looking at wider generalisations, we consider another simplified model where the fractures can only be oriented parallel to the axis directions as shown in Figure 5.25.

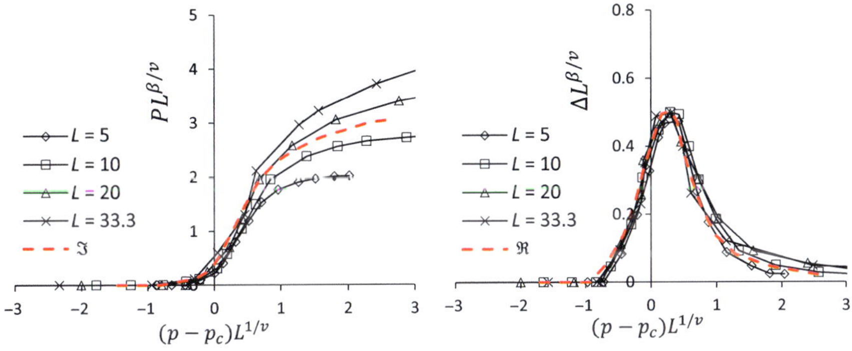

Figure 5.24. Illustration of finite-size scaling for (a) the mean of connectivity curves and (b) the standard deviation of connectivity results in 3D randomly oriented fracture systems.

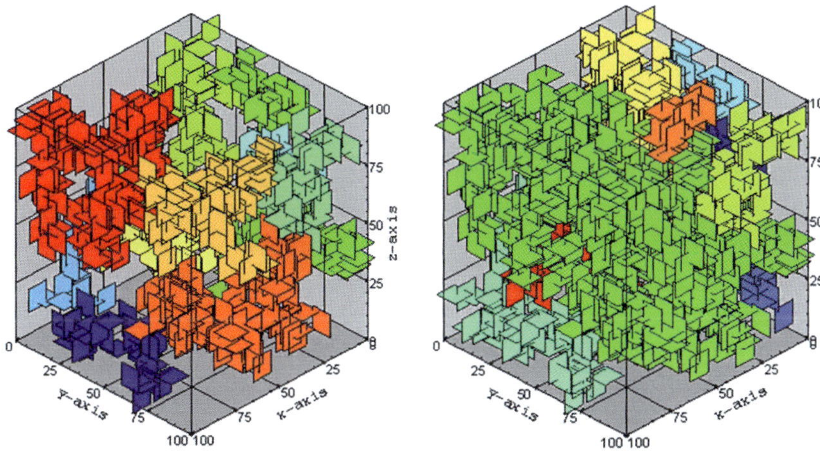

Figure 5.25. Realisations of the fracture models with constant-length fractures of length $l = 10$ in a cube of size $L = 100$ showing the 10 largest clusters in different colours. In this case, the fractures are restricted to orientations parallel to the axes: (a) below the threshold with 1,600 fractures and (b) at the threshold with 1,836 fractures, and the spanning cluster shown in green consists of 682 fractures.

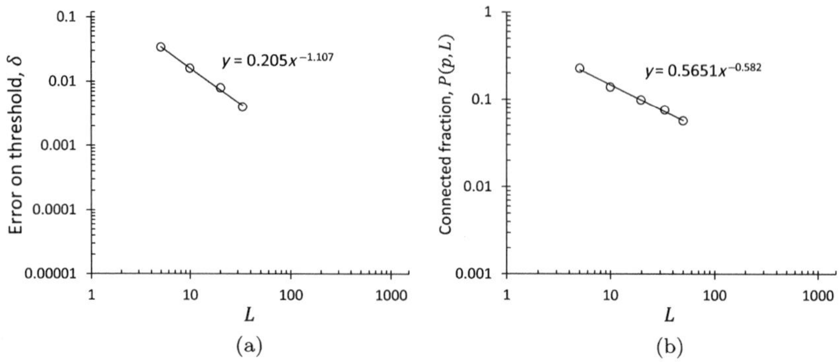

Figure 5.26. Estimation of critical exponents for the model of three orthogonal families of fractures in 3D by using the scale dependency of (a) the error obtained on the threshold and (b) the percolation probability P.

Since the excluded volume is zero for parallel squares and is $V_{\text{ex}} = 2l^3$ for orthogonal squares of length l, the average excluded volume per square in this model is $\langle V_{\text{ex}} \rangle = 4l^3/3$. Numerical investigation shows that for this orthogonal fracture model, the threshold for the different occupancy parameter definitions is $\rho_c = 0.187$, $I_c = 2$, $p_c = 0.220$ which is slightly less than the value obtained for the randomly oriented fracture model (Robinson, 1984; Adler and Thovert, 1999; Masihi *et al.*, 2007). However, the estimated critical exponents shown in Figure 5.26 are in agreement with the literature values in Table 2.2 emphasising that they belong to the same universality class.

Hence, the same scaling laws (Eqs. (2.11) and (2.14)) can be used to estimate the connectivity behaviour of this model. Figure 5.27 compares the scaled curves for these basic fracture models (the randomly oriented and aligned models). This shows that the connectivity master curves are very similar for both models.

5.6.2 *More complex fracture models*

There are other factors including the fracture size, orientation and spatial distributions, fracture shape, anisotropy and well patterns which need to be accounted for to estimate connectivity or conductivity in real systems. We shall discuss some of these effects in this section. However, this is still an area for active research.

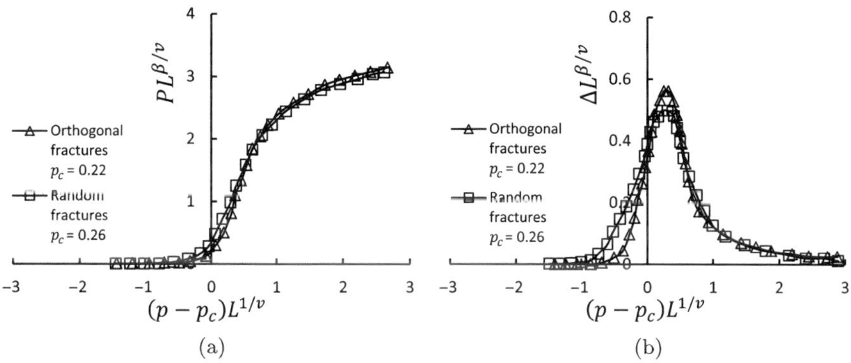

Figure 5.27. Comparison of master connectivity curves for the randomly oriented fracture model and orthogonal fracture models in 3D. (a) Master curve for mean connectivity results and (b) master curve for the standard deviation in connectivity results.

We first need an appropriate definition of the occupancy parameter which accounts for the effects of the size, shape and orientation distributions of the fractures. We have already introduced four measures of fracture density in the previous section. Adler and Thovert (1999) used a general dimensionless density $\rho' = \rho\langle V_{\text{ex}}\rangle$ in Eq. (5.14) where $\rho = N/L^3$ is the number density of fractures. The angular brackets denote averages over the fracture population. The excluded volume between two randomly oriented and distributed fractures, represented by convex polygons of areas A and perimeters P, is (Mourzenko *et al.*, 2005)

$$V_{\text{ex}} = \frac{1}{4}(A_1P_2 + A_2P_1). \tag{5.15}$$

For identical fractures, Eq. (5.15) reduces to $V_{\text{ex}} = AP/2$. Hence, the excluded volume for randomly oriented identical squares of size l becomes $V_{\text{ex}} = 2l^3$. So the percolation threshold based on this is around $\rho'_c = 2.36$ (Mourzenko *et al.*, 2011).

5.6.2.1 *Anisotropy effects*

We now investigate the effects of anisotropy on the connectivity of the orthogonal fractures model. The anisotropy arises as a consequence of using either rectangles with size (b, c), (a, c) and (a, b) perpendicular

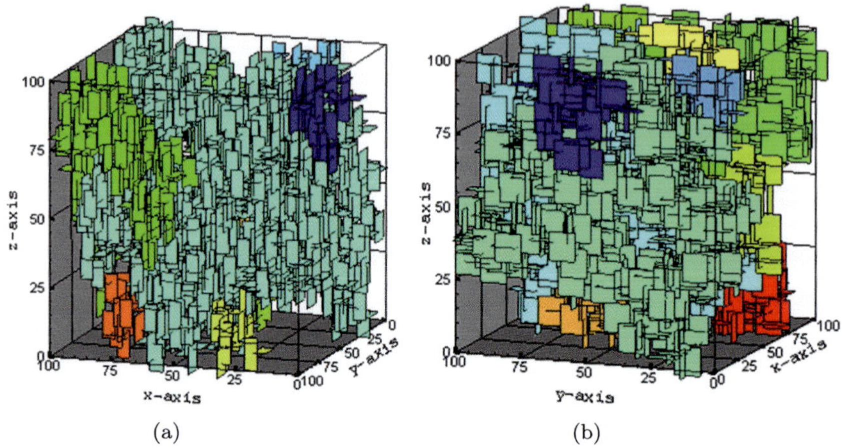

Figure 5.28. Two realisations of anisotropic orthogonal fracture model with $a = 5$, $b = c = 10$ ($\omega_1 = \omega_2 = 2$) at the percolation threshold showing (a) the difficult direction for connection along the x axis with 3,497 fractures where the spanning cluster consists of 1,706 fractures and (b) the easy direction for connection along the y axis with 3,412 fractures where the spanning cluster consists of 581 fractures.

to the x, y and z directions repectively or different fracture numbers aligned with each direction. We use the first in this section and assume a particular case $b = c$ which enables us to compare the x direction connectivity to the y direction connectivity. The three dimensionless lengths are $L_x = L/a$, $L_y = L/b$ and $L_z = L/c$. The average excluded volume for this case model is $\langle V_{\text{ex}} \rangle = 4abc/3$. The framework to account for the anisotropy effects which uses two aspect ratios $\omega_1 = L_x/L_y$ and $\omega_2 = L_x/L_z$ is similar to that explained in Section 4.5.2. Two realisations of this anisotropic fracture model emphasising different connectivity in the x and y (or z) directions are shown in Figure 5.28.

We now carry out the usual finite-size scaling and find the characteristic data collapse for anisotropic systems (Figure 5.29).

As before, we can further collapse these onto a single curve by accounting for the shift in the apparent finite-size threshold (Eq. (4.30)). Using the same argument from Section 4.5.2 we assume that the same correction formula (Eq. (4.39)) can be used here. As a

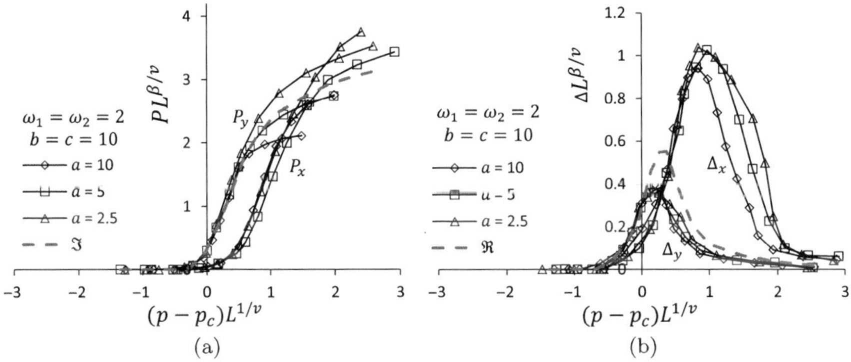

Figure 5.29. Finite-size scaling for the connectivity of the anisotropic orthogonal fracture model with $\omega_1 = \omega_2 = 2$: (a) rescaled mean (P) of connectivity curves and (b) rescaled standard deviation of connectivity curves (Δ).

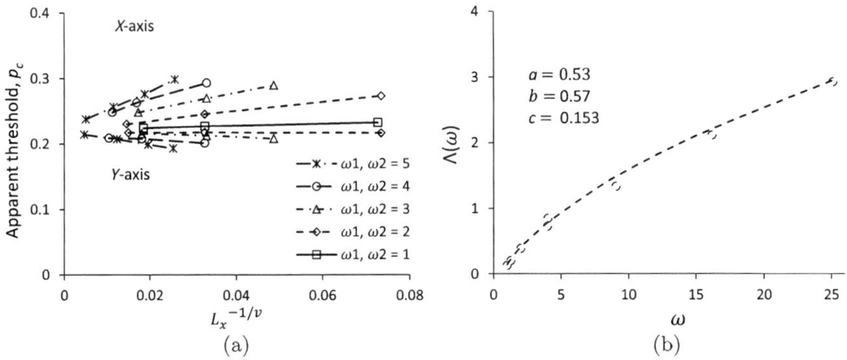

Figure 5.30. (a) The variation of the apparent threshold of the anisotropic orthogonal fracture models in the x and y directions against $L_x^{-1/\nu}$ for different aspect ratios and (b) the shift in the apparent threshold Λ as a function of ω (the product of ω_1 and ω_2).

reminder, the correction term is $\Lambda_x(\omega_1, \omega_2) = a[(\omega_1\omega_2)^b - 1] + c$. The numerical results obtained fit this form with coefficients $a = 0.53$, $b = 0.57$, $c = 0.153$ as shown in Figure 5.30.

Using the corrected finite-size percolation threshold in the finite-size-scaling law (Eqs. (4.41) and (4.42)) gives the usual collapse of all the connectivity curves onto the same universal connectivity curves (see Figure 5.31).

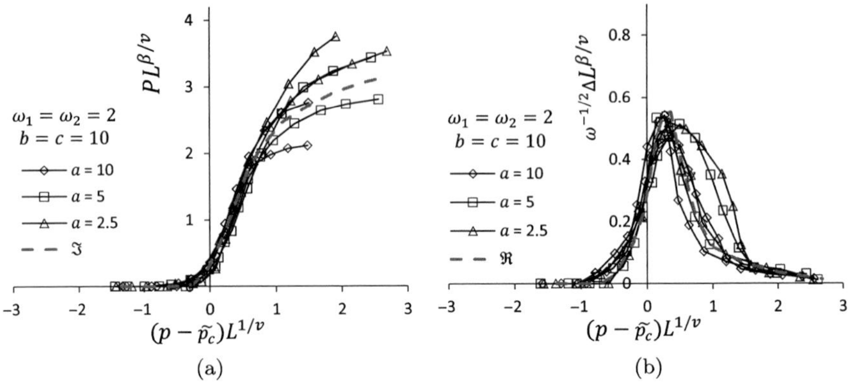

Figure 5.31. Illustration of corrected finite-size scaling for (a) the mean and (b) the standard deviation of connectivity for the anisotropic orthogonal fracture system with aspect ratios $\omega_1 = \omega_2 = 2$. The isotropic connectivity curve is shown as a dotted line.

5.6.2.2 *Fracture size variation*

To take into account the effects of fracture size variation, we use the same approach as that in 2D. For example, in 3D, the effective length based on the average excluded volume can be defined by the product of the first and the second moment (Charlaix *et al.*, 1984; Koudina *et al.*, 1998; Adler and Thovert, 1999),

$$l_{\text{eff}} = \sqrt[3]{\langle l^2 \rangle \langle l \rangle}, \tag{5.16}$$

which is consistent with the mean number of intersections per fracture. Or we could use the third moment of size distribution (Balberg *et al.*, 1984; Bour and Davy, 1998; Mourzenko *et al.*, 2005),

$$l_{\text{eff}} = \sqrt[3]{\langle l^3 \rangle}. \tag{5.17}$$

As pointed out by Mourzenko *et al.* (2005), the former is a measure of local connectivity, whereas the latter controls the global network percolation. The effective length is again assumed to control the connectivity of the fracture network which enables us to use the standard connectivity master curve for systems with a distribution of fracture sizes.

Bogdanov *et al.* (2007) studied the effects of a power law size distribution for fracture networks above and below the percolation threshold in 3D. They used a more general fracture density based on the dimensionless density in Eq. (5.14) for polydisperse fractures to describe the percolation properties,

$$\rho' = \rho\langle v_{ex}\rangle\langle l^3\rangle, \quad \text{where} \quad V_{ex} = v_{ex}\frac{l_1 l_2^2 + l_2 l_1^2}{2}, \tag{5.18}$$

where v_{ex} is a dimensionless shape factor. Mourzenko *et al.*, (2005) showed that the critical density is almost constant ($\rho'_c = 2.3 \pm 0.1$) for a wide range of fracture shapes and size distributions.

5.6.2.3 *Fracture shape*

We also need to consider more realistic fracture shapes. Huseby *et al.* (1997) analysed the geometrical and topological properties of networks of random fractures with similar sizes but having different shapes. They used the dimensionless fracture density parameter defined in Eq. (5.14). A unique form of the percolation threshold was reported to apply for a wide range of fracture shapes (such as discs, squares, regular hexagons, equilateral triangles and rectangular fractures) or a mixture of shapes (Cacas *et al.*, 1990; Koudina *et al.*, 1998; Bogdanov *et al.*, 2003; Charlaix, 1986; Gawlinski and Stanley, 1981; Mourzenko *et al.*, 2011). Mourzenko *et al.* (2005) proposed a shape factor in the definition of the dimensionless percolation parameter ρ' (Eq. (5.18)) and found the threshold values of 2.36, 2.24, and 1.90 for squares, triangles and rectangles respectively. Bogdanov *et al.* (2003, 2007) studied the effects of different fracture shapes including polygons (triangles, squares, hexagons, icosagons), rectangles and ellipses with different aspect ratios and showed that the results for networks of fractures with various shapes obey a single law when they used Eq. (5.18) for the fracture density with appropriate shape factor v_{ex}. Moreover, Mourzenko *et al.* (2011) reported a unique threshold value of $\rho' = 2.3 \pm 0.1$ for a variety of fracture shapes.

5.6.2.4 *Well pattern*

So far, we have defined wells as planes in 3D. In reality, wells are lines with specific trajectory. In such case, the spanning cluster is defined by the connection between two lines at the two different locations of the system instead of two opposite faces of the system. We expect that the connectivity between two lines would be much lower than the connectivity between two faces. Hence, we expect the finite-size threshold to be changed. In such cases, the percolation properties such as the cluster size distribution, the structure of the spanning cluster, backbone or shortest path are also expected to change models (Soares *et al.*, 2004). However, it is unclear how to transform face-to-face connectivity results to line-to-line connectivity in order to use the same master connectivity curves as before. In more realistic cases, the properties of clusters connecting three or more wells may be needed (Geiger-Boschung *et al.*, 2009). Figure 5.32 shows a typical fracture pattern in 3D with connectivity to multiple wells. This is an area which requires further research.

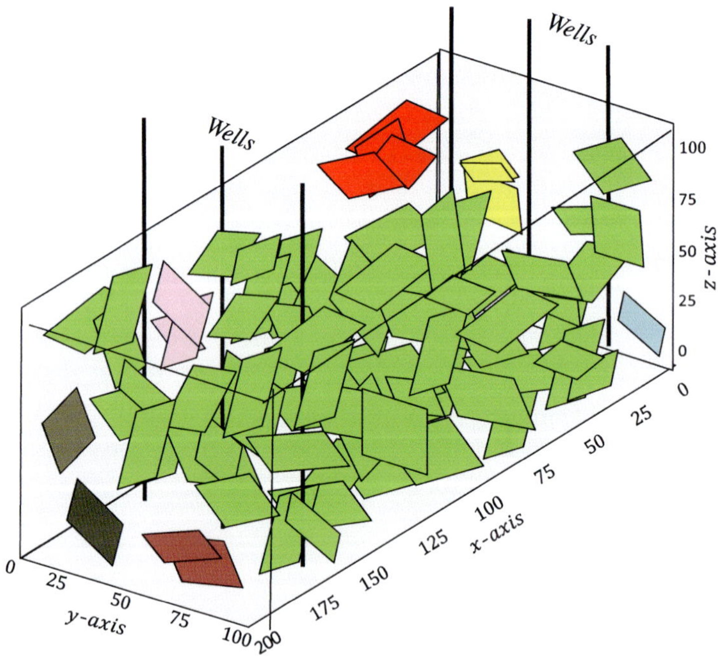

Figure 5.32. Illustration of well patterns and fractures distribution in 3D.

5.6.3 *Effective permeability*

In principal, it is straightforward to extend the methods described in Section 5.5 to determine the effective permeability of fracture networks to 3D. We restrict ourselves to the case where the matrix is completely impermeable and the fractures all have the same permeability of one. We again solve the flow equations on fracture network numerically to find the equivalent permeability. The normalised effective permeability given in Mourzenko *et al.* (2011) for randomly oriented and distributed squares in system of size $L = 20$ is shown in Figure 5.33(a). To perform the finite-size scaling as before, we need the critical exponents (μ, ν). These were estimated by using the scale dependency (Eq. (5.12)) of the effective permeability of fracture systems in 3D and gave $\frac{\mu}{\nu} = 2.22$ which is consistent with the literature value in Table 2.2 (Mourzenko *et al.*, 2011). The normalised effective permeability is shown in Figure 5.33(a) and in rescaled form (using finite-size scaling with $p_c = 0.26$ and estimated critical exponents ν and μ) in Figure 5.33(b).

We expect that orientation disorder can also cause anisotropy in the effective permeability as was found with the connectivity.

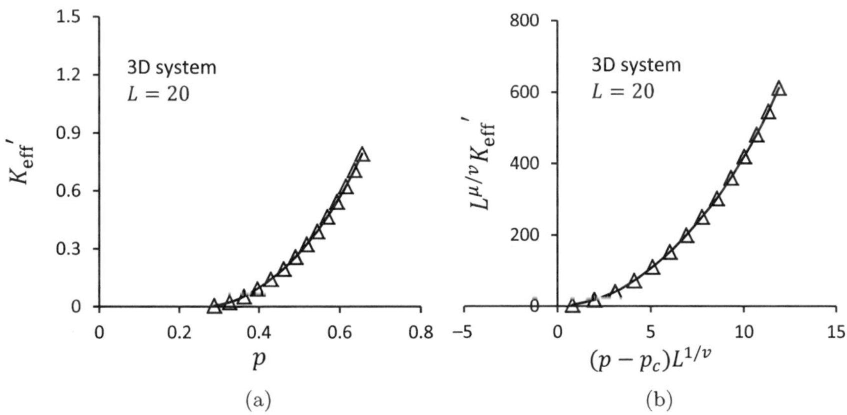

Figure 5.33. Illustration of the normalised effective permeability of 3D fracture system of size $L = 20$ made of randomly oriented squares. (a) The normalised effective permeability against percolation parameter and (b) the rescaled effective permeability using finite-size-scaling laws.

However, the detailed description of the effective permeability needs further investigation. Savik *et al.* (2013) studied the effects of anisotropy due to the fracture sets with different relative orientation and used effective medium methods to determine the effective permeability. Mourzenko *et al.* (2011) modified the definition of the percolation parameter (Eq. (5.14)) to include the effect of the anisotropy due to the orientational effects by using a Fisher distribution for orientations.

Lang *et al.* (2014) proposed a full equivalent permeability tensor to approximate flow in 3D when only 2D analyses were available. They observed numerically that 3D fracture networks exhibit percolation in the range of $2.2 < \rho' < 4.2$ which is less than half of the value observed for 2D trace maps. This emphasises that fracture connectivity in 3D is not accurately captured by studies in the plane. Mourzenko *et al.* (2011) suggested an expression for the effective permeability based on curve fitting of the permeability data calculated numerically,

$$K' = \frac{0.037\Delta\rho'^2}{1 + 0.155\Delta\rho'}, \tag{5.19}$$

which has a similar power law behaviour in the critical region. The permeability, calculated numerically, for a variety of fracture shapes is shown in Table 5.1.

So far, we have assumed that only the fractures are open to the flow. However, in reality, the rock matrix is also permeable (Hamzehpour *et al.*, 2009). Hence, the effective permeability of the fractured porous medium (K_{eff}) is not the same as the upscaled permeability K_n of the fracture network. The case where the matrix

Table 5.1. The percolation threshold and the power law scaling for the permeability of fracture networks with a variety of fracture shapes (Mourzenko *et al.*, 2011).

Range of density parameter	**Power law scaling**	**Percolation threshold ($\Delta\rho'$)**
$0 < \Delta\rho' \leq 10$	$K' = 0.045\Delta\rho'^{1.57}$	2.3 ± 0.1
$0 < \Delta\rho' \leq 1.5$	$K' = 0.032\Delta\rho'^{1.9}$	
$0 < \Delta\rho' \leq 5$	$K' = 0.031\Delta\rho'^{1.8}$	

permeability is negligible corresponds to the steady-state effective permeability in dual-porosity models. The effective permeability for the more complex case can be computed using unstructured grids and a finite volume discretisation of steady-state single-phase flow equations (Koudina *et al.*, 1998).

The interaction of fracture flow with the permeable matrix can also be approximately accounted for by a 1D dual-porosity approach (Dershowitz and Miller, 1995). The relation between the effective permeability of the fractured porous medium (K_{eff}) and the upscaled permeability of the fracture network (K_n) was discussed in Bogdanov *et al.* (2007). They showed that the increment of the normalised effective permeability K_{eff}/K_m is significant at lower of fracture density (ρ') and higher normalised fracture conductivity (σ'). The same approach was used by Mourzenko *et al.* (2011) to propose a model to account for the effect of matrix permeability on the effective permeability of the fractured porous medium.

5.7 Case studies

The study area used by Belayneh *et al.* (2006) is a fractured limestone bed exposed on the southern margin of the Bristol Channel Basin (150 km long and 30 km across) in north Somerset, UK on the limb of the rollover fold formed during opening of the basin (Figure 5.34).

A mosaic of photographs for the 150 m^2 study area was used to produce line drawings of two vein sets. The vein sets were formed during basin subsidence. The relative ages of the two vein sets (V_1 and V_2) were determined from the cross-cutting relationships. The older vein set V_1 was formed on the limb of a rollover fold in response to the outer arc extension, and it is parallel to the fold axis. The vein set V_2 was formed in response to the inhomogeneous settling of the bed which may be caused by fluidisation and withdrawal of shale or a partial compaction and is oblique to the fold axis (40° in the clockwise sense). The spatial distribution and the aperture of the veins were recorded along eight scan lines as shown in Figure 5.35. Statistical analysis showed that the majority of these scan lines show lognormal spacing distributions for V_1 (with an average spacing that

Figure 5.34. Illustration of the study area that is a fractured limestone bed exposed on the southern margin of the Bristol Channel Basin, the north Somerset, UK (Belayneh *et al.*, 2006).

ranges between 9.22 and 20.26 cm) and a negative exponential for V_2 (with an average spacing ranging between 10.29 and 25.45 cm). To check the scale dependency, the network is sub-divided into eight domain areas ranging in size from 9 to 144 m^2.

We used the standard algorithm (Figure 3.3) to identify the various clusters in each sub-domain (Hoshen and Kopelman, 1976). Then the statistics of the fracture length distributions in the eight sub-domains and in the whole network were obtained and the vein connectivities calculated. The results are summarised in Table 5.2.

The percolation occupancy p and the connected fraction of fractures P of the eight sub-domains and the whole network were rescaled using finite-size scaling (Eqs. (2.39) and (2.40)). Figure 5.36 compares the computed connectivity results with the universal connectivity curves (Figure 3.21).

Most of the connectivities in the area under study are within one standard deviation of the mean predicted value. This still holds

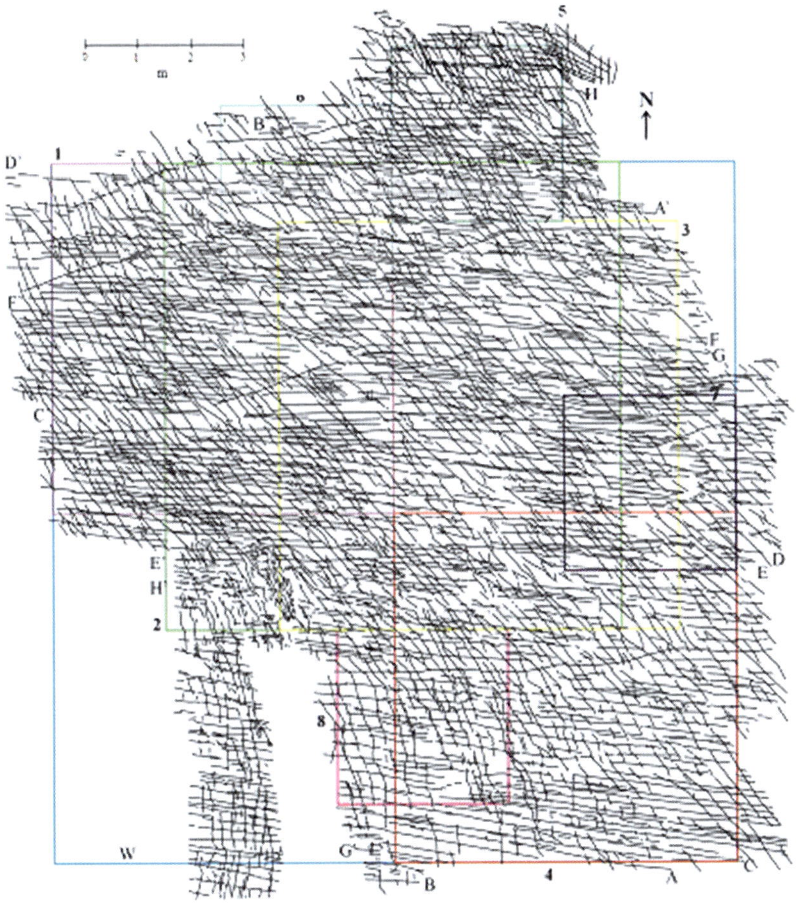

Figure 5.35. Illustration of line drawings of fractures under study and eight sub-domain areas (Belayneh *et al.*, 2006).

even when the system is above the percolation threshold, which demonstrates the applicability of this approach for real systems.

5.8 Problems

5.1. Show that the excluded volume between two randomly oriented fractures with areas A and perimeters P can be given by Eq. (5.15).

Table 5.2. Statistics of the fracture lengths and connectivity in various domains.

Domain no.	Domain size	Number of fractures (N)	Lognormal length distribution with ($\bar{l}, \sigma$)	$P(p, L)$
Whole (w)	12×12	3,387	(0.479, 0.603)	0.930
1	6×6	951	(0.558, 0.710)	0.987
2	8×8	1,886	(0.477, 0.584)	0.957
3	7×7	1,366	(0.481, 0.563)	0.942
4	6×6	949	(0.436, 0.525)	0.903
5	3×3	361	(0.423, 0.446)	0.987
6	3×3	282	(0.436, 0.529)	0.970
7	3×3	294	(0.402, 0.466)	0.799
8	3×3	275	(0.385, 0.376)	0.972

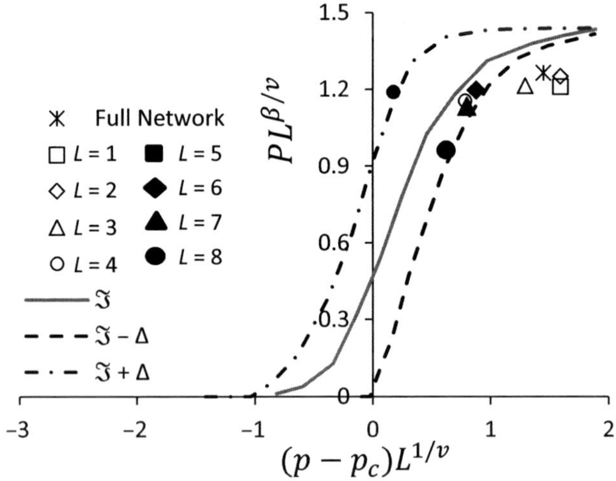

Figure 5.36. Comparison of the rescaled connected fraction of fractures in eight sub-domains and the whole network with the universal connectivity curves (Figure 3.21). The solid line is the mean value and the dashed lines are plus and minus one standard deviation of universal connected fraction curves.

5.2. The dimensionless density ρ' based on the average excluded volume of fractures has been defined in Eq. (5.14) with which we can characterise percolation properties in 3D. Find the ρ' for identical squares of size l with a uniform distribution of orientation.

5.3. Find the average excluded volume for randomly distributed fractures with these geometrical shapes: discs, ellipses, squares, regular hexagons, equilateral triangles and rectangular fractures.

5.4. Find the dimensionless density ρ' for polydisperse fractures such as identical squares with a power law distribution of size l.

5.5. Four percolation parameters (ρ, ρ', I, p) have been introduced for fracture systems in 3D. Work out the possible relationship between these measures.

5.6. Find an appropriate shape factor v_{ex} based on Eq. (5.18) to be considered in the dimensionless density ρ' for fracture systems with polygons (such as triangles, squares, hexagons), rectangles, discs and ellipse shapes.

5.7. Compare the percolation properties of two fracture models made of (a) constant-length randomly oriented squares and (b)three orthogonal sets of squares of the same size and density and conclude.

5.8. Use the master connectivity curves of the 2D isotropic randomly oriented fracture model shown in Figure 3.21 to reconstruct the fracture connectivity curves for an anisotropic case when fractures have uniform orientation distribution in the range of $-\theta_1 \leq \theta \leq \theta_1$ for various system sizes given in Figure 3.21.

5.9. Consider a simple 2D isotropic constant-unit-size orthogonal fracture model (two orientations are $\theta_1 = 0$, $\theta_2 = \pi/2$) with the same fracture density $\rho = 0.5$ for each fracture set. Assume the connectivity of $P = 0.58$ at occupation fraction $p = 0.61$. Then let us make anisotropic models by assuming that the density of fractures with orientation $\theta_1 = 0$ to be $\rho = 0.6, 0.65, 0.7, 0.75$ and 0.8. Use the finite-size scaling to determine the connectivity of these anisotropic models and plot the fracture connectivity against aspect ratio ω.

5.10. An engineer has tried to apply the connectivity master curve for a 2D fracture in an outcrop region of size 15 m. Statistical analysis shows that the 4,000 fractures analysed in the region follow a lognormal size distribution (mean of 0.479 m and standard deviation of 0.603 m). Estimate the fraction of connected fractures. Using this information, estimate the fraction of connected fractures of this region in 3D.

References

Adler, P.M. and Thovert, J.F., *Fractures and Fracture Networks*. Kluwer Academic Publishers, London, pp. 163–218, 1999.

Adler, P.M. and Berkowitz, B., Effective medium analysis of random lattices. *Transport in Porous Media*, 40: 145–151, 2000.

Baker, R., Paul, G., Sreenivasan, S. and Stanley, H.E., Continuum percolation threshold for interpenetrating squares and cubes. *Physical Review E*, 66: 046136.1–046136.5, 2002.

Balberg, I., Universal percolation threshold limits in the continuum. *Physical Review B*, 31(6): 4053–4055, 1985.

Balberg, I. and Binenbaum, N., Computer study of the percolation threshold in a two dimensional anisotropic system of conducting sticks. *Physical Review B*, 28(7): 3799–3811, 1983.

Balberg, I., Anderson, C.H., Alexander, S. and Wagner, N., Excluded volume and its relation to the onset of percolation. *Physical Review B*, 30(7): 3933–3943, 1984.

Bear, J., Tsang, C.F. and de Marsily, G., Flow and contaminant transport in fractured rock. Academic Press, San Diego, CA, pp. 169–231, 1993.

Belayneh, M., Masihi, M., King, P.R. and Matthäi, S.K., Prediction of fracture uncertainty using percolation approach: model test with field data. *Journal of Geophysics and Engieering*, 3: 219–229, 2006.

Berkowitz, B., Analysis of fracture network connectivity using percolation theory. *Mathematical Geology*, 27(4): 467–483, 1995.

Berkowitz, B., Characterizing flow and transport in fractured geological media: a review. *Advances in Water Resources*, 25: 861–884, 2002.

Berkowitz, B. and Balberg, I., Percolation theory and its application to ground water hydrology. *Water Resources Research*, 29(4): 775–794, 1993.

Berkowitz, B. and Adler, P.M., Stereological analysis of fracture network structure in geological formations. *Journal of Geophysical Research*, 103(B7): 15339–15360, 1998.

Berkowitz, B., Bour, O., Davy, P. and Odling, N., Scaling of fracture connectivity in geological formations. *Geophysical Research Letters*, 27(14): 2061–2064, 2000.

Bogdanov, I.I., Mourzenko, V.V., Thovert, J.-F. and Adler, P.M., Effective permeability of fractured porous media in steady state flow. *Water Resources Research*, 39(1): 1023, doi: 10.1029/2001WR000756, 2003.

Bogdanov, I.I., Mourzenko, V.V., Thovert, J.-F. and Adler, P.M., Effective permeability of fractured porous media with power-law distribution of fracture sizes. *Physical Review E*, 76: 036309, 2007.

Bonnet, E., Bour, O., Odling, N.E., Davy, P., Main, I., Cowie, P. and Berkowitz, B., Scaling of fracture systems in geological media. *Reviews of Geophysics*, 39(3): 347–383, 2001.

Bour, O. and Davy, P., Connectivity of random fault networks following a power law fault length distribution. *Water Resources Research*, 33(7): 1567–1583, 1997.

Bour, O. and Davy, P., On the connectivity of three dimensional fault networks. *Water Resources Research*, 34(10): 2611–2622, 1998.

Bour, O. and Davy, P., Clustering and size distributions of fault pattern: theory and measurements. *Geophysical Research Letters*, 26: 2001–2004, 1999.

Cacas, M.C., Ledoux, E., de Marsily, G., Tillie, B., Barbreau, A., Durand, E., Feuga, B. and Peaudecerf, P., Modeling fracture flow with stochastic discrete fracture network: calibration and validation: 1. The flow model. *Water Resources Research*, 26(3): 479–489, 1990.

Charlaix, E., Percolation threshold of random array of discs: a numerical simulation. *Journal of Physics A: Mathematical and General*, 19(9): L533–L536, 1986.

Charlaix, E., Guyon, E. and Rivier, N., A criterion for percolation threshold in a random array of plates. *Solid State Communications*, 50(11): 999–1002, 1984.

Darcel, C., Bour, O. and Davy, P., Cross-correlation between length and position in real fracture networks. *Geophysical Research Letters*, 30(12): 52.1–52.4, 2003a.

Darcel, C., Bour, O., Davy, P. and de Dreuzy, J.R., Connectivity properties of two dimensional fracture networks with stochastic fractal correlation. *Water Resources Research*, 39(10): SBH1.1–SBH1.13, 2003b.

de Dreuzy, J.R., Darcel, C., Davy, P. and Bour, O., Influence of spatial correlation of fracture centres on the permeability of two-dimensional fracture networks following a power law length distribution. *Water Resources Research*, 40: W01502.1–W01502.11, 2004.

Dershowitz, W.S. and Miller, I., Dual porosity fracture flow and transport. *Geophysical Research Letters*, 22: 1441–1444, 1995.

Desbarats, A.J., Spatial averaging of hydraulic conductivity in three-dimensional heterogeneous porous media. *Mathematical Geology*, 24: 249–267, 1992.

Gawlinski, E.T. and Stanley, H.E., Continuum percolation in two dimensions: Monte Carlo tests of scaling and universality for non-interacting discs. *Journal of Physics A: Mathematical and General*, 14: L291–L299, 1981.

Geiger-Boschung, S., Huangfu, Q., Reid, F. *et al.*, Massively parallel sector scale discrete fracture and matrix simulations. Presented at the SPE Reservoir Simulation Symposium, 2–4 February, The Woodlands, Texas (SPE 118924), 2009.

Gillespie, P.A., Howard, C.B., Walsh, J.J. and Watterson, J., Measurement and characterization of spatial distributions of fractures. *Tectonophysics*, 226: 113–141, 1993.

Hamzehpour, H., Mourzenko, V.V., Thovert, J.-F. and Adler, P.M., Percolation and permeability of networks of heterogeneous fractures. *Physical Review E*, 79: 036302, 2009.

Harris, C.K., Effective medium treatment of flow through anisotropic fracture system — improved permeability estimates using a new lattice mapping. *Transport in Porous Media*, 9: 287–295, 1992.

Harter, T., Finite size scaling analysis of percolation in three dimensional correlated binary Markov chain random fields. *Physical Review E*, 72: 026120.1–026120.7, 2005.

Heffer, K.J. and Bevan, T.G., Scaling relationships in natural fractures: data, theory and applications. Paper SPE 20981, Proceedings of the Europec 90, The Hague, Netherlands, 22–24 October, 1990.

Heffer, K.J. and King, P.R., Spatial scaling of effective modulus and correlation of deformation near the critical point of fracturing. *Pure and Applied Geophysics*, 163(10): 2223–2242, 2006.

Herrmann, H.J. and Stanley, H.E., Building blocks of percolation clusters: volatile fractals. *Physical Review Letters*, 53: 1121, 1984.

Herrmann, H.J., Hong, D.C. and Stanley, H.E., Backbone and elastic backbone of percolation clusters obtained by the new method of 'burning'. *Journal of Physics A: Mathematical and General*, 17: L261–L266, 1984.

Hestir, K. and Long, J.C.S., Analytical expressions for the permeability of random two-dimensional Poisson fracture networks based on regular lattice percolation and equivalent media theories. *Journal of Geophysical Research: Solid Earth*, 95(B13): 21565–21581, 1990.

Hoshen, J. and Kopelman, R., Percolation and cluster distribution. I. Cluster multiple labelling technique and critical concentration algorithm. *Physical Review B*, 14(8): 3438–3445, 1976.

Hudson, J.A. and Priest, S.D., Discontinuities and rock mass geometry. *International Journal of Rock Mechanics and Mining Sciences & Geomechanics Abstracts*, 16: 339–362, 1979.

Huseby, O., Thovert, J.-F. and Adler, P.M., Geometry and topology of fracture systems. *Journal of Physics A: Mathematical and General*, 30(5): 1415, 1997.

Khamforoush, M., Shams, K., Thovert, J.F. and Adler, P.M., Permeability and percolation of anisotropic three-dimensional fracture networks. *Physical Review E*, 77: 056307, 2008.

Koudina, N., Garcia, R.G., Thovert, J.-F. and Adler, P.M., Permeability of three-dimensional fracture networks. *Physical Review E*, 57: 4466–4479, 1998.

Lang, P.S., Paluszny, A. and Zimmerman, R.W., Permeability tensor of three-dimensional fractured porous rock and a comparison to tracemap predictions. *Journal of Geophysical Research: Soild Earth*, 119(8): 6288–6307, doi: 10.1002/2014JB011027, 2014.

Leung, C.T.O. and Zimmerman, R.W., Estimating the hydraulic conductivity of two-dimensional fracture networks using network geometric properties. *Transport in Porous Media*, 93: 777–797, 2012.

Long, J.C.S. and Witherspoon, P.A., The relationship of the degree of interconnection to permeability in fracture networks. *Journal of Geophysical Research: Solid Earth*, 90: 3087–3098, 1985.

Mandelbrot, B.B., *The Fractal Geometry of Nature*. W. H. Freeman, New York, pp. 468, 1982.

Masihi, M., King, P.R. and Nurafza, P., Connectivity prediction in fractured reservoirs with variable fracture size: analysis and validation. Paper SPE 100229, Proceedings of the the SPE Europec, Vienna, Austria, 12–15 June, 2006a.

Masihi, M., King, P.R. and Nurafza, P., The effect of anisotropy on finite size scaling in percolation theory. *Physical Review E*, 74: 042102, 2006b.

Masihi, M. and King, P.R., A correlated fracture network: modeling and percolation properties. *Water Resources Research*, 43, doi: 10.1029/2006WR005331, 2007a.

Masihi, M. and King, P.R., Connectivity of spatially correlated fractures: simulation and field studies. Paper SPE 107132, Proceedings of the the SPE Europec, London, UK, 11–14 June, 2007b.

Masihi, M., King, P.R. and Nurafza, P., Fast estimation of connectivity in fractured reservoirs using percolation theory. *SPE Journal*, 12(2): 167–178, 2007.

Masihi, M., King, P.R. and Nurafza, P., Connectivity prediction in fractured reservoirs with variable fracture size: analysis and validation. *SPE Journal*, 13(1): 88–98, 2008.

Matthäi, S.K. and Belayneh, M., Fluid flow partitioning between fractures and a permeable rock matrix. *Geophysical Research Letters*, 31: L07602.1–L07602.5, 2004.

Mourzenko, V.V., Thovert, J.-F. and Adler, P.M., Macroscopic permeability of three-dimensional fracture networks with power-law size distribution. *Physical Review E*, 69: 066307, 2004.

Mourzenko, V.V., Thovert, J.F. and Adler, P.M., Percolation of three dimensional fracture networks with power law size distribution. *Physical Review E*, 72: 036103.1–036103.14, 2005.

Mourzenko, V.V., Thovert, J.F. and Adler, P.M., Permeability of isotropic and anisotropic fracture networks, from the percolation threshold to very large densities. *Physical Review E*, 84(3): 036307, doi: 10.1103/PhysRevE.84.036307, 2011.

Narr, W. and Suppe, J., Joint spacing in sedimentary rocks. *Journal of Structural Geology*, 13(9): 1037–1048, 1991.

Odling, N.E., Scaling and connectivity of joint systems in sandstones from western Norway. *Journal of Structural Geology*, 19(10): 1257–1271, 1997.

Odling, N.E., Gillespie, P., Bourgine, B., Castaing, C., Chiles, J-P., Christensen, N.P., Fillion, E., Genter, A., Olsen, C., Thrane, L., Trice, R., Aarseth, E., Walsh, J.J. and Watterson, J., Variations in fracture system geometry and their implications for fluid flow in fractured hydrocarbon reservoirs. *Petroleum Geosciences*, 5: 373–384, 1999.

Olson, J.E., Sublinear scaling of fracture aperture versus length: an exception or the rule? *Journal of Geophysical Research*, 108(B9): ETG3.1–ETG3.11, 2003.

Prakash, S., Havlin, S., Schwartz, M. and Stanley, H.E., Structural and long-range correlated percolation. *Physical Review A*, 46(4): R1724–R1727, 1992.

Priest, S.D. and Hudson, J.A., Discontinuity spacings in rock. *International Journal of Rock Mechanics and Mining Sciences & Geomechanics Abstracts*, 13: 135–148, 1976.

Priest, S.D. and Hudson, J.A., Estimation of discontinuity spacings and trace length using scan line surveys. *International Journal of Rock Mechanics and Mining Sciences & Geomechanics Abstracts*, 18: 185–197, 1981.

Renshaw, C.E., Connectivity of joint networks with power law length distributions. *Water Resources Research*, 35(9): 2661–2670, 1999.

Renshaw, C.E. and Pollard, D.D., Numerical simulation of fracture set formation: a fracture mechanics model consistent with experimental observations. *Journal of Geophysical Research: Solid Earth*, 99(B5): 9359–9372, 1994.

Rives, T., Razack, M., Pett, J.P. and Rawnsley, K.D., Joint spacing: analogue and numerical simulation. *Journal of Structural Geology*, 14(8–9): 925–937, 1992.

Robinson, P.C., Connectivity of fracture systems: a percolation theory approach. *Journal of Physics A: Mathematical and General*, 16: 605–614, 1983.

Robinson, P.C., Numerical calculations of critical densities for lines and planes. *Journal of Physics A: Mathematical and General*, 17(14): 2823–2830, 1984.

Rossen, W.R., Gu, Y. and Lake, L.W., Connectivity and permeability in fracture networks obeying power law statistics. Paper SPE 59720, Proceedings of the SPE Permian Basin Oil and Gas Recovery Conference, Midland, Texas, 21–23 March, 2000.

Rouleau, A. and Gale, J.E., Statistical characterization of the fracture system in the Stripa Granite, Sweden. *International Journal of Rock Mechanics and Mining Sciences & Geomechanics Abstracts*, 22(6): 353–367, 1985.

Sadeghnejad, S., Masihi, M. and King, P.R., Dependency of percolation critical exponents on the exponent of power law size distribution. *Physica A: Statistical Mechanics and its Applications*, 392(24): 6189–6197, 2013.

Sahimi, M., *Flow and Transport in Porous Media and Fractured Rock: From Classical Models to Modern Approaches*. Wiley-VCH, New York, pp. 103–157, 1995.

Saidi, A.M., *Reservoir Engineering of Fractured Reservoirs*. Total Edition Presse, Paris, pp. 864, 1987.

Savik, P.N., Berre, I., Jakobsen, M. and Lien, M., A 3D computational study of effective medium methods applied to fractured media. *Transport in Porous Media*, 100: 115–142, 2013.

Schmittbuhl, J., Vilotte, J.P. and Roux, S., Percolation through self affine surfaces. *Journal of Physics A: Mathematical and General*, 26: 6115–6133, 1993.

Segall, P. and Pollard, D.D., Joint formation in granitic rock in the Sierra Nevada. *Geological Society of America Bulletin*, 94: 563–575, 1983.

Sen, Z. and Kazi, A., Discontinuity spacing and RQD estimates from finite length scan lines. *International Journal of Rock Mechanics and Mining Sciences & Geomechanics Abstracts*, 21(4): 203–212, 1984.

Sisavath, S., Mourzenko, V.V., Genthon, P., Thovert, J.F. and Adler, P.M. Geometry, percolation and transport properties of fracture networks derived from line data. *Geophysics Journal International: Tectonics and Geodynamics*, 157: 917–934, 2004.

Snow, D.T., Anisotropic permeability of fractured media. *Water Resources Research*, 5: 1273–1289, 1969.

Soares, R.F., Corso, G., Lucena, L.S., Freitas, J.E., da Silva, L.R., Paul, G. and Stanley, H.E., Distribution of shortest paths at percolation threshold: application to oil recovery with multiple wells. *Physica A*, 343: 739–747, 2004.

Taylor, W.L., Pollard, D. and Aydin, A., Fluid flow in discrete joint sets: field observations and numerical simulations. *Journal of Geophysical Research: Solid Earth*, 104(B12): 28983–29006, 1999.

Thovert, J.-F. and Adler, P.M., Trace analysis for fracture networks of any convex shape. *Journal of Geophysical Research: Solid Earth*, 31: L22502.1–L22502.5, 2004.

van Dijk, J.P., Bello, M., Toscano, C., Bersani, A. and Nardon, S., Tectonic model and three-dimensional fracture analysis of Monte Alpi (southern Apennines). *Tectonophysics*, 324: 203–237, 2000.

van Golf-Racht, T.D., *Fundamentals of Fractured Reservoir Engineering*. Developments in Petroleum Science, vol. 12. Elsevier Scientific Publishing Company, Amsterdam, 1982.

Vermilye, J.M. and Scholz, C.H., Relationship between vein length and aperture. *Journal of Structural Geology*, 17(3): 423–434, 1995.

Watanabe, K. and Takahashi, H., Fractal geometry characterization of geothermal reservoir fracture networks. *Journal of Geophysical Research: Solid Earth*, 100(B1): 521–528, 1995.

Wu, H. and Pollard, D.D., An experimental study of the relationship between joint spacing and layer thickness. *Journal of Structural Geology*, 17(6): 887–905, 1995.

Zhang, X. and Sanderson, D.J., *Numerical Modelling and Analysis of Fluid Flow and Deformation of Fractured Rock Masses, First Edition*. Pergamon, Elsevier Science Ltd., Oxford, pp. 288, 2002.

Zimmerman, R. and Bodvarsson, G.S. Effective transmissivity of two-dimensional fracture networks. *International Journal of Rock Mechanics and Mining Sciences & Geomechanics Abstracts*, 33(4): 433–438, 1996.

Chapter 6

Backbone, dangling ends and effective permeability

So far, we have concentrated on the prediction of the static connectivity of complex reservoirs using percolation theory. We now turn our attention to how the cluster structure influences fluid flow. We find that the percolating cluster consists of two parts. These are "dead ends" that contribute verly little to the flow and the "backbone" where most of the flow is concentrated. These two parts of the cluster have distinct statistics and scale differently. However, the distinction is very useful. For example, in a waterflood, only the oil in the backbone will be swept. Also, the different scaling of the two parts of the cluster accounts for the difference between the effective permeability exponent and the connectivity exponent. This chapter describes the appropriate percolation scaling for backbone, dangling ends and effective permeability. We shall concentrate on the overlapping sandbody problem. Similar scaling behaviour is found for the fracture models.

6.1 Backbones and dangling ends

So far, we have shown how percolation theory can be used to make rapid predictions of the connected sand. The percolating cluster has a tortuous conducting path with many dead ends that do not contribute to the flow. These dead ends contribute to the fraction

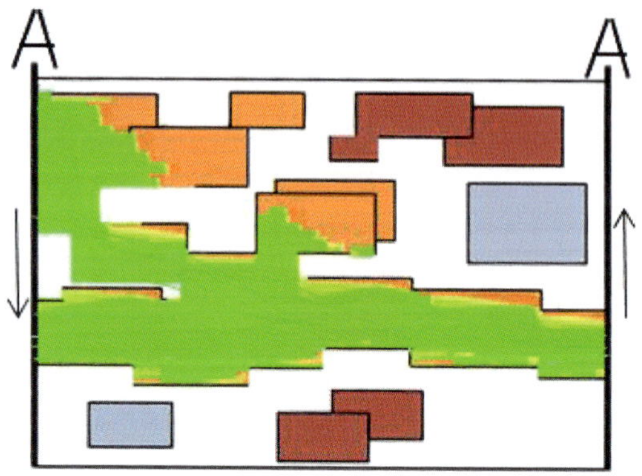

Figure 6.1. Illustration of the flow through backbone (shown in green) with tortuous conducting path with dangling ends attached to it and the other non-connecting sands (non-accessible oil).

of sands that are connected but not to the conductivity as there is no significant pressure drop to drive flow out of them. These regions are unlikely to be swept during secondary displacement. The conducting path which is the only part of the connected cluster that contributes to the flow is called the backbone (Figure 6.1). It should be noted that the backbone is a much more direct pathway between two wells than the whole cluster and so has a lower fractal dimension (Table 2.3). The dead ends are also sometimes known as dangling ends.

The precise structure of the backbone and dangling ends is not fully known (Porto *et al.*, 1999). The backbone can be thought of as being made up of "blobs" (multiple connected parts of the backbone) and "red links" (Stanley, 1977), which are single "bonds" that must carry all the flow. These are the links between two blobs. A schematic of this decomposition of the percolating cluster is shown in Figure 6.2.

The scaling of the backbone and dead ends is distinct. They are important for reservoir engineering purposes because the backbone represents swept oil and the dead ends unswept oil only recoverable by infill drilling, for example. The size structure of the dead ends can give an indication of the appropriate well spacing to use. Wen *et al.* (2014) showed numerically that a significant volume of oil

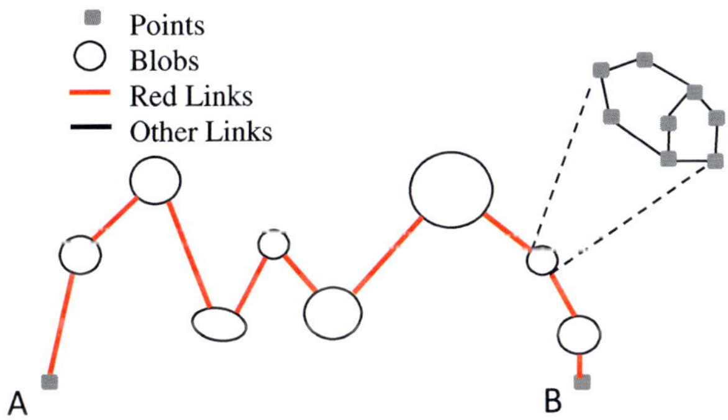

Figure 6.2. Illustration of the connection between the red links and blobs between two points on the backbone.

(up to 50% original oil in place (OOIP)) could be bypassed in these systems. In this chapter, we shall use the simple sandbody model described in Chapter 3 to develop the general finite-size-scaling laws for these percolation quantities. As before, we can then predict the mean and standard deviation in unswept oil, e.g., using simple scaling.

Consider the simple model of overlapping aligned sandbodies represented by squares in 2D as described in Chapter 3. Again, for simplicity we consider constant-size sandbodies and neglect spatial correlation between the sands. In each realisation, we identified the clusters connecting to the injection and production wells (placed on the left and right sides of the system as shown in Figure 6.1) as before and then subdivided the connected cluster into the backbone and the dangling ends (Figure 6.3).

The backbone fraction, B, is defined as the ratio of the area of this path to the total area of the sand in the system. Moreover, the dangling-ends fraction, D, is defined as the ratio of remaining part of the connected cluster once the backbone is removed to the total area of the sand in the system. Hence, the relation between three percolation quantities P, B and D is

$$D = P - B. \tag{6.1}$$

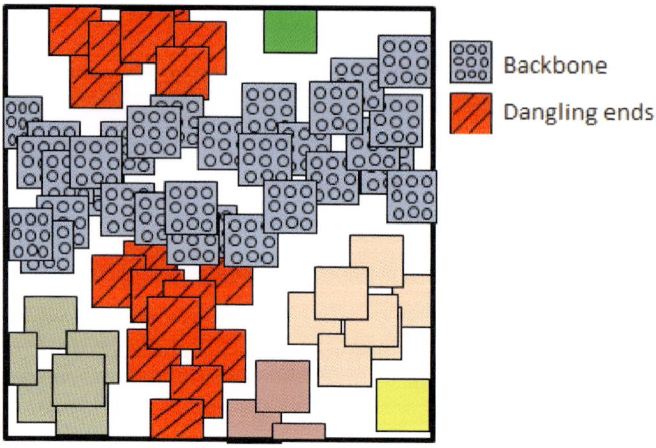

Figure 6.3. Illustration of connected cluster connecting the left and right sides of the system which is divided into the backbone (in blue) and dangling ends (in red) and the other non-connected finite clusters.

The backbone can be found geometrically by finding those parts of the connected cluster that are only connected by one sandbody to the cluster. Or it can be found by using a flow-based criterion (Wen *et al.*, 2014). In particular, the various versions of the burning algorithm, Tarjan algorithm or other recursive algorithms (Tarjan, 1972; Herrmann *et al.*, 1984; Roux and Hansen, 1987; Rintoul and Nakanishi, 1992; Grassberger, 1999; Danziger *et al.*, 2015), can be used to identify the backbone.

In the flow-based method, a skeleton is formed that consists of lines connecting sandbody centres, treating each line as a resistor with its resistance inversely related to the length of the line. Then the current flowing through each part of the resulting resistor network is calculated. A sandbody is considered to be part of a dangling end if the current flowing through it is negligible (e.g., less than a few percent of the main current). In the overlapping sandbody model where the connections is between two wells in two opposite sides of the system, the backbone structure formed is similar to the so-called *bus bar backbone* structure (Grassberger, 1992) in which the two opposite edges of a random resistor network are kept at different potentials. However, we could use alternative definitions to account for different well configurations, such as the *point-to-point backbone*.

As with the percolation probability P, there is a finite-size dependency for the backbone at the threshold from which the backbone exponents can be estimated as $\frac{\beta_B}{\nu} = 0.365 \pm 0.005$ (Rintoul and Nakanishi, 1992),

$$B(p, L) = L^{-\beta_B/\nu}. \tag{6.2}$$

The numerical value for the backbone exponent in 2D is found to be $\beta_B = 0.4758$ (Table 2.2). It is related to its fractal dimension $D_B = 1.6432$ (Table 2.3) through Eq. (2.45). The dangling ends are known to have a similar fractal dimension to the percolating cluster (Porto *et al.*, 1999).

To study the behaviour of these new percolation quantities, we generate various realisations of the sandbody model by using different square sizes (a) and at various net-to-gross ratios (p). Then we collect the statistics of the backbone fraction B, the dangling-ends fraction D and isolated clusters in each realisation. We also calculate the the standard deviation of these quantities over many realisations. The mean fractional volume of the backbones $B(p, L)$ and its fluctuations $\Delta_B(p, L)$ around the mean (as measured by the standard deviation) as a function of the net-to-gross ratio at different system sizes are shown in Figure 6.4.

The mean fractional volume of the dangling ends $D(p, L)$ and its fluctuation $\Delta_D(p, L)$ around the mean (the standard deviation) as a function of the net-to-gross ratio at different system sizes are shown in Figure 6.5.

As expected, the mean backbone fraction increases rapidly with net-to-gross ratio at the percolation threshold and there is a system size dependence. The standard deviation of the area of these sub-networks is highest near the percolation threshold. However, for dangling ends, the peak is shifted due to finite-size effects. The standard deviation of the connected cluster (Figure 3.10) and the backbone (Figure 6.4) both decrease as the system size increases, whereas the standard deviation of the volume of the dangling ends (Figure 6.5) increases. The mean dangling-ends fraction increases

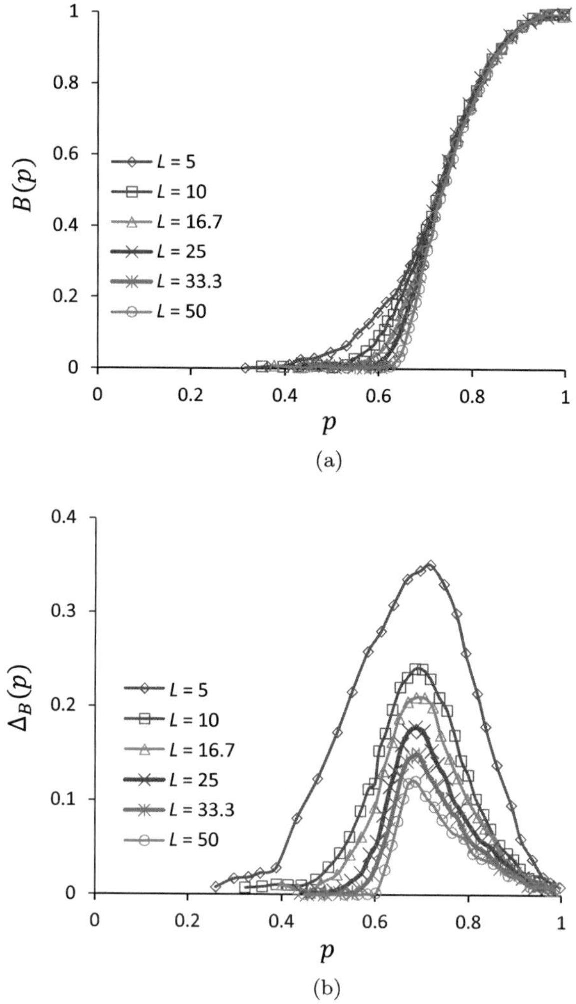

Figure 6.4. (a) The mean fractional volume of the backbones $B(p, L)$ and (b) its standard deviation $\Delta_B(p, L)$ as a function of the net-to-gross ratio for different sandbody sizes in 2D.

as the sandbody size decreases, approaching almost half the connected sand volume at the percolation threshold (Figure 6.5). This implies that a significant volume of oil could be bypassed during a conventional waterflood through a moderately low net-to-gross fluvial or turbidite reservoir.

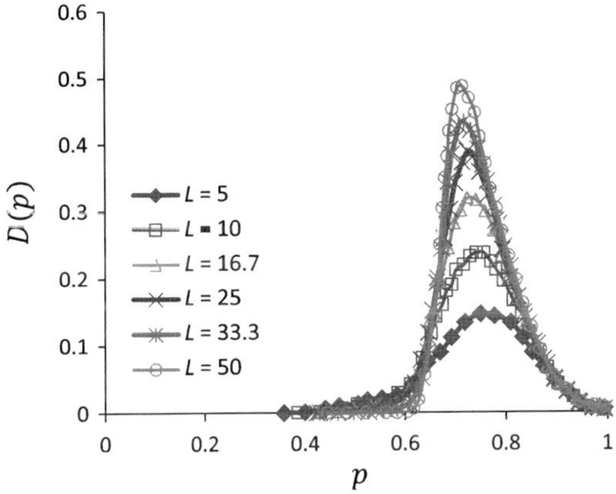

(a)

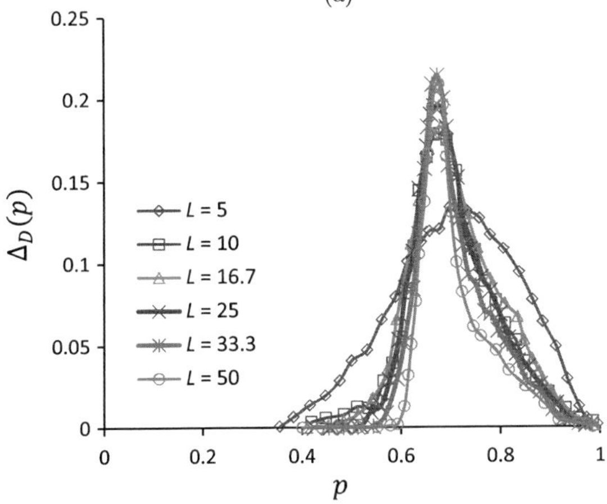

(b)

Figure 6.5. (a) The mean fractional volume of the dangling ends $D(p, L)$ and (b) its standard deviation $\Delta_D(p, L)$ as a function of the net-to-gross ratio for different sandbody sizes in 2D.

The connected sand fraction, the backbone fraction and the dangling-ends fraction are related to each other through Eq. (6.1). Using the property of variance of sum of two variables, the uncertainty in these percolation quantities has the following relation also: $\Delta^2 = \Delta_B^2 + \Delta_D^2 + 2\sigma_{BD}$, where $\sigma_{BD} = \langle BD \rangle - \langle B \rangle \langle D \rangle$ is the covariance between the backbone and dangling-ends fractions. Using the numerical results shown above for $\Delta_D(p, L)$ and $\Delta_B(p, L)$,

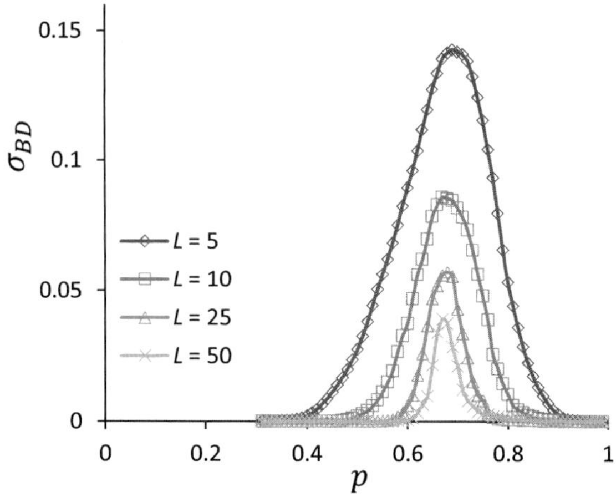

Figure 6.6. The covariance σ_{BD} between the backbone and the dangling-ends fractions as a function of the net-to-gross ratio for different sandbody sizes in 2D.

the covariance function σ_{BD} as a function of the occupation fraction and the system size can be calculated. This is shown in Figure 6.6.

As can be seen in Figure 6.6, the covariance (which is a measure of similarity in value) is highest around the threshold.

Finally we used finite-size-scaling laws to develop a simple mathematical scaling for these quantities. We expect that the backbones and dangling-ends fractions follow the same scaling (Eq. (2.6)) for large systems. The estimates for the backbone fraction exponents β_B and the dangling-ends fraction β_D by using scaling Eq. (2.6) are shown in Figure 6.7. The critical exponent of the backbone fraction is found empirically by finite-size scaling Eq. (2.6) to be $\beta_B = 0.5$, which is in agreement with the published values in Table (2.2). A similar power law scaling to Eq. (2.6) was used to estimate the critical exponent of the dangling ends as $\beta_D = -0.45$.

The same arguments described in Chapter 2 for developing the finite-size scaling for connectivity (Eqs. (2.11) and (2.14)) can be applied with these new critical exponents to determine the scaling

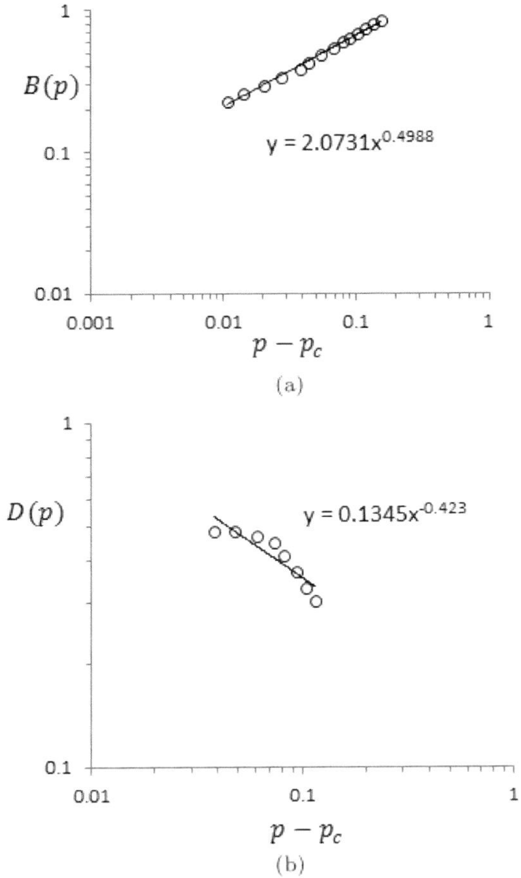

Figure 6.7. The scaling Eq. (2.6) for the mean fractional volume of (a) the backbones $B(p, L)$ and (b) the dangling ends $D(p, L)$ as a function of $p - p_c$ used to estimate the critical exponents.

for the mean and standard deviation of the backbone fraction as

$$B(p, L)L^{\beta_B/\nu} = \Im_B[(p - p_c)L^{1/\nu}], \tag{6.3}$$

$$\Delta_B(p, L)L^{\beta_B/\nu} = \Re_B[(p - p_c)L^{1/\nu}], \tag{6.4}$$

and for the dangling-ends fraction as

$$D(p, L)L^{\beta_D/\nu} = \Im_D[(p - p_c)L^{1/\nu}], \tag{6.5}$$

$$\Delta_D(p, L)L^{\beta_D/\nu} = \Re_D[(p - p_c)L^{1/\nu}], \tag{6.6}$$

where $\Im$ and $\Re$ are the universal mean and its associated uncertainty for the percolation quantities and subscripts B and D denote the backbone and dangling-ends fractions respectively.

Now if we rescale the data shown in Figures (6.4) and (6.5) for the fractional volumes of backbones and dangling ends and plot $XL^{\beta/\nu}$ (where X represents B or D) against $(p - p_c)L^{1/\nu}$, we see that all the curves for different system sizes collapse onto each other, and from these we find the universal master curve for the mean and its associated uncertainty of the fractional volumes of backbones and dangling ends, which are $\Im_B$, $\Im_D$, $\Re_B$, and $\Re_D$ (Figures 6.8 and 6.9).

The master curves for the backbone and dangling ends can then be encoded in a spreadsheet from which the prediction of these impotant quantities can be made very quickly. In the case of the dangling ends, the data collapse is not so good, indicating that there are corrections to the simple forms used which require further investigation. It is also required to test whether the methods described in Chapter 4 to account for more complicated models with size and orientation distributions still apply.

6.2 Effective (single-phase) permeability

Percolation theory has been used to study the conductivity behaviour on both lattices and continuum percolation problems (Watson and Leath, 1974; Pike and Seager, 1974; Masihi *et al.*, 2008; Sadeghnejad *et al.*, 2011a, 2011b). Understanding the dynamic behaviour of percolating systems is a complicated problem, and we shall first consider a simple problem here which is the prediction of the effective (single-phase) permeability.

We start with the usual simple overlapping sandbody model of squares or rectangles in 2D and assume perfect hydraulic connection between the intersecting sandbodies. The background is assumed to be impermeable, and all the sandbodies have the same permeability equal to 1. It is reasonable to assume that the sands have much higher permeability than the background shale so the flow will be confined to the sandbodies. Also the variation of permeability within the sandbodies is much less than that between the sand and the

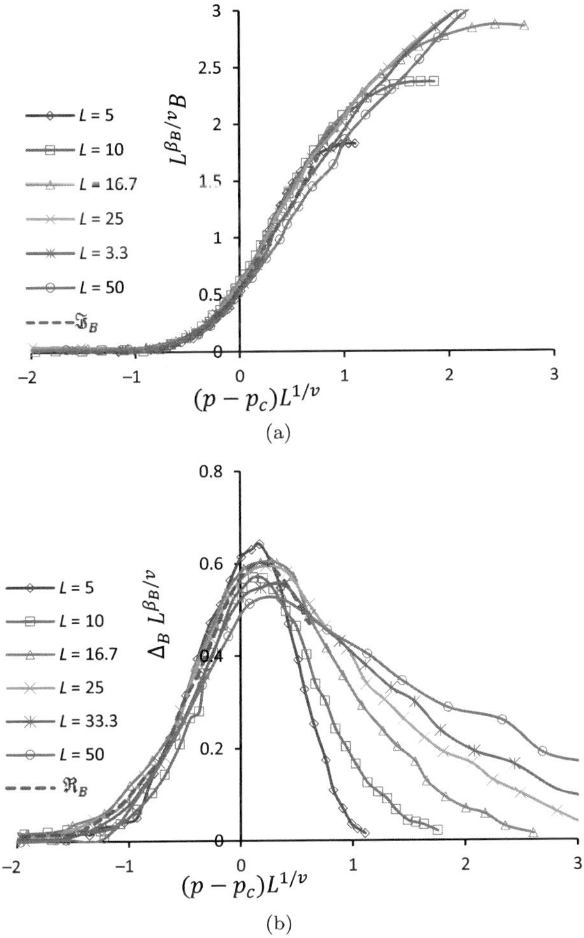

Figure 6.8. The finite-size scaling for (a) the mean of the backbone fraction and (b) its standard deviations near the percolation threshold.

background. We will treat the impact of a continuous permeability distribution in Chapter 7.

We consider the case of single-phase flow which is similar to the problem of the electrical conductivity of percolating networks of resistors. The effective permeability (K_{eff}) of single-phase flow is that permeability which gives the same overall flow rate for a fixed pressure drop as in the heterogeneous system. We only look

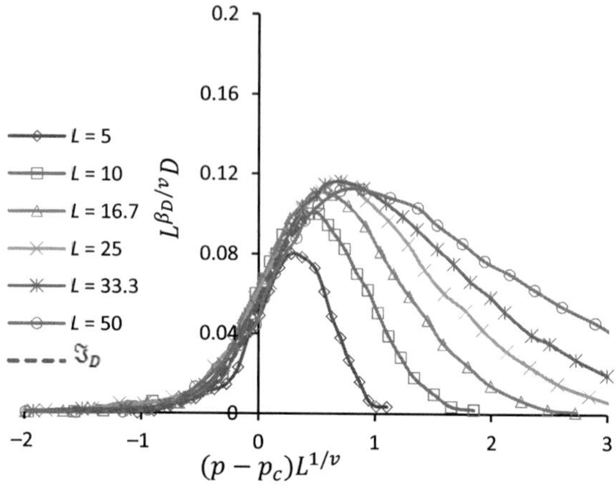

(a)

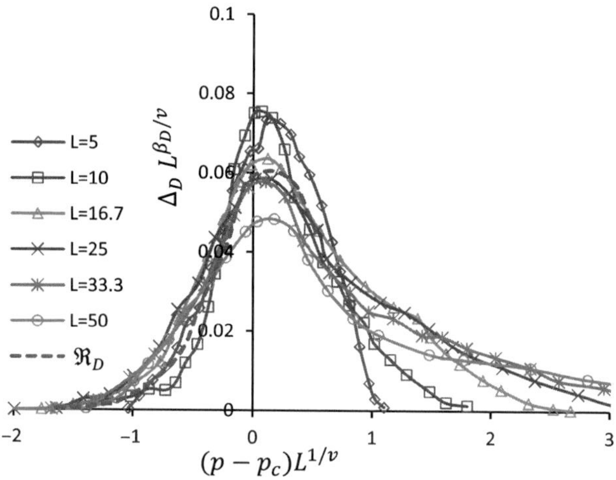

(b)

Figure 6.9. The finite-size scaling for (a) the mean dangling-ends fraction and (b) its standard deviation near the percolation threshold.

at low flow rates controlled by Darcy's law $\left(\mathbf{v} = -\frac{K}{\mu}\nabla P\right)$ and incompressible steady-state flow where the continuity equation is $\nabla \cdot \mathbf{v} = 0$ which leads to the single-phase pressure equation (we assume that the viscosity can be set to 1):

$$\nabla \cdot K\nabla P = 0. \tag{6.7}$$

This equation is solved numerically by covering the system with a fine-grid giving sites within the sands a permeability of 1 and within the background an essentially zero permeability (setting it to exactly zero can cause problems with many numerical schemes). Then applying finite-difference or finite-volume methods leads to a matrix representation of Eq. (6.7). These are solved to calculate the total flow rate, and then by equating fluxes with those from an equivalent homogenous system, the effective permeability is found. Although there are various numerical methods including direct and iterative methods for solving the resulting equations (Ertekin *et al.*, 2001), these are slow and need huge computational efforts (CPU) if it is done for many realisations. Hence, we may need alternative much faster approaches to get reasonable statistics on the effective permeability results. For example, we could use real-space renormalisation (King, 1989) to calculate the effective permeability. This considerably reduces the computation time, but there are known errors associated with this approximation.

Below the threshold, there cannot be any overall conductance as there is no connected flow pathway from inlet to outlet. On the other hand, when the net-to-gross ratio is equal to 1, then the permeability is also 1. Between the threshold and full occupancy ($p = 1$), the effective permeability must rise continuously and monotonically. We may expect that sand conductivity is proportional to the connected sand fraction, $K_{\text{eff}} \propto P(p)$. However, the numerical results show that this is not the case and the conductivity raises much slowly (Last and Thouless, 1971) although again it has a power law equation for infinite-size systems (Eq. (2.7)) with the numerical value for the conductivity exponent in Table 2.2. The reason for this slower rise of the conductivity behaviour is that the percolating cluster has a tortuous conducting path with many dead ends as discussed in Section 6.1. As the dead-ends carry no flux, the rise in conductivity is slower than the rise in the connected volume. A realisation of overlapping sandbody model of size $L = 8$ above the threshold point with the calculated effective permeability ($K_{\text{eff}} = 0.1193$) and connected sand fraction ($P = 0.731$) is shown in Figure 6.10.

Figure 6.10. A realisation showing the connected path in brown in an overlapping sandbody model of size $L = 8$ at $p = 0.6518$, $P = 0.7310$, $K = 0.1193$, $N = 68$.

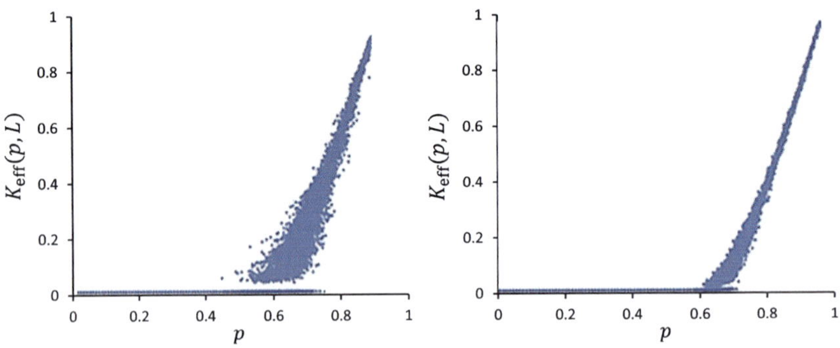

(a) (b)

Figure 6.11. The typical scatter plots of the calculated effective permeability values versus net-to-gross ratio at system size (a) $L = 16$ and (b) $L = 64$.

As before, we collect the statistics of the effective permeability K_{eff} and net-to-gross ratio p for a given system size and plot the effective permeability against the net-to-gross ratio. This gives a scatter of points as shown for two system sizes in Figure 6.11. The range of scatter depends on the system size.

We take the average (K_{eff}) and standard deviation (Δ_k) about the average for the effective permeability results. The variation of

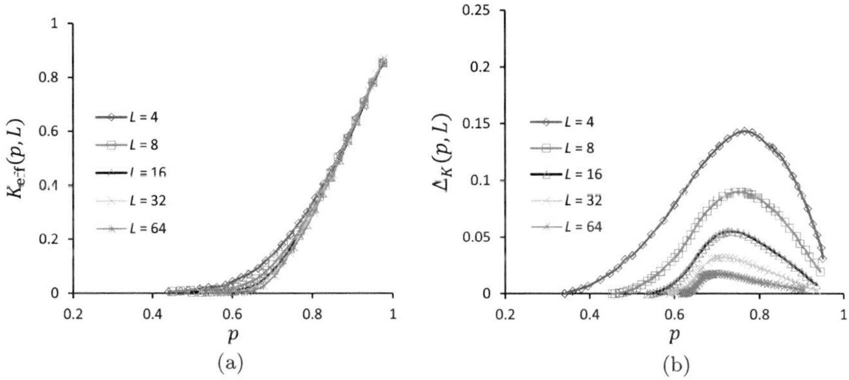

Figure 6.12. The variation of (a) the mean effective permeability (K_{eff}) and (b) its associated error (Δ_k) against net-to-gross ratio for a squared sandbody model at various system sizes.

the calculated mean effective permeability values and its associated error as a function of net-to-gross ratio for variety of system sizes is shown in Figure 6.12.

In principal, it is not necessary to determine the critical exponents as they are universal. Previous estimations of conductivity exponent (King, 1990; Sadeghnejad *et al.*, 2013a, 2013b) were very close to those from lattice percolation in Table 2.2. To estimate *the conductivity exponent* (μ), one can take a log–log plot of the K_{eff} against $p - p_c$ for a very large system size (Eq. (2.7)). Figure 6.13 shows an example of extrapolating the permeability data (K_{eff}) against ($p - p_c$) to find the slope which is the *conductivity exponent* μ.

The same arguments described in Chapter 2 for developing the finite-size scaling for connectivity (Eqs. (2.11) and (2.14)) can be applied with new critical exponents to determine the scaling for the mean and standard deviation for the effective permeability results as

$$K_{\text{eff}}(p, L)L^{\mu/\nu} = \Im_k[(p - p_c)L^{1/\nu}], \tag{6.8}$$

$$\Delta_k(p, L)L^{\mu/\nu} = \Re_k[(p - p_c)L^{1/\nu}], \tag{6.9}$$

where $\Im_k$ and $\Re_k$ are the master curves respectively for the mean and standard deviation of effective permeability results in isotropic overlapping sandbody model. Now by using these scaling laws, we

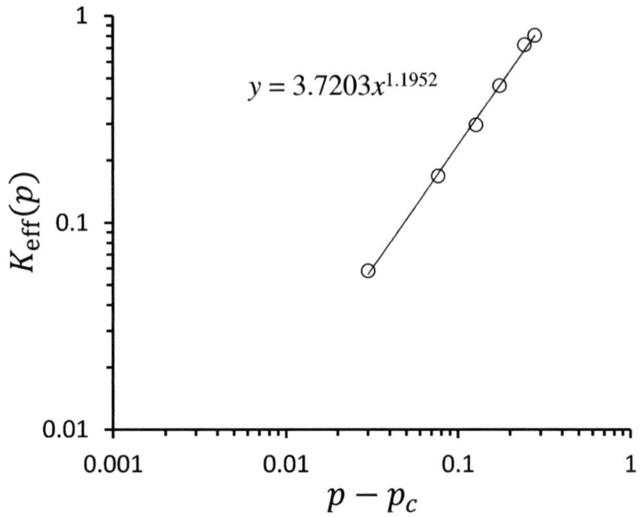

Figure 6.13. The use of the scaling law (Eq. (2.7)) to estimate the conductivity exponent μ for the overlapping sandbodies model in 2D.

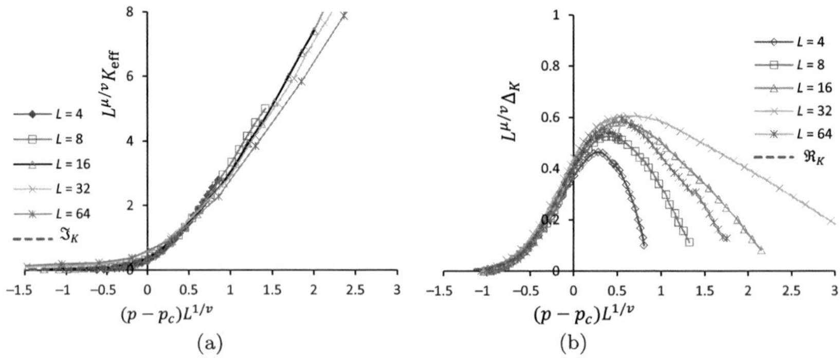

Figure 6.14. Illustration of finite-size-scaling transformations for (a) the average and (b) the standard deviation of the effective permeability results shown in Figure 6.12.

can rescale the effective permeability results in Figure 6.12 to find the data collapse and determine the relevant master curves as shown in Figure 6.14.

These results enable us to use the master curves for the average effective permeability ($\Im_k$) and its associated uncertainty ($\Re_k$) to predict the effective permeability of any system size very quickly. Moreover, these can be used to find the effective permeability as a function of well spacing at different net-to-gross ratios. As expected, the scaling is valid for the region close to the threshold; for other regimes far from the threshold, other techniques such as effective medium approximation (see Chapter 7) can be used.

6.2.1 *Anisotropic effects in 2D*

We consider the effects of anisotropy on the effective permeability results in 2D. The details of how to generate the anisotropic models has already been described in Section 4.2. We determine the effective permeability values of various realisations of sandbody models. As was shown for the connectivity results (Masihi *et al.*, 2006), we see data collapses for the effective permeability results in both the horizontal and vertical directions. They are clearly different from the isotropic effective permeability curves (see Figure 6.15 for the anisotropic system with aspect ratio $\omega = 2$).

The same arguments described in Section 4.2 for incorporating the anisotropic effects into the finite-size scaling by using an effective apparent threshold ($\tilde{p}_c$) and rescaling the standard deviation results by a factor ($\omega^{1/2}$) can be applied with new critical exponents to determine the appropriate scaling for the mean and standard

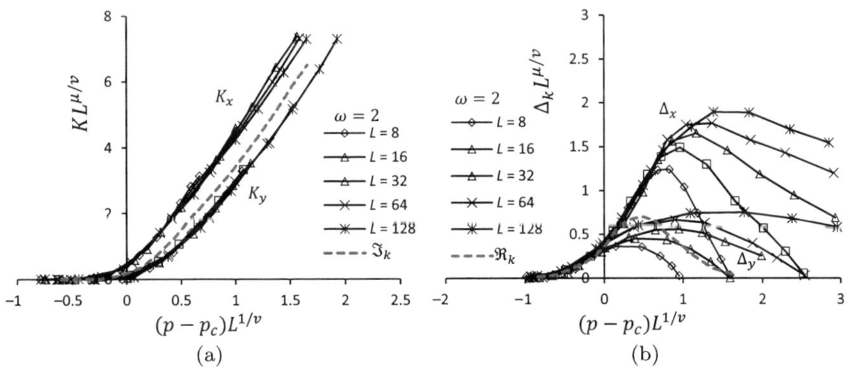

Figure 6.15. Illustration of usual finite-size scaling of effective permeability results (at a fixed aspect ratio $\omega = 2$) by using (a) Eq. (6.8) and (b) Eq. (6.9) which shows two sets of data collapse in two directions.

deviation for the effective permeability results as

$$K_{\text{eff}}(p, L_x, \omega) = L_x^{-\mu/\nu} \Im_k[(p - \tilde{p}_c)L_x^{1/\nu}], \tag{6.10}$$

$$\Delta_k(p, L_x, \omega) = \omega^{1/2} L_x^{-\mu/\nu} \Re_k[(p - \tilde{p}_c)L_x^{1/\nu}]. \tag{6.11}$$

The rescaled results of Figure 6.15 for the anisotropic overlapping sandbody system with aspect ratio $\omega = 2$ is shown in Figure 6.16. The

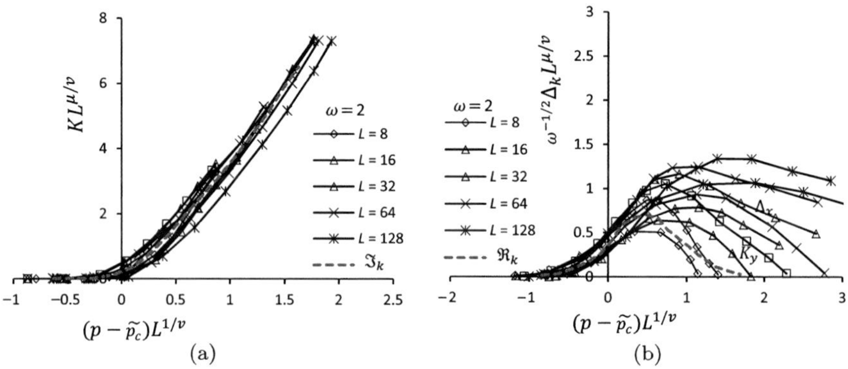

Figure 6.16. Illustration of the corrected finite-size scaling of effective permeability results (at a fixed aspect ratio $\omega = 2$) by using (a) Eq. (6.10) and (b) Eq. (6.11).

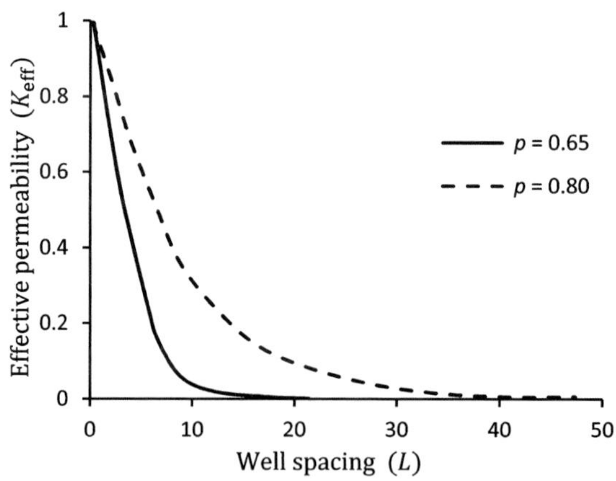

Figure 6.17. Illustration of the effective permeability results against well spacing at various net-to-gross values in an anisotropic system.

data collapse in the critical region is reasonable, especially for approximations.

With this analysis, the variation of the effective permeability against well spacing for a variety of net-to-gross ratios of 2D cross-sectional models can be obtained by simply inverting the scaling laws (Figure 6.17).

The model used for calculating the effective permeability was very simple. In reality, we need to consider the effects of sandbody size and orientation distribution as well as the sandbody shapes.

6.2.2 *Effects of sandbody orientation disorder*

In reality, sandbodies have different orientations which will have an impact on the effective permeability. If the sands are restricted to the range between $-\theta$ and θ (with respect to the horizontal axis), the main effect is to make the system anisotropic. The same framework described in Section 4.3 can be used to account for this anisotropy on the effective permeability. So the effective sandbody sizes (Eq. (4.25)) and the corrected threshold (Eq. (4.23)) can be used to include the effects of the orientation disorder. A comparison of the results of the mean effective permeability for squares of size 1×1 ($\omega = 1$) with uniform angular dispersion between $-15°$ and $15°$ (and $-30°$ and $30°$) distributed in a region of size 16×16 in 2D is shown in Figure 6.18. For such a system, the average length along the x and y directions are $L = 1.118$ (and 1.21 for $\pm 30°$). The corrected threshold is $p_c = 0.641$ (and 0.626 for $\pm 30°$). As the aspect ratio is equal to 1 ($\omega = 1$), there is no effective finite-size shift in this case. The results show that the methodology given to consider the effect of sandbody orientations is appropriate and again enables us to use the isotropic connectivity master curves (Figure 6.14).

6.2.3 *Effects of sandbody size distribution*

A similar treatment to consider the effects of sandbody size distribution as discussed in Section 4.1 based on the appropriate definition of effective size can be used to estimate the effective permeability. This

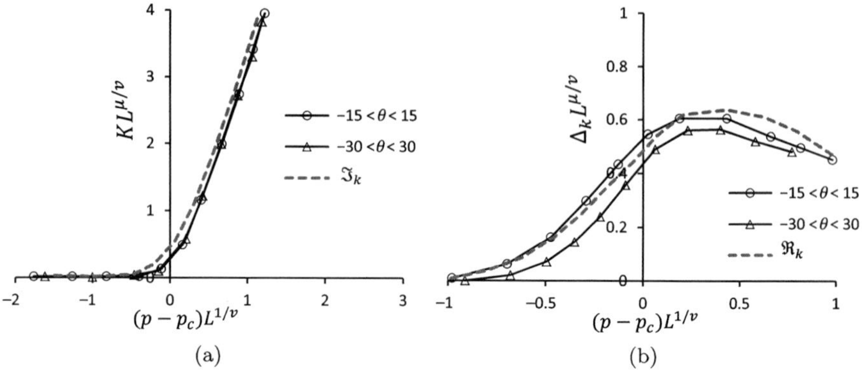

Figure 6.18. Comparison of the computed effective permeability results for sandbodies model with uniform orientation between −15° and 15° and between −30° and 30° with the corrected scaling rules. (a) Mean and (b) standard deviation.

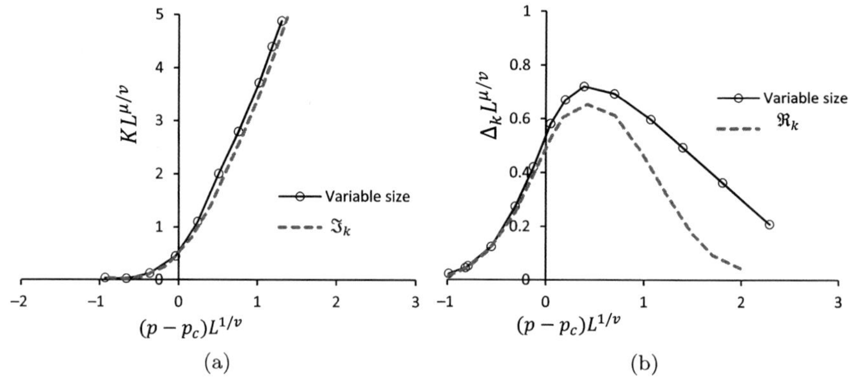

Figure 6.19. Illustration of finite-size scaling of (a) mean and (b) standard deviation in effective permeability for variable sizes (solid line) and sand size l_{eff} (dotted line) in a sandbody model with uniform size distribution of sands in the range 0.5–1.5.

hypothesis allows us to use the finite-size-scaling laws (Eqs. (6.8) and (6.9)) to estimate the mean effective permeability and the standard deviation. For example, the results for a sandbody model with uniform size distribution of sands in the range 0.5–1.5 with the effective length ($l_{\text{eff}} = 1.04$) in a system of dimensionless size $L_x = 24/1.041 = 23.08$ are shown in Figure 6.19.

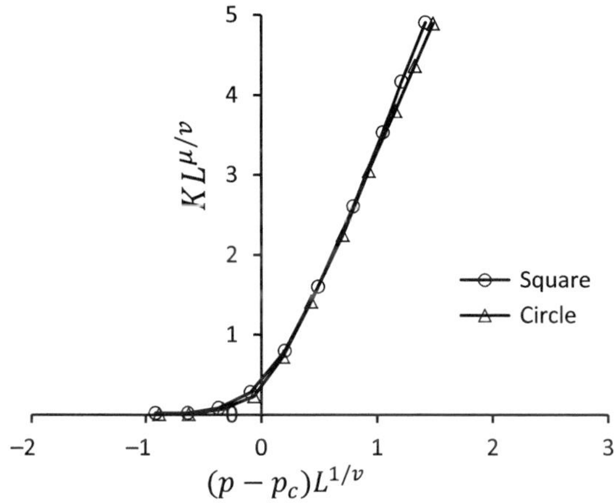

Figure 6.20. A comparison of the effective permeability master curves for the sandbody models made by circles and squares.

6.2.4 *Effects of sandbody shape*

As before, the main effect of sandbody shape is to change the threshold value (Table 3.1). This is true providing the shape of the sandbody is not so extreme as to change the effective dimensionality of the system, which would change the critical exponents. To illustrate this, we compare the effective permeability results in rescaled form for sandbodies represented by circles and squares in 2D. The results are shown in Figure 6.20.

6.3 Extensions to 3D

So far, the behaviour of the effective permeability of sandbodies in 2D has been discussed. It can be assumed that similar arguments apply in 3D. We first restrict ourselves to the simple cubic model and for computational efficiency use renormalisation to calculate the effective permeability. A realisation of cubic sandbody model near the threshold in a region of size $L = 100/25 = 4$ is shown in Figure 6.21. The computed effective permeability is $K = 0.0302$ and the connected sand fraction $P = 0.511$.

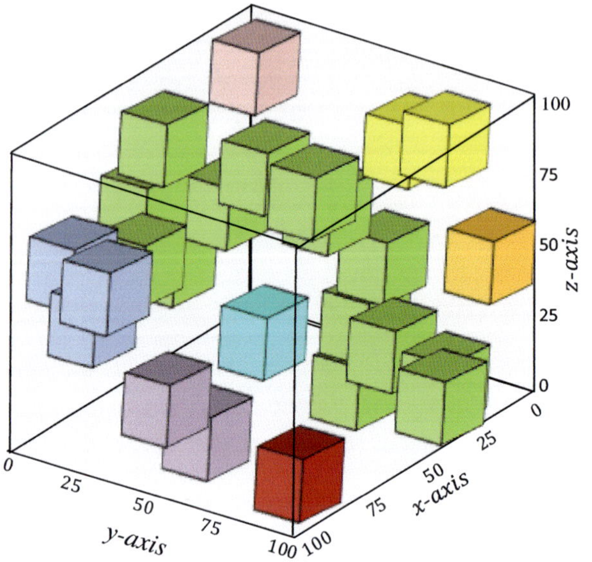

Figure 6.21. A realisation of the sandbody model in 3D with $N = 27$, $p = 0.2933$, $P = 0.5110$ and $K = 0.0302$ where percolation cluster is in green.

We plot the effective permeability $K_{\text{eff}}(p, L)$ against the net-to-gross ratio p and then extract the average effective permeability $K_{\text{eff}}(p, L)$ and its standard deviation $\Delta_k(p, L)$. The effective permeability results in 3D at five different system sizes are shown in Figure 6.22. The results clearly show that the errors in the effective permeability strongly depend on the dimensionless size of the system. It is interesting to note that the errors are much lower than in 2D (Figure 6.12).

To perform the finite-size scaling, we need the threshold and the conductivity exponent in 3D. The estimation of the percolation threshold has been explained in Section 4.5. To obtain an estimate of the exponent μ, one can take a log–log plot of $K_{\text{eff}}(p, L)$ versus $p - p_c^{\infty}$ for a relatively large system; then by using Eq. (2.7), the gradient on the data gives an estimate for the exponent μ. The slope shows an agreement with the literature value of 2 in 3D percolating systems demonstrating that universality holds (shown in Figure 6.23).

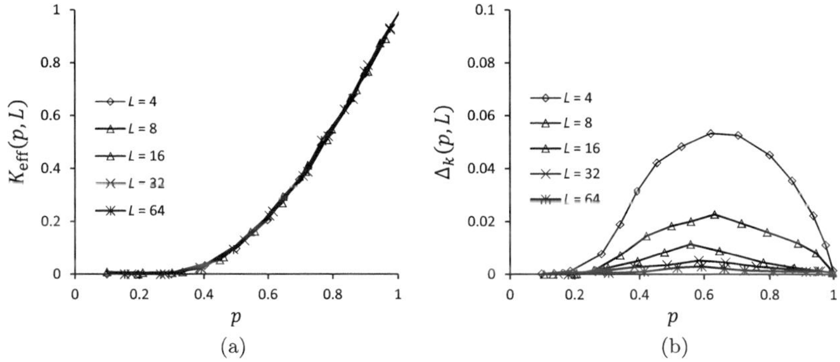

Figure 6.22. Illustration of finite-size effects by plotting the (a) mean and (b) standard deviation in the effective permeability results versus occupancy of cubic sandbody at five different system sizes.

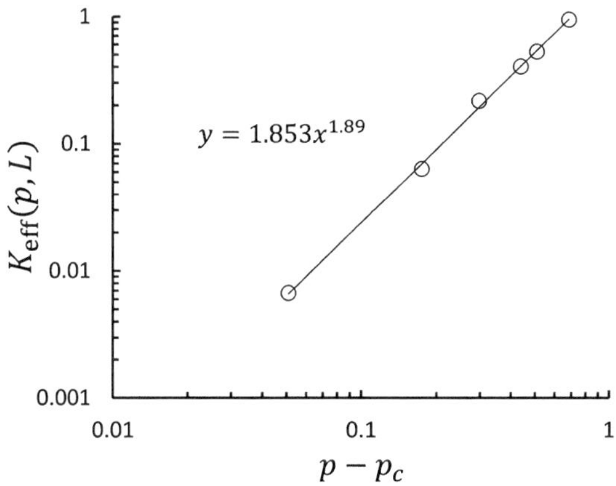

Figure 6.23. Plot of the computed effective permeability versus $(p - p_c)$ whose slope gives an estimate for the exponent μ in 3D.

The finite-size-scaling laws (Eqs. (6.8) and (6.9)) can be used to find the universal master curves for the mean effective permeability ($\Im_k$) and its associated uncertainty ($\Re_k$) of the overlapping sandbody model in 3D. These are shown in Figure 6.24.

6.3.1 *Anisotropy effects in 3D*

The same analysis of the effects of anisotropy as discussed in Section 4.5 can be used. We use the corrected threshold given by

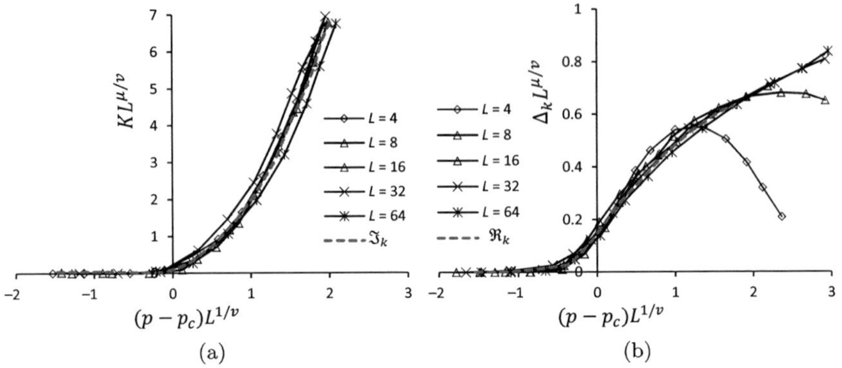

Figure 6.24. The Illustration of data rescaling of the effective permeability results shown in Figure 6.22 by using finite-size-scaling law in 3D. (a) Mean and (b) standard deviation.

Eqs. (6.10) and (6.11) to develop the finite-size scaling for the effective permeability and its standard deviation in the presence of anisotropy. In 3D, we note that the aspect ratio is the product of two independent aspect ratios $\omega = \omega_1\omega_2$. As an example to show the use of these scaling laws, we used 500 realisations to generate various effective permeability curves for different system sizes. In Figure 6.25, we plot $KL_x^{\mu/\nu}$ against $(p - \tilde{p}_c)L_x^{1/\nu}$ for a fixed value of the aspect ratio of 4 which shows the data collapse for the mean and the standard deviation in the effective permeability results. The results enable us to use the same isotropic universal curves shown in Figure 6.24 for predicting the effective permeability of anisotropic sandbody systems.

6.3.2 *Effects of sandbody orientation disorder*

The analysis given in Section 6.2 can be used to account for the effects of sandbody orientation disorder on the effective permeability results. For example, sandbody sizes (Eq. (4.25)) and corrected infinite threshold (Eq. (4.23)) can be used to include the effects of the orientation disorder in 3D. A comparison of the computed results of the mean effective permeability for a set of cubes of size $1 \times 1 \times 1$ with uniform angular dispersion between $-15°$ and $15°$ and $-30°$ and $30°$ along the x direction in a cubic region of size $16 \times 16 \times 16$ with

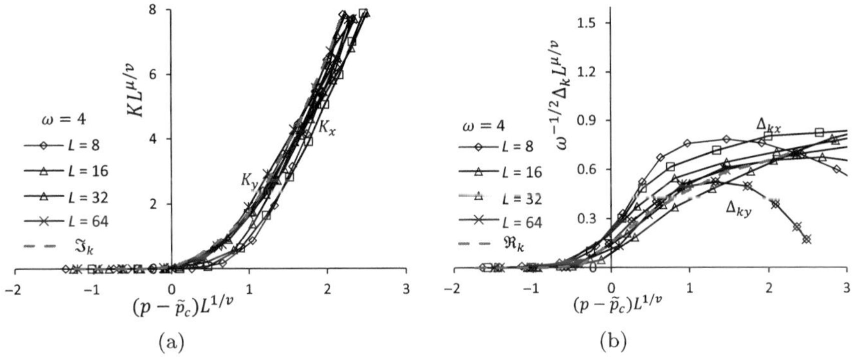

Figure 6.25. Illustration of data collapse of (a) the effective permeability and (b) its associated standard deviation using the corrected finite-size apparent threshold (Eqs. (6.10) and (6.11)) at the aspect ratio of 4; the dotted curves show the isotropic effective permeability master curves in 3D (Figure 6.24).

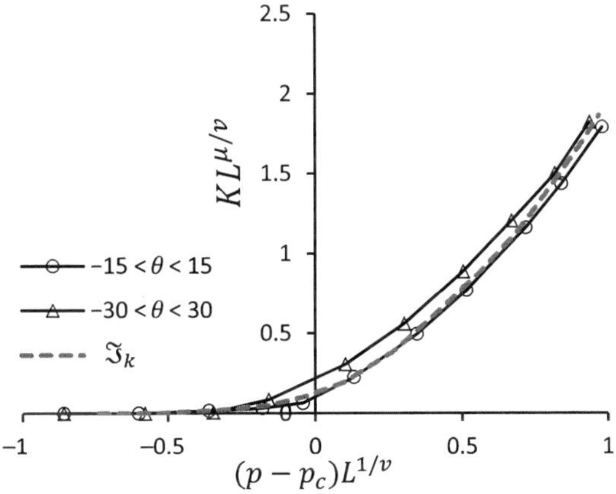

Figure 6.26. Comparison of the computed effective permeability results for sandbodies model with uniform orientation between -15° and 15° and between -30° and 30° with the corrected scaling rules.

the predictions from finite-size-scaling rules which uses the effective lengths and the corrected threshold is shown in Figure 6.26.

6.3.3 *Effects of sandbody size distribution*

We use the same analysis as in Section 6.2 to account for the effects of sandbody size distribution on the effective permeability results.

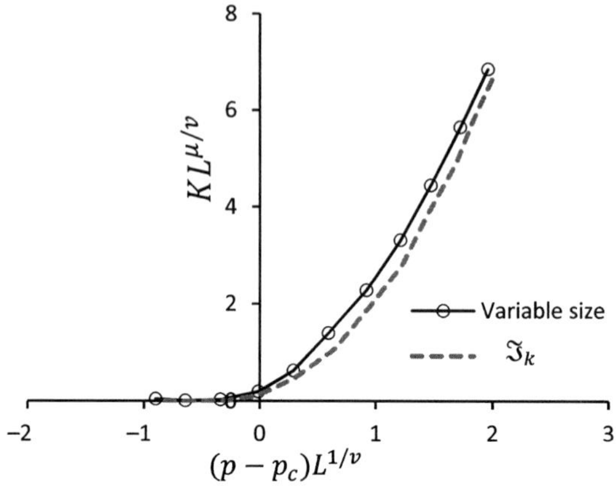

Figure 6.27. Illustration of finite-size scaling of mean effective permeability for variable sizes (solid line) and sand size l_{eff} (dotted line) in a sandbody model with uniform size distribution of sands in the range 0.5–1.5.

For example, the rescaled mean effective permeability results for a sandbody model with uniform size distribution of sands in the range 0.5–1.5 with the effective length ($l_{\text{eff}} = \sqrt[3]{1.041^2 \times 1} = 1.027$) in a system of size $L = 24$ (the system dimensionless size $L_x = \frac{24}{1.027} = 23.37$) are shown in Figure 6.27.

6.4 Case studies

6.4.1 *Case study 1*

As described in Section 4.6, Sadeghnejad *et al.* (2011a) applied the finite-size scaling on a fine-grid model of an offshore oil field. We have already described the details of method for the connectivity calculations in 2D and 3D. In this section, we present the validation of the effective permeability calculations with the predictions from the finite-size-scaling laws for the effective permeability (Eqs. (6.8) and (6.9)) both in 2D and 3D. For permeability computations of each extracted permeability map, we used the renormalisation method (King, 1989). Then the computed values can be compared with the predictions from the finite-size scaling of the effective permeability (Figures 6.14 and 6.24). As an example, Figure 6.28 compares the computed effective permeability values (shown in points) for 1,000 2D areal models (Figure 6.28(a)) and the cross-sectional

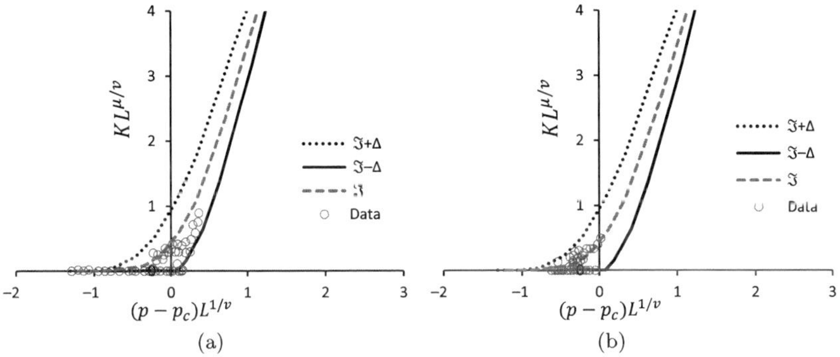

Figure 6.28. A comparison of the computed effective permeability values (points) for (a) 1,000 2D areal models and (b) cross-sectional maps with the effective permeability master curves (Figure 6.14). The average size of sandbodies is $1{,}200 \times 1{,}300\,\mathrm{m}^2$ and the critical value of permeability is 2,930 md.

maps (Figure 6.28(b)) with the effective permeability master curves (Figure 6.14). Clearly there is a reasonable agreement between the effective permeability calculated from field data and the predictions from the percolation approach shown by the universal effective permeability curves.

Similarly, we can validate the effective permeability predictions from percolation theory approach against the field data in 3D. Figure 6.29 compares the computed effective permeability values (points) for 20 3D maps with the effective permeability master curves (Figure 6.24).

6.4.2 *Case study 2*

In another case study, Sadeghnejad *et al.* (2011b) applied percolation theory method to the data set of a gas condensate in the south of Iran to evaluate the applicability of the effective permeability scaling laws. The reservoir was described in Section 4.6, and the details for the calculations are as before. The size of sandbodies in the scaling laws was $4{,}500 \times 800\,\mathrm{m}^2$ in the X and Y directions. The results for the effective permeability is shown in Figure 6.30 which indicates a reasonable match. The fit may not be perfect for a small number of the cross sections, but the CPU time required to use the scaling laws is a fraction of a second on a spreadsheet whereas the conventional calculations is computationally very intensive.

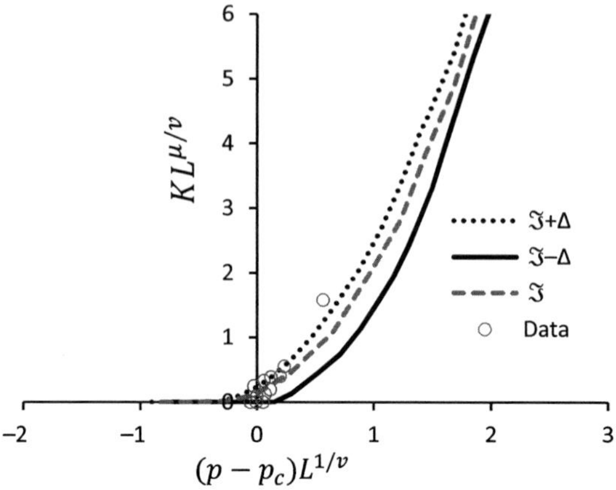

Figure 6.29. A comparison of the computed effective permeability values (points) for 20 3D maps with the effective permeability master curves (Figure 6.24). The average size of sandbodies is 1,200 × 1,300 × 15 m^3 and the critical value of permeability is 1,800 md.

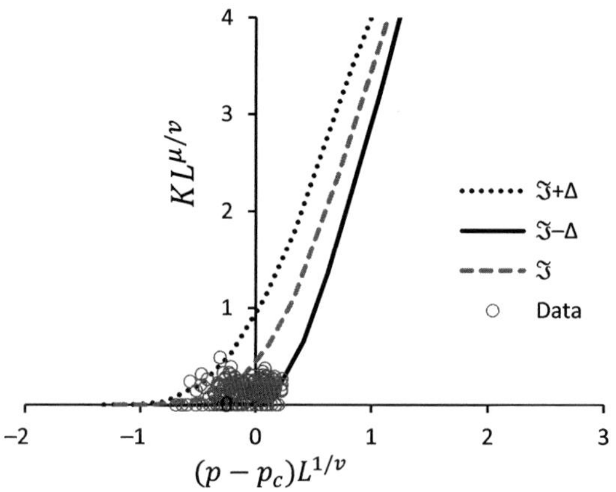

Figure 6.30. Reservoir's computed effective permeability results from 2D cross-sectional maps of the reservoir C compared with the percolation predictions of the effective permeability universal master curves (Figure 6.14).

6.5 Problems

6.1. Use the master curves for the average backbone fraction ($\Im_B$) and its associated uncertainty ($\Re_B$) shown in Figure 6.8 to reconstruct the backbone fraction curve of the overlapping sandbody systems with sizes $L = 10$, 50, 100 and 500 for the region close to the finite-size threshold.

6.2. Use the master curves for the average dead-ends fraction ($\Im_D$) and its associated uncertainty ($\Re_D$) shown in Figure 6.9 to reconstruct the dead-ends fraction curve of the overlapping sandbody systems with sizes $L = 10$, 50, 100 and 500 for the region close to the finite-size threshold.

6.3. Use the master curves for the average effective permeability ($\Im_k$) in 2D and its associated uncertainty ($\Re_k$) shown in Figure 6.14 to reconstruct the effective permeability curve of the overlapping sandbody systems with sizes $L = 10$, 50, 100 and 500 for the region close to the finite-size threshold.

6.4. Use the master curves of effective permeability for the 2D isotropic overlapping sandbody model shown in Figure 6.14 to reconstruct the effective permeability curves for an anisotropic model with rectangular shape sands with aspect ratios of $\omega = 5$ and 10 and for various system sizes given in Figure 6.15.

6.5. Discuss the effect of using two different definitions of the effective lengths (based on the various moments of length distribution) on the effective permeability estimation for the overlapping sandbody model with a distribution of sizes in 3D.

6.6. The sandbody size used in percolation analysis of an overlapping sandbody model is estimated within a 20% tolerance. How does this assumption affect the estimation of the effective permeability in the model? You may assess this effect numerically by using an arbitrary point on the effective permeability master curve.

6.7. Use the master curves for the average connected sand fraction, backbone fraction and effective permeability to reconstruct the percolation quantities for a specific sandbody dimensionless size. Then compare the behaviour of these three percolation quantity curves and conclude.

References

Danziger, M.M., Bashan, A. and Havlin, S., Interdependent resistor networks with process-based dependency. *New Journal of Physics*, 17: 043046, 2015.

Ertekin, T., Abou-Kassem, J. and King, G.R., *Basic Applied Reservoir Simulations.* SPE Textbook Series, vol. 7. Society of Petroleum Engineers, Richardson, TX, 2001.

Grassberger, P., Spreading and backbone dimensions of 2D percolation. *Journal of Physics A: Mathematical and General*, 25: 5475, 1992.

Grassberger, P., Conductivity exponent and backbone dimension in 2-D percolation. *Physica A: Statistical Mechanics and its Applications*, 262: 251–263, 1999.

Herrmann, H.J., Hong, D.C. and Stanley, H.E., Backbone and elastic backbone of percolation clusters obtained by the new method of 'burning'. *Journal of Physics A: Mathematical and General*, 17: L261–L266, 1984.

King, P.R., The use of renormalization for effective permeability. *Transport in Porous Media*, 4: 37–58, 1989.

King, P.R., The connectivity and conductivity of overlapping sandbodies. In: Buller, A.T. (ed.), *North Sea Oil and Gas Reservoirs — II.* Graham and Trotman, London, pp. 353–361, 1990.

Last, B.J. and Thouless, D.J., Percolation theory and electrical conductivity. *Physical Review Letters*, 27: 1719, 1971.

Masihi, M., King, P.R. and Nurafza, P., The effect of anisotropy on finite size scaling in percolation theory. *Physical Review E*, 74: 042102, 2006.

Masihi, M., Sadeghnejad, S., Bazrafkan, S., Forouzanfar, F., Jahangiri, H.R. and King, P.R., Facies connectivity and conductivity modeling. Proceedings of the XVII International Conference on Computational Methods in Water Resources, San Francisco, CA, July 6–10, 2008.

Pike, G.E. and Seager, C.H., Percolation and conductivity: a computer study. I. *Physical Review B*, 10(4): 1421–1434, 1974.

Porto, M., Bunde, A., and Havlin, S., Distribution of dangling ends on the incipient percolation cluster. *Physica A: Statistical Mechanics and its Applications*, 266(1–4): 96–99, 1999.

Rintoul, M.D. and Nakanishi, H., A precise determination of the backbone fractal dimension on two-dimensional percolation clusters. *Journal of Physics A: Mathematical and General*, 25: L945–L948, 1992.

Roux, S. and Hansen, A., A new algorithm to extract the backbone in a random resistor network. *Journal of Physics A: Mathematical and General*, 20: L1281, 1987.

Sadeghnejad, S., Masihi, M., King, P.R., Shojaei, A. and Pishvaei, M., A reservoir conductivity evaluation using percolation theory. *Petroleum Science and Technology*, 29(10): 1041–1051, 2011a.

Sadeghnejad, S., Masihi, M., Pishvaei, M., Shojaei, A. and King, P.R., Utilization of percolation approach to evaluate the reservoir connectivity and effective permeability: a case study on North Pars Gas Field. *Scientia Iranica C*, 18(6): 1391–1396, 2011b.

Sadeghnejad, S., Masihi, M., Pishvaie, M. and King, P.R., Rock type connectivity estimation using percolation theory. *Mathematical Geosciences*, 45: 321–340, 2013a.

Sadeghnejad, S., Masihi, M. and King, P.R., Dependency of percolation critical exponents on the exponent of power law size distribution. *Physica A: Statistical Mechanics and its Applications*, 392(24): 6189–6197, 2013b.

Stanley, H.E., Cluster shapes at the percolation threshold: an effective cluster dimensionality and its connection with critical-point exponents. *Journal of Physics A: Mathematical and General*, 10(11): L211, 1977.

Tarjan, R., Depth first search and linear graph algorithms. *SIAM Journal on Computing*, 1: 146, 1972.

Watson, B.P. and Leath, P.L., Conductivity in the two-dimensional-site percolation problem. *Physical Review B*, 9: 48–93, 1974.

Wen, H., King, P.R., Muggeridge, H. and Vittoratos, E.S., Using percolation theory to estimate recovery from poorly connected sands using pressure depletion. Paper presented at the ECMOR XIV — 14th European Conference on the Mathematics of Oil Recovery Catania, Sicily, Italy, 8–11 September, 2014.

Chapter 7

Beyond simple percolation

The scaling laws presented so far are strictly only valid for the critical region close to the percolation threshold, although they work as reasonable approximations away from this region. Also they are only strictly correct for binary systems with two media: one permeable and the other impermeable. In this chapter, we address how to extend these concepts to more general systems without these limitations. As before, we are primarily interested in the connected sand fraction, the backbone fraction, the dangling-ends fraction and the effective permeability. For the systems discussed so far, there are four distinct regions: (1) well below the percolation threshold where the percolation quantities are almost zero, (2) the critical region described by percolation theory scaling laws, (3) close to full occupancy $p = 1$ which is described by the homogenisation methods like effective medium theory and (4) the region in between regions (2) and (3). This regime is more complex as it contains corrections to the power law scaling theories. There is no complete formulation for this regime, so we use interpolation between the critical region and the nearly fully occupied region. In this chapter, we first describe this interpolation approach and then go on to describe how to extend percolation to systems with a continuous permeability distribution.

7.1 Prediction of the percolation quantities in binary permeability media

In previous chapters, we have described how the various quantities of interest (the mean connected sand fraction (P), the mean backbone fraction (B), the mean dangling-ends fraction (D) and the mean effective permeability (K_{eff})) can be determined from some simple universal curves by using finite-size scaling. Furthermore, the associated uncertainties related to these can also be derived from simple curves. However, the scaling laws that were used to derive the finite-size formulae are strictly only valid in the region close to the percolation threshold. For the region very close to the full occupancy, when $p \to 1$, we expect that $P \to 1, B \to 1, D \to 0$ and $K_{\text{eff}} \to 1$. For the region in between (region III in Figure 7.1), we use interpolation to derive approximately some mathematical functional forms. We will discuss each of the percolation quantities in turn.

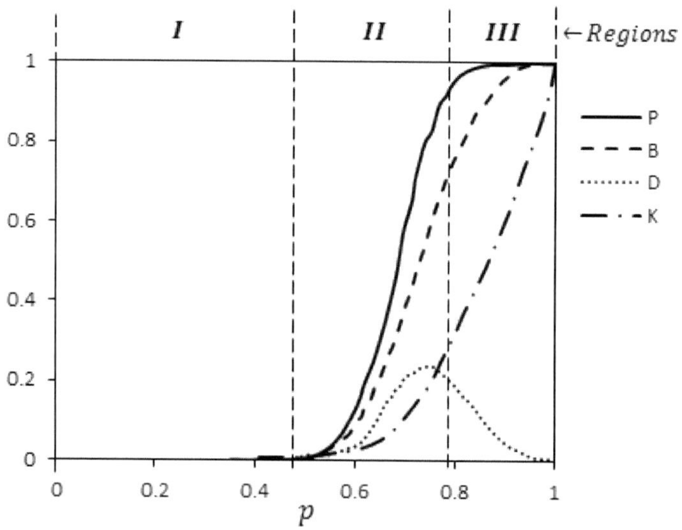

Figure 7.1. Illustration of three regions for percolation quantities: region I (no percolation), region II (critical region around the threshold where finite-size-scaling laws apply) and region III (above the critical region) which extends to just before full occupancy $p = 1$.

7.1.1 *The connected sand fraction*

First consider the connected sand fraction. The results of data collapse for the connected sand fractions of overlapping sandbody model in 2D are shown in Figures 3.11 and 3.12. We can try to fit a simple functional form to the universal mean connected sand fraction curve ($\Im$). A suitable function that fits the empirical data for regions I and II is

$$P(p, L) = 0 \qquad \text{if } (p - p_c)L^{\frac{1}{\nu}} < -1.5, \tag{7.1a}$$

$$P(p, L) = \frac{1.5L^{-\beta/\nu}}{1 + 1.9e^{-3.2(p-p_c)L^{1/\nu}}} \qquad \text{if } -1.5 < (p - p_c)L^{1/\nu} < 0.5. \tag{7.1b}$$

This has an intercept on the *ordinate* axis (at $p = p_c$) as $L^{\beta/\nu}P = 0.52$ which is observed in the numerical curves in Figure 3.11.

For the region far above the percolation threshold (labelled region III in Figure 7.1) when $(p - p_c)L^{1/\nu} > 0.5$, the mean connected sand fraction in Figure 3.11 curves for different system sizes do not all collapse onto the same universal curve. This is because of the corrections to scaling mentioned before, so Eq. (7.1b) cannot be valid for region III. Also we know that at full occupancy, when $p = 1$, the connected sand fraction must also equal 1 for any system size, $P(p = 1, L) = 1$. In region III, we are able to collapse all the mean connected sand fractions data of different system sizes on top of each other by using a suitable scheme shown in Figure 7.2. The rescaling of the axes was found empirically.

This data collapse leads to the following functional form for the mean connected sand fraction in the region III,

$$P(p, L) = L^{-0.01}[1.02 - 0.67e^{-8.4(p-p_c)L^{0.35}}] \quad \text{if } (p - p_c)L^{1/\nu} > 0.5, \tag{7.1c}$$

$$P(p, L) = 1 \qquad \text{if } p = 1. \tag{7.1d}$$

This enables us to determine the mean connected sand fraction curve ($P(p, L)$ versus p) for any system size without need for any further

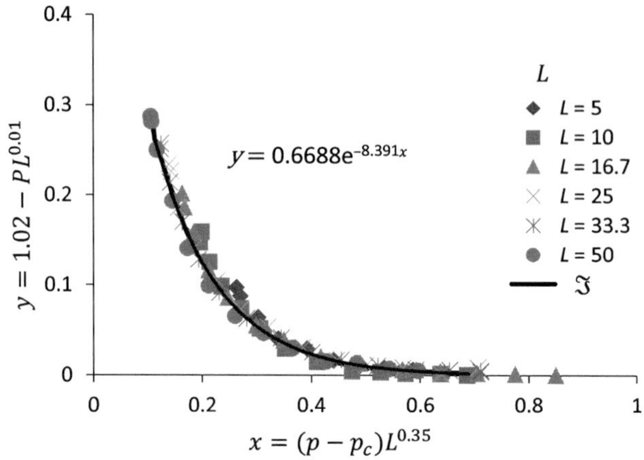

Figure 7.2. Illustration of data collapse for the mean connected sand fraction results (shown in Figure 3.11) of the region III that is far above the threshold $(p - p_c)L^{1/\nu} > 0.5$.

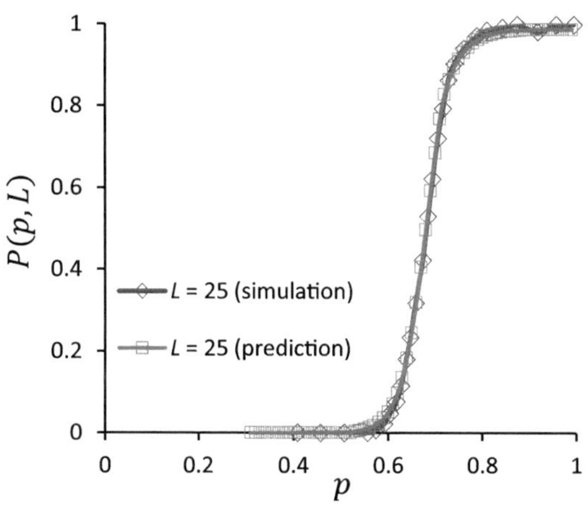

Figure 7.3. Illustration of the computed connected sand fraction by using Eq. (7.1) for the system size $L = 25$ to compare with results shown in Figure 3.10.

simulations. In Figure 7.3, we show the mean connected sand fraction curves for the system size $L = 25$ as found from numerical simulation and from Eq. (7.1).

We can also find fits to the uncertainty in the connected sand fraction results. The results of data collapse are shown in Figure 3.12.

By trial and error, we find that the following formulae give good fits to the data:

$$\Delta(p, L) = 0 \quad \text{if } (p - p_c)L^{1/\nu} < -1.5, \tag{7.2a}$$

$$\Delta(p, L) = 0.48L^{-\frac{\beta}{\nu}} e^{-0.9\left[1.4(p-p_c)L^{\frac{1}{\nu}}-0.1\right]^2}$$
$$\text{if } -1.5 < (p - p_c)L^{1/\nu} < 0.5. \tag{7.2b}$$

As before, for net-to-gross ratio far above the percolation threshold, when $(p-p_c)L^{1/\nu} > 0.5$, the uncertainty does not fit this form. Again, we use an empirical fit as shown in Figure 7.4.

The functional form obtained for this curve is

$$\Delta(p, L) = 0.48L^{-\beta/\nu} e^{-4[1.4(p-p_c)L^{0.55}-0.1]^2} \quad \text{if } (p - p_c)L^{1/\nu} > 0.5, \tag{7.2c}$$

$$\Delta(p, L) = 0 \quad \text{if } p = 1. \tag{7.2d}$$

This enables us to determine the standard deviation of the connected sand fraction for any system size without the need for any further

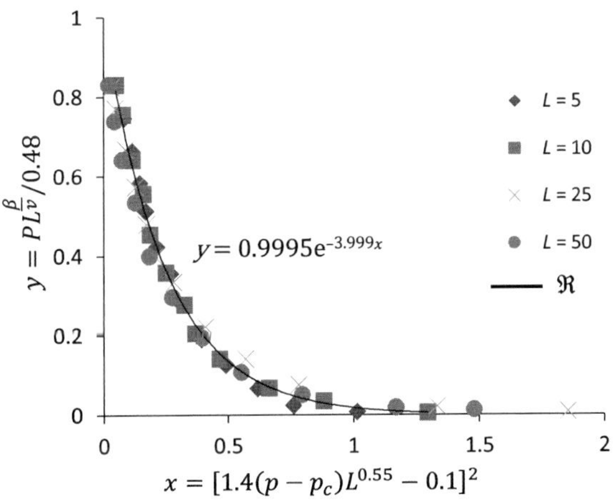

Figure 7.4. Illustration of data collapse for the uncertainty in the connected sand fraction results (shown in Figure 3.12) of the region III that is far above the threshold $(p - p_c)L^{1/\nu} > 0.5$.

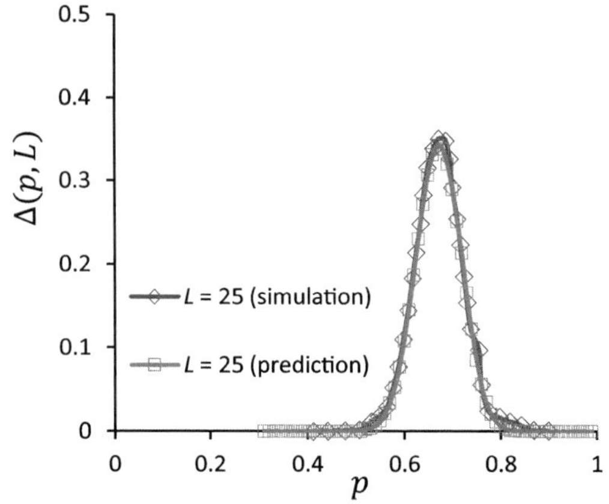

Figure 7.5. Illustration of the computed uncertainty in the connected sand fraction by using Eq. (7.2) for system size $L = 25$ to compare with the results shown in Figure 3.10.

simulations. The uncertainty in the connected sand fraction for $L = 25$ from numerical studies and from these formulae is shown in Figure 7.5.

7.1.2 *The backbone fraction*

We now turn to the backbone fraction of the connected cluster. Again we find empirical formulae for regions I and II,

$$B(p, L) = 0 \quad \text{if } (p - p_c)L^{1/\nu} < -1.5, \tag{7.3a}$$

$$B(p, L) = 0.2L^{-\beta_B/\nu}[1.36 + (p - p_c)L^{1/\nu}]^3$$
$$\text{if } -1.5 < (p - p_c)L^{1/\nu} < 0.5. \tag{7.3b}$$

The backbone fraction appears to be independent of the system size above the critical regime as shown in Figure 7.6.

This allows us to find the following functional form for the mean backbone fraction in region III,

$$B(p, L) = 1.1 - e^{5.6-8.3p} \quad \text{if } (p - p_c)L^{1/\nu} > 0.5, \tag{7.3c}$$

$$B(p, L) = 1 \quad \text{if } p = 1. \tag{7.3d}$$

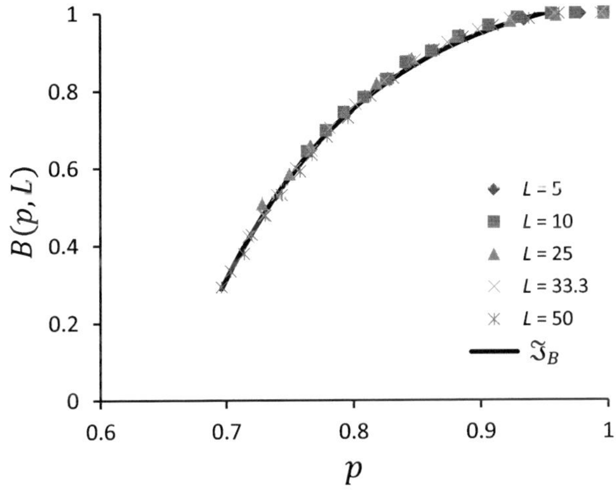

Figure 7.6. Illustration of data collapse for the mean backbone fraction results (shown in Figure 6.4(a)) in the region far above the threshold $(p - p_c)L^{1/\nu} > 0.5$ indicating that the results are approximately independent of the system size.

A comparison of this prediction with numerical data is shown in Figure 7.7.

We now follow the same process as before for the uncertainty in the backbone fraction results. The results of data collapse for the uncertainty in the backbone fraction of the overlapping sandbody model in 2D are shown in Figure 6.4(b). Extracting the standard deviation in the backbone fraction curve ($\Re_B$) based on the data collapse shown in Figure 6.4(b) gives

$$\Delta_B(p, L) = 0 \quad \text{if } (p - p_c)L^{\frac{1}{\nu}} < -1.5, \tag{7.4a}$$

$$\Delta_B(p, L) = 0.6L^{-\frac{\beta_B}{\nu}} e^{-0.8\left[1.3(p-p_c)L^{\frac{1}{\nu}} - 0.3\right]^2}$$
$$\text{if } -1.5 < (p - p_c)L^{1/\nu} < 0.5. \tag{7.4b}$$

Again far above the percolation threshold, when $(p - p_c)L^{1/\nu} > 0.5$, we look for an empirical rescaling to collapse the data onto a single curve shown in Figure 7.8.

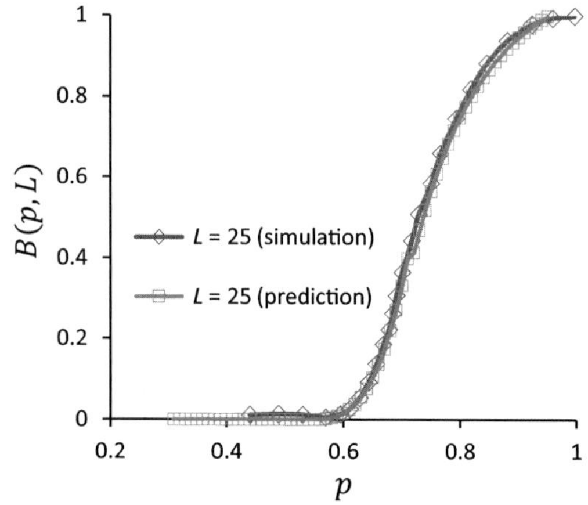

Figure 7.7. Illustration of the computed mean backbone fraction by using Eq. (7.3) for the system size $L = 25$ to compare with results shown in Figure 6.4(a).

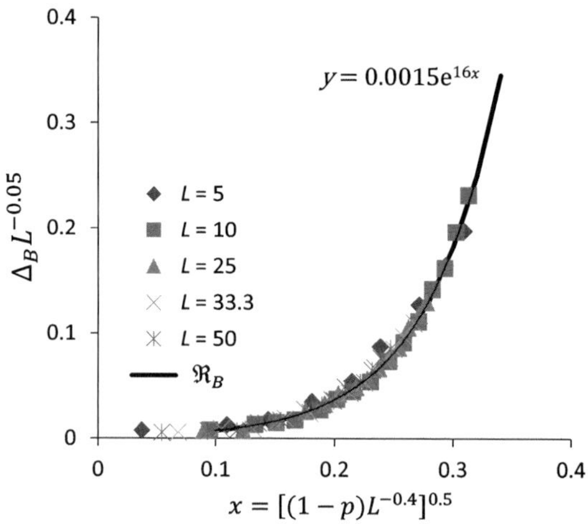

Figure 7.8. Illustration of data collapse for the uncertainty in the backbone fraction results (shown in Figure 6.4(b)) of the region III that is far above the threshold $(p - p_c)L^{1/\nu} > 0.5$.

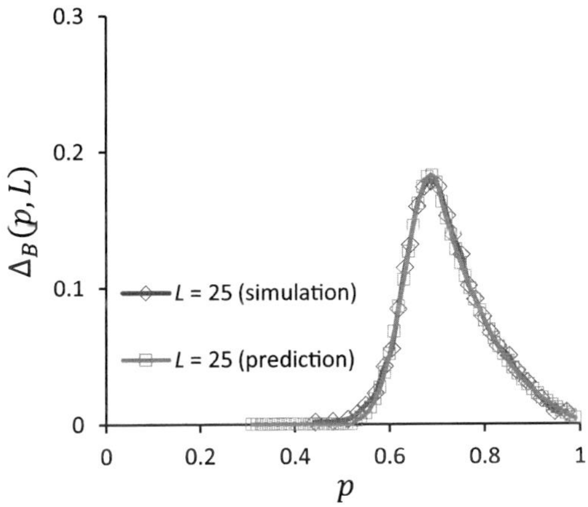

Figure 7.9. Illustration of the computed uncertainty in the backbone fraction by using Eq. (7.4) for the system size $L = 25$ to compare with the results shown in Figure 6.4(b).

The simple functional form that best fits this is

$$\Delta_B(p, L) = 0.0015L^{0.05}e^{16[(1-p)L^{-0.4}]^{0.5}} \quad \text{if } (p - p_c)L^{1/\nu} > 0.5, \tag{7.4c}$$

$$\Delta_B(p, L) = 0 \quad \text{if } p = 1. \tag{7.4d}$$

Once again we show a comparison of numerical results with these formulae in Figure 7.9.

7.1.3 *The dangling-ends fraction*

The procedure used here is identical to that described in the previous section. We look for simple analytic formulae to rescale the curves for regime III and then for simple formulae that fit these curves. So we shall not repeat the details but simply summarise the results.

The mean dead-end fraction is fit by the following formulae for regions I and II:

$$D(p, L) = 0 \quad \text{if } (p - p_c)L^{\frac{1}{\nu}} < -1.5, \tag{7.5a}$$

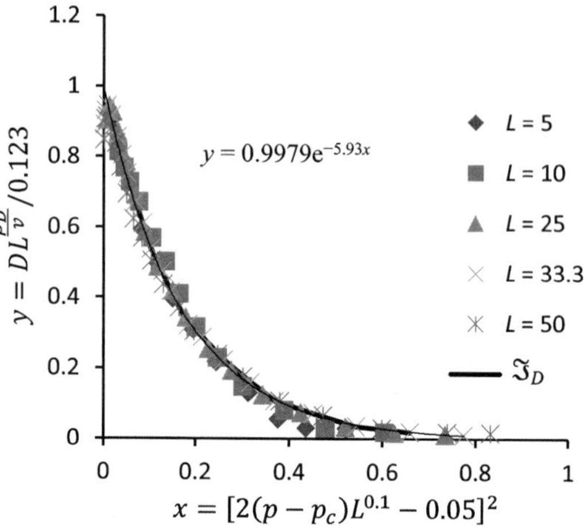

Figure 7.10. Illustration of data collapse for the mean dangling-ends fraction results of the region III that is far above the threshold $(p - p_c)L^{1/\nu} > 0.5$.

$$D(p, L) = P - B = \frac{0.15L^{\frac{-\beta_D}{\nu}}}{1 + 1.6e^{-3.5(p-p_c)L^{\frac{1}{\nu}}}}$$

$$\text{if } -1.5 < (p - p_c)L^{1/\nu} < 0.5. \tag{7.5b}$$

For region III, a suitable rescaling of the axes is shown in Figure 7.10.

This data collapse leads to the following functional form for the mean dangling-ends fraction results shown in the region III as

$$D = P - B = 0.123L^{-\beta_D/\nu}e^{-5.93[2(p-p_c)L^{0.1}-0.05]^2}$$

$$\text{if } (p - p_c)L^{1/\nu} > 0.5, \tag{7.5c}$$

$$D(p, L) = P - B = 0 \quad \text{if } p = 1. \tag{7.5d}$$

The comparison between prediction and numerical results is shown in Figure 7.11.

The standard deviation in the dead-end fraction is fit by

$$\Delta_D(p, L) = 0 \quad \text{if } (p - p_c)L^{\frac{1}{\nu}} < -1.5, \tag{7.6a}$$

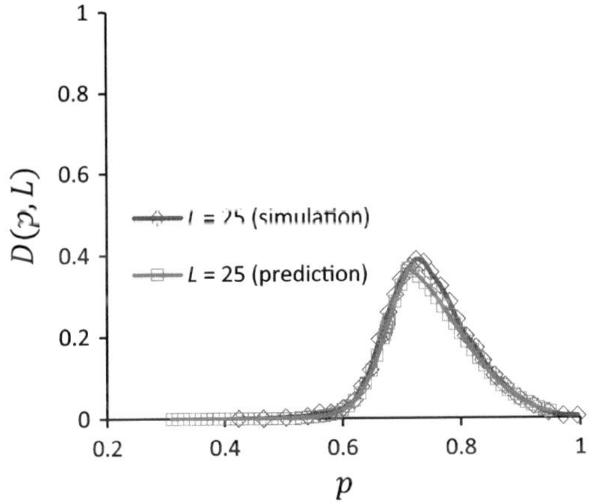

Figure 7.11. Illustration of the computed mean dangling-ends fraction by using Eq. (7.5) for system sizes $L = 25$ to compare with results shown in Figure 6.5(a).

$$\Delta_D(p, L) = 0.063L^{-\beta_D/\nu}e^{-[1.4(p-p_c)L^{1/\nu}-0.2]^2}$$
$$\text{if } -1.5 < (p-p_c)L^{1/\nu} < 0.5. \qquad (7.6\text{b})$$

Above the critical region the rescaling is shown in Figure 7.12.

This can be fit by the following:

$$\Delta_D(p, L) = 0.26L^{-0.2}e^{-16L^{0.3}(p-p_c)^2} \quad \text{if } (p-p_c)L^{1/\nu} > 0.5, \qquad (7.6\text{c})$$
$$\Delta_D(p, L) = 0 \quad \text{if } p = 1. \qquad (7.6\text{d})$$

The standard comparison between simulation and the formulae is shown in Figure 7.13.

7.1.4 *The effective permeability*

Finally we consider the effective permeability. Empirical fits to the mean effective permeability in regions I and II are

$$K_{\text{eff}}(p, L) = 0 \quad \text{if } (p-p_c)L^{\frac{1}{\nu}} < -1.5, \qquad (7.7\text{a})$$
$$K_{\text{eff}} = L^{-\mu/\nu}[-0.04 + 0.41e^{2.8(p-p_c)L^{1/\nu}}]$$
$$\text{if } -1.5 < (p-p_c)L^{1/\nu} < 0.6. \qquad (7.7\text{b})$$

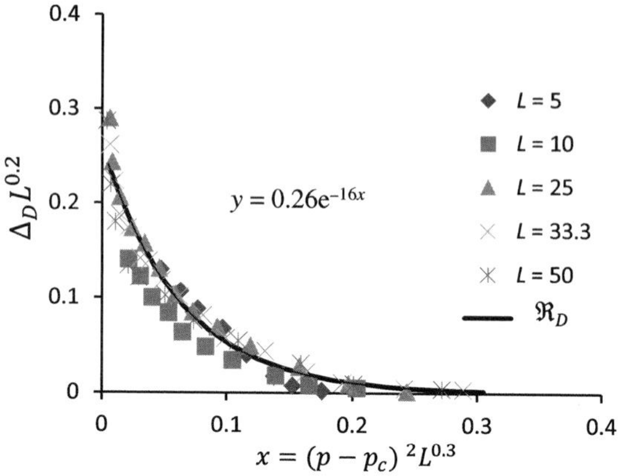

Figure 7.12. Illustration of data collapse for the uncertainty in the dangling-ends fraction results (shown in Figure 6.5(b)) of the region III that is far above the threshold $(p - p_c)L^{1/\nu} > 0.5$.

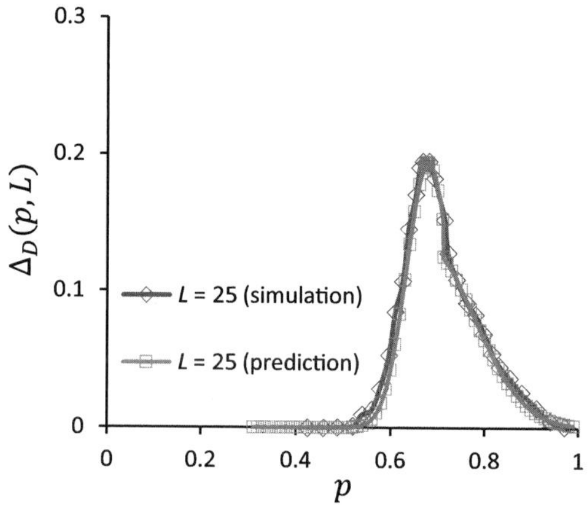

Figure 7.13. Illustration of the computed uncertainty in the dangling-ends fractions by using Eq. (7.6) for the system size $L = 25$ to compare with results shown in Figure 6.5(b).

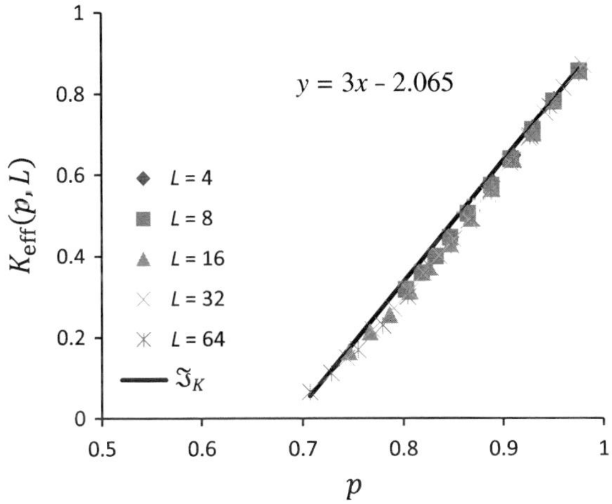

Figure 7.14. Illustration of data collapse for the mean effective permeability results for region III that is far above the threshold $(p - p_c)L^{1/\nu} > 0.6$.

Above the critical region, it appears that the effective permeability becomes independent of system size as shown in Figure 7.14.

This data collapse leads to the following functional form for the mean effective permeability in the region III,

$$K_{\text{eff}}(p, L) = 3p - 2.065 \quad \text{if } (p - p_c)L^{1/\nu} > 0.6, \tag{7.7c}$$

$$K_{\text{eff}}(p, L) = 1 \quad \text{if } p = 1. \tag{7.7d}$$

The comparison for the mean effective permeability curves for the system size $L = 8$ is shown in Figure 7.15.

The standard deviations in the effective permeability below and in the critical region are found to fit.

$$\Delta_K(p, L) = 0 \quad \text{if } (p - p_c)L^{\frac{1}{\nu}} < -1.5, \tag{7.8a}$$

$$\Delta_K(p, L) = 0.55L^{-\frac{\mu}{\nu}} e^{-2\left[1.1(p-p_c)L^{\frac{1}{\nu}} - 0.39\right]^2}$$

$$\text{if } -1.5 < (p - p_c)L^{1/\nu} < 0.5. \tag{7.8b}$$

A suitable rescaling above the critical region is shown in Figure 7.16.

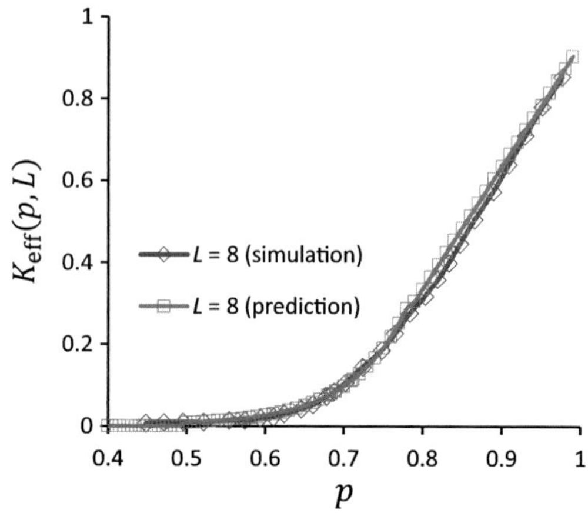

Figure 7.15. Illustration of the computed mean effective permeability by using Eq. (7.7) for the system size $L = 8$ to compare with the results shown in Figure 6.12(a).

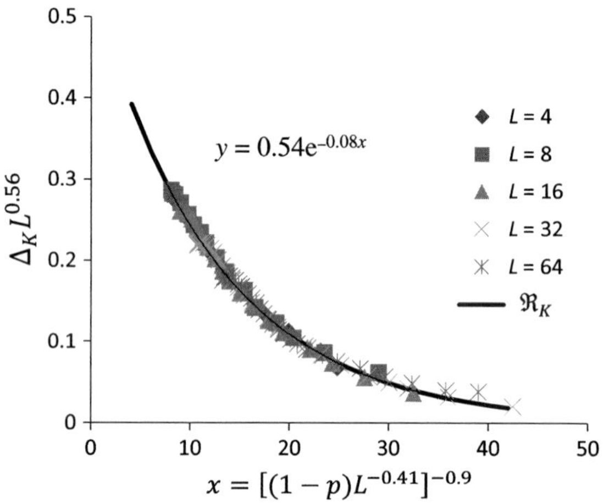

Figure 7.16. Illustration of data collapse for the uncertainty in the effective permeability results (shown in Figure 6.12(b)) of the region III that is far above the threshold $(p - p_c)L^{1/\nu} > 0.5$.

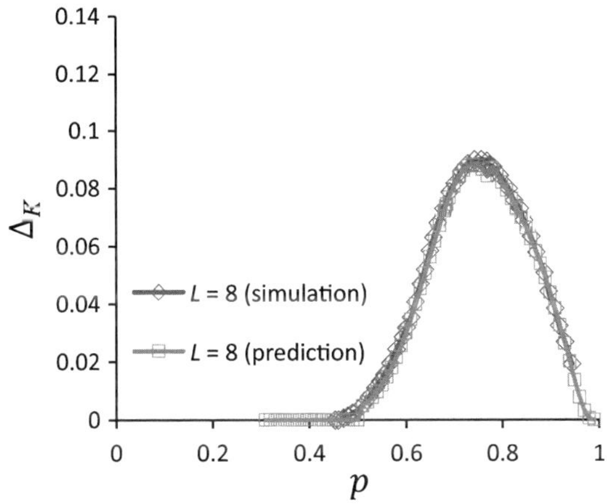

Figure 7.17. Illustration of the computed uncertainty in the effective permeability by using Eq. (7.8) for the system size $L = 8$ to compare with results shown in Figure 6.12(b).

This leads to the following functional form:

$$\Delta_K(p, L) = 0.54L^{-0.56}e^{-0.08[(1-p)L^{-0.41}]^{-0.9}} \quad \text{if } (p - p_c)L^{1/\nu} > 0.5, \tag{7.8c}$$

$$\Delta_K(p, L) = 0 \quad \text{if } p = 1. \tag{7.8d}$$

Comparisons for $L = 8$ are shown in Figure 7.17.

We now give an example to demonstrate the application of the formalism developed to predict the various percolation quantities. Consider two cross-sectional reservoir models with well spacing of $L = 800$ m and thickness of 160 m made by overlapping rectangular sands of side length $a = 80$ m and width $b = 20$ m as shown in Figure 7.18. The sands have permeability $K_{ss} = 1$ and the background permeability is zero. The values of the percolation quantities calculated by the conventional reservoir engineering methods are summarised in Table 7.1.

We use the formulae developed in this chapter (Eqs. (7.1)–(7.8)) to predict the sand fraction, backbone fraction, dangling-ends fraction and effective permeability and compare with values in

Table 7.1. The system is isotropic and the effective size is $L = 8$. We first check which regime each of the cross-sectional models belongs to. As $(p_1 - p_c)L^{1/\nu} = -0.41$ and $(p_2 - p_c)L^{1/\nu} = 0.28$, both models belong to region II. From the formulae above, the various percolation quantities can be estimated as summarised in Table 7.2.

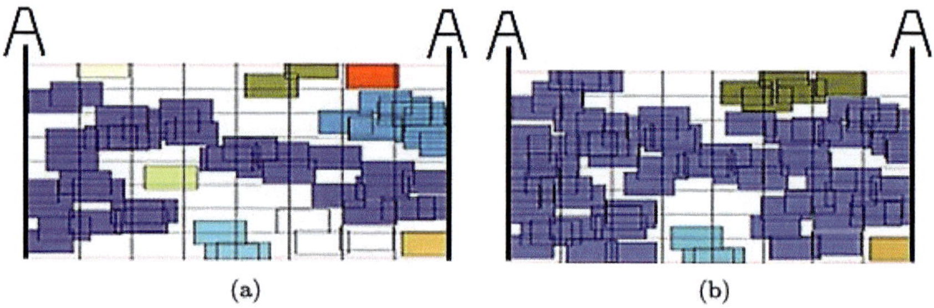

Figure 7.18. Two cross-sectional reservoir models made of overlapping rectangular sandbodies at net-to-gross ratios (a) $p = 0.579$ and (b) $p = 0.726$.

Table 7.1. The numerical values of percolation quantities calculated conventionally for two cross-sectional models shown in Figure 7.18.

Model no.	No. of sands	Sand fraction (p)	Connected sand fraction (P)	Backbone fraction (B)	Dangling-ends fraction (D)	Effective perme-ability (K_{eff})
1	58	0.579	0.625	0.473	0.152	0.0002
2	88	0.726	0.867	0.606	0.261	0.1476

Table 7.2. The numerical values of percolation quantities calculated from Eqs. (7.1)–(7.8) for two cross-sectional models.

Model no.	Sand fraction (p)	Connected sand fraction (P)	Backbone fraction (B)	Dangling-ends fraction (D)	Effective permeability (K_{eff})
1	0.579	0.15(±0.24)	0.08(±0.16)	0.07(±0.07)	0.014(±0.021)
2	0.726	0.68(±0.35)	0.42(±0.28)	0.26(±0.12)	0.14(±0.09)

7.2 Estimation of effective permeability in heterogeneous media

We now turn to the problem of estimating effective permeability when there is a continuous permeability distribution, i.e., when the system can no longer be divided into two components, one in which fluids flow and one where they do not. Hydrocarbon reservoirs are very heterogeneous due to the complex sedimentary processes and post-sedimentary events. These heterogeneities occur on a very wide range of length scales and have a big impact on the flow and reservoir performance. Conventionally this is modelled with a finite-difference or finite-volume simulation of the flow. This requires a set of grid cells constructed to represent the geology (Koltermann and Gorelick, 1996). In each block, it is generally assumed that rock properties are constant. In typical models, the grid cells can be tens to hundreds of metres in length. We need an appropriate estimate of the "effective" grid-block values from data which are typically gathered from core plugs that are tens of centimetres in size. For example, the fine-grid reservoir model of a typical reservoir can have as many as 10^9 grids to account for the small-scale heterogeneities. To be manageable by most flow simulators, this needs to be reduced to around 10^5 grid blocks. This process of coarse graining is also called upscaling and is an important subject in reservoir engineering. It is also important in hydrology, hydrogeology and sequestration of carbon dioxide for example.

There are a variety of upscaling techniques in the literature from flow-based (or numerical) methods to stochastic methods (Durlofsky, 1991; Pickup and Sorbie, 1994). To understand how percolation concepts can be used in upscaling, we focus on the effective (single-phase) permeability.

First we need a definition of what is meant by effective permeability, K_{eff}. The usual definition is based on the relation between average flow rate in a heterogeneous system $\langle q \rangle$ and the average flow potential $\langle \nabla \Phi \rangle$:

$$\langle q \rangle = -K_{\text{eff}} \langle \nabla \Phi \rangle. \tag{7.9}$$

There are numerous different approaches to how this can be calculated numerically or estimated theoretically. It is not appropriate to review all these methods in this book as there are many subtleties not relevant to percolation. The interested reader should refer to the excellent reviews in the literature (such as Renard and de Marsily, 1997) or the other references to be found at the end of this chapter.

What is clear is that connectivity of flow paths plays a very significant role for the effective permeability of the medium. So we expect that percolation concepts are a useful tool in its calculation. This is particularly true when the permeability variation is very broad when the flow tends to get "localised" into a smaller domain corresponding to the connected higher permeability regions (Ambegaokar *et al.*, 1971). Katz and Thompson (1986) conjectured that single-phase flow in disordered porous medium at the pore scale is largely controlled over a connected *critical path* consisting of pores of sufficiently high permeability that span the region. Moreno and Tsang (1994) show this numerically at the larger scale by calculating the tracer breakthrough curves for a pulse injection into a lognormal permeability distribution and see that at a high standard deviation, two peaks appear in the breakthrough plot. One emerges at a much earlier time showing the channelling effect. This is very reminiscent of the backbone of percolation clusters. We make use of this in the following section.

7.2.1 *Methods based on the connectivity in the medium*

As we have shown, percolation theory can be used to estimate the effective permeability of random binary permeability fields (Berkowitz and Balberg, 1993; Sahimi, 1993). The fundamental observation is that the permeability scales with the net-to-gross ratio as

$$K_{\text{eff}} = K_1(p - p_c^{\infty})^{\mu}. \tag{7.10}$$

One approach to the continuous permeability distribution was briefly discussed in Chapter 6. We first transformed the continuous permeability map into a binary medium by choosing a suitable threshold

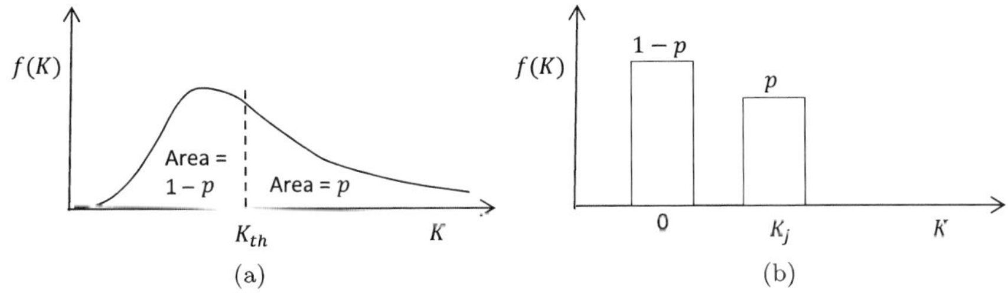

Figure 7.19. Illustration of choosing a suitable threshold value (K_{th}) to change (a) the real permeability distribution to (b) an equivalent binary system.

value (K_{th}) as shown in Figure 7.19. Then the fraction of sites with permeabilities $K > K_{\text{th}}$ is equivalent to the usual occupancy fraction (p),

$$p = \int_{K_{\text{th}}}^{\infty} f(K)dK, \tag{7.11}$$

where $f(K)$ is the probability density function for the permeability.

Then we can easily use the effective permeability scaling laws (Eq. (7.7)) to estimate the effective permeability of the equivalent binary medium. For example, one may use

$$K_{\text{eff}} = K_{\text{th}}(p - p_c^{\infty})^{\mu}. \tag{7.12}$$

This is amended by finite-size and anisotropy effects as described in previous chapters. The exponent μ in Eq. (7.12) may not be the universal value as it may depend on the spatial correlation and the details of permeability distribution.

We then have to scale back appropriately to find the effective permeability of the original medium. However, the problem with this approach is that it is unclear how to choose an appropriate threshold value and a suitable scaling factor to represent the effective permeability.

The concept of a high-permeability path is consistent with the fact that in highly disordered media, only a small subset (similar to the backbone in percolation theory) of highly permeable elements contributes to the overall conductance. This path will be almost 1D topologically, so one could treat the smaller resistances in the system as though they were in series along the path which gives an approximation (Hunt and Idriss, 2009):

$$K_{\text{eff}} = \frac{K_{\text{ar}} K_{h+}}{K_{\max}}, \tag{7.13}$$

where K_{ar} is the arithmetic average of the permeability distribution, K_{h+} is the harmonic average of the high permeability values (those above the threshold) and $K_{\max}$ is the highest permeability value. The choice of the normalisation factor in the denominator of Eq. (7.13) works in monomodal distributions but may fail in bimodal distributions when the higher permeability mode does not percolate.

The concept of high-permeability pathways has been extended to estimate the effective permeability for systems with lognormal distributions (Shah and Yortsos 1996; Hunt and Idriss 2009). For example, Shah and Yortsos (1996) proposed an implicit method to estimate the effective permeability that depends on the standard deviation (σ) of the lognormal permeability distribution as

$$K_{\text{eff}} \sim \sigma^{-\mu} \exp\left(\ln\left(\frac{K_{\text{th}}}{K_g}\right) + \mu\left[\frac{\ln(K_{\text{th}}/K_g)}{\sigma\sqrt{2}}\right]^2\right), \tag{7.14}$$

$$\operatorname{erfc}\left(\frac{\ln(K_{\text{th}}/K_g)}{\sigma\sqrt{2}}\right) = 2p_c + \frac{\mu e^{-\left[\frac{\ln(K_{\text{th}}/K_g)}{\sigma\sqrt{2}}\right]^2}}{\sigma\sqrt{\pi/2}}. \tag{7.15}$$

Hunt and Ewing (2005) proposed an approximate effective permeability of a medium with a bimodal permeability distribution (for sand and mud deposits) as

$$K_{\text{eff}} = \frac{L}{\frac{L-\xi}{K_{\text{th}}} + \frac{\xi}{K_{g+}}}, \tag{7.16}$$

where L is the length of the sample, K_{th} the permeability threshold, K_{g+} the geometric mean for the high permeability values and ξ is the correlation length of high permeability values.

Hunt and Idriss (2009) used a power law interpolation scheme similar to that of Scheibe and Yabusaki (1998) for estimating the effective permeability of a medium with a bimodal permeability distribution with two distributions and proposed

$$K_{\text{eff}} = K_g(1 - p_{\text{ss}}) \qquad \text{for } p_{\text{ss}} < p_c, \tag{7.17}$$

$$K_{\text{eff}} = K_g(1 - p_{\text{ss}}) + K_{g+}\left(\frac{p_{\text{ss}} - p_c}{1 - p_c}\right)^2 \qquad \text{for } p_{\text{ss}} > p_c, \tag{7.18}$$

where p_{ss} is the fraction of cells with permeability greater than K_{th}. When the upper mode of the permeability distribution does not percolate, then Eq. (7.17) can be used instead of Eq. (7.13). In general, a hybrid between the two estimations (Eqs. (7.13) and (7.17)) may be used.

Since this approach which uses the concept of high-permeability pathways covers a wide range of practical cases, the rest of this chapter concentrates on developing methods based on this approach to estimate the effective permeability of heterogeneous porous media. The proposed methods do not need fitting to the experimental data of permeability measurements to estimate the model parameters as was necessary in the previous empirical methods. We examine these methods on different layers of the 10th SPE model (Christie and Blunt, 2001) and compare the effective permeability results with the values determined from the numerical flow simulators for validation.

Consider a general 2D reservoir permeability map with any spatial structure and a given probability density distribution with a schematic cumulative distribution function shown in Figure 7.20. We wish to estimate the effective permeability parallel to the y axis in Figure 7.21. We use the argument that it is the grid blocks with high permeability values that contribute most to the flow, while the flow in grid blocks with very low permeability values may be neglected. It is essential that there is connectivity across the system parallel to the y axis through the high permeabilities. In percolation terminology, this means that we must have a connected cluster of highly permeable cells that connect the two opposite sides of the system.

To find the connected cluster of cells with high permeability values on a given permeability map we choose a threshold value

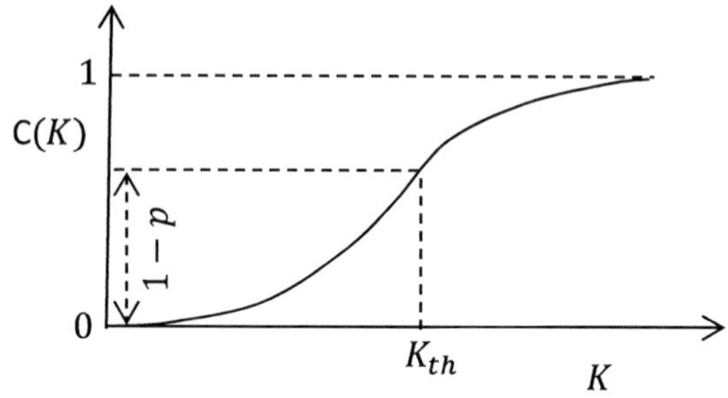

Figure 7.20. A schematic permeability cumulative probability function (CPF) to illustrate the threshold permeability value (K_{th}) and the fraction of cells (p) occupied by high permeability values ($K > K_{\mathrm{th}}$).

(K_{th}) and check that the cells with permeabilities above this do connect. This is done by using, e.g., the algorithm of Hoshen and Kopelman (1976). The threshold value is chosen by starting with a high value and then lowering until connection first appears (as shown in Figure 7.21). Then the fraction of cells occupied by high permeability values ($K > K_{\mathrm{th}}$) is called the occupancy fraction (p), and its corresponding permeability value is called the threshold permeability value (K_{th}). So K_{th} is the threshold value at which the first percolating cluster appears. Now we assume that this spanning cluster is the main path that controls the flow. There will be other finite disconnected and smaller clusters of high permeability values that have just local conductivity effects (these can be seen in Figure 7.21).

To see the effects of applying the permeability threshold visually, two different layers of 10th SPE model are used. The SPE model is a geostatistical model based on the Jurassic Upper Brent formation that consists of $60 \times 220 \times 85$ grid cells, each of size $20\,\mathrm{ft} \times 10\,\mathrm{ft} \times 2\,\mathrm{ft}$ (Christie and Blunt, 2001). The top 35 layers (70 ft) represent the shallow marine Tarbert formation which consists of sandstone, siltstone and shales, and the lower 50 layers (100 ft) represents the fluvial Ness formation formed from well-sorted high-permeability sandstone channels imposed on a low-permeability background of

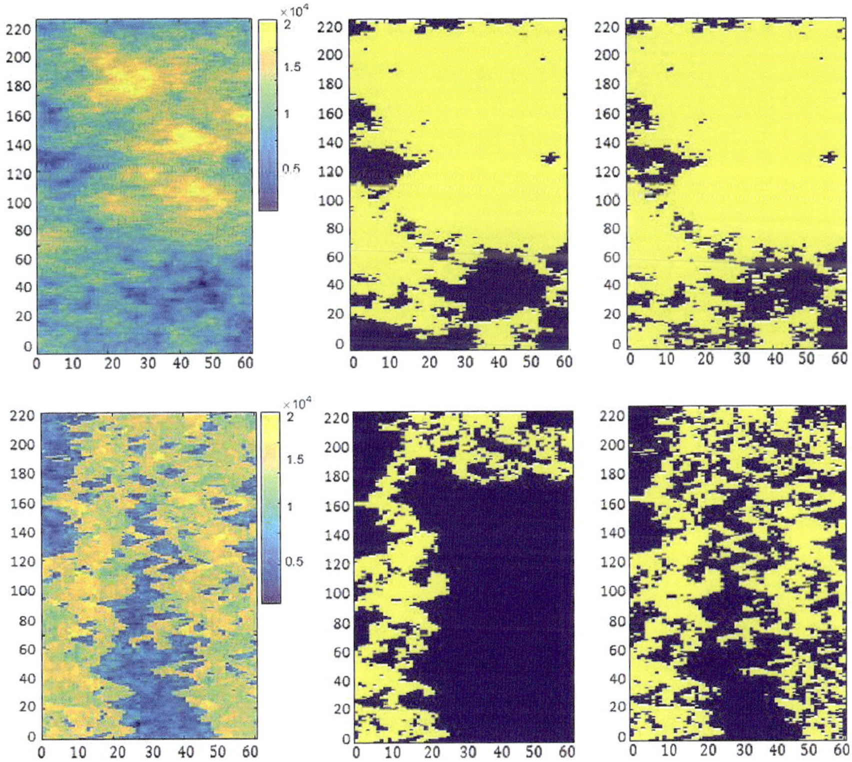

Figure 7.21. Examples of (left) the original permeability map, (middle) only the connected cluster of high permeability values along the y axis and (right) the connected cluster along with other finite smaller clusters for (top) layer 14 (at $p = 0.798$, $P = 0.958$ and $K_{\mathrm{th}} = 6.889\,\mathrm{md}$) and (bottom) layer 57 (at $p = 0.468$, $P = 0.540$ and $K_{\mathrm{th}} = 106.84\,\mathrm{md}$) of 10th SPE model. Yellow shows the area occupied by high permeability values ($K > K_{\mathrm{th}}$).

shales and coal. There is a large permeability variation (8–12 orders of magnitude) in the model with significant differences in the types of permeability distributions in the upper and lower layers.

Figure 7.21 shows the original permeability map, the structure of the connected cluster and the connected cluster along with other finite smaller clusters for layer 14 (in the upper structure) and layer 57 (in the lower channelised structure).

The shape of the connected cluster depends on the spatial correlation of permeability. For example, we can use the same

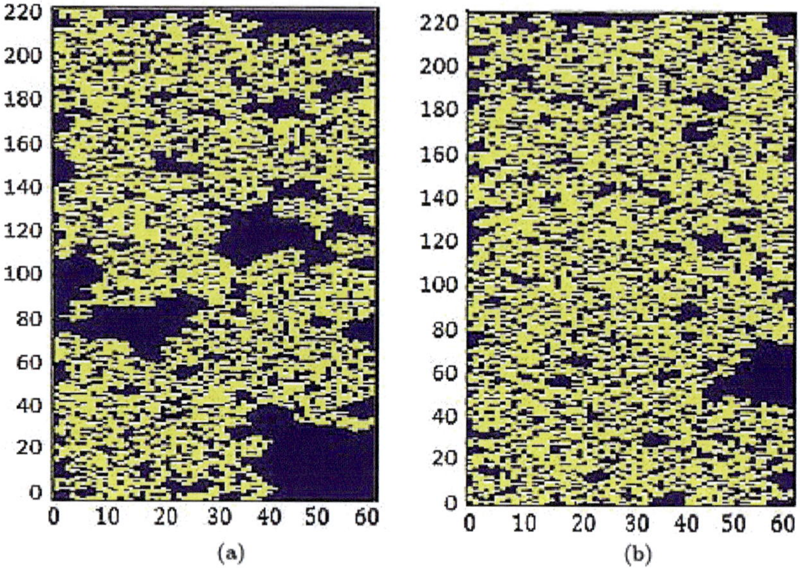

Figure 7.22. Two realisations of spatially uncorrelated permeability maps with the same original permeability distributions of 10th SPE model shown in Figure 7.21 for (a) layer 14 at $p = 0.6256$, $P = 0.794$ with $K_{\text{th-rand}} = 20.93\,\text{md}$ and (b) layer 57 at $p = 0.6442$, $P = 0.915$ with $K_{\text{th-rand}} = 23.70\,\text{md}$ showing the connected clusters of cells with high permeability values ($K > K_{\text{th}}$) in yellow.

permeability values from layers 14 and 57 and distribute them without correlation to generate corresponding uncorrelated random fields. We can then generate many realisations and determine the permeability threshold value ($K_{\text{th-rand}}$). Two realisations are shown in Figure 7.22 and clearly have a very different structure from the clusters in Figure 7.21.

The average threshold value for connectivity in the y direction for the 10th SPE model over all realisations is about $p = 0.62$. The difference between this value and the infinite threshold value for site percolation ($p_c = 0.593$) is due to the anisotropy of the system (Masihi *et al.*, 2006) arising from the different dimensionless lengths in the x and y directions. This results in a shift in the threshold. We will use this permeability threshold value ($K_{\text{th-rand}} = 20.93\,\text{md}$) of $p = 0.62$ as the representative permeability threshold of uncorrelated random fields in subsequent sections.

Masihi *et al.* (2016) presented four different connectivity-based approaches to estimate the effective permeability of continuous permeability distributions. We use the notation K_{est} to represent the estimated value for the effective permeability from the proposed percolation-based methods. We shall compare these with the effective permeability computed numerically from a commercial flow simulator (K_{sim}). The estimates are as follows:

1. *The threshold value* (K_{th})

   In this approach, we just use the threshold permeability to estimate the effective permeability:

   $$K_{\text{est}} = K_{\text{th}}. \tag{7.19}$$

   This can be thought of as a power average of the component permeabilities (Scheibe and Yabusaki, 1998) in a binary medium made by high-permeability (K_{th}) and low-permeability cells ($K = 0$) with occupation fraction p and $1 - p$ respectively.

2. *A weighted power average between the threshold value and geometric mean of the permeability distribution*

   To estimate the effective permeability in this approach, we hypothesise a weighted power average between the permeability threshold value (K_{th}) and the geometric mean of the permeability distribution (K_g) defined as

   $$K_{\text{est}} = (pK_{\text{th}}^m + (1-p)K_g^m)^{1/m}. \tag{7.20}$$

   This considers both the effects of the permeability distribution and the topology of the high-conductance pathways. An appropriate value for the exponent m can be considered in following different ways:

   (A) The exponent m is assumed to be one that is consistent with the assumption of parallel flow in layers with permeability of K_{th} and K_g,

   $$m = 1. \tag{7.21}$$

(B) The exponent m is assumed to be the fractal dimension of the sub-percolation network called the minimal path in percolation theory:

$$m = 1.13. \tag{7.22}$$

(C) The exponent m is assumed to be the ratio of the fractal dimension of the connected clusters in two cases. The first is the connected clusters of the real permeability map. The second case is the connected clusters of the uncorrelated random field with the same permeability probability density function as the real one,

$$m = \frac{1.89}{D}, \tag{7.23}$$

where 1.89 is the fractal dimension of percolating cluster in 2D.

3. *A power law function of the occupancy probability* (p) *scaled by the threshold value* (K_{th})

To estimate the effective permeability in this approach, we hypothesise a power law function of the high-permeability occupancy scaled by the permeability threshold value (K_{th}),

$$K_{\text{est}} = K_{\text{th}} \cdot p^n. \tag{7.24}$$

This is a generalised percolation-type power law function with exponent n as a characteristic parameter. Note that it is not related to the power in the previous section. McCarthy (1991) also emphasised that this exponent depends on the shape of the inclusions and whether they are oriented, partially oriented or random in the binary percolation medium. This method considers the topology of the highly conductive pathways. An estimate for the characteristic parameter (n) can be inferred in the following different ways:

(A) The exponent n is inferred from the equality of the permeability threshold value (K_{th}) and the weighted power average between the geometric mean (K_{g-}) of the permeability values below the threshold ($K < K_{\text{th}}$) and the geometric mean (K_{g+}) of permeability values above the threshold ($K > K_{\text{th}}$),

$$K_{\text{th}} = (pK_{g+}^n + (1-p)K_{g-}^n)^{1/n}. \tag{7.25}$$

(B) n is inferred from equality of the permeability threshold value (K_{th}) and the power law average between the geometric mean (K_{g-}) of permeability values below the threshold ($K < K_{\text{th}}$) and the geometric mean (K_{g+}) of permeability values above the threshold ($K > K_{\text{th}}$),

$$K_{\text{th}} = K_{g+}^n \cdot K_{g-}^{1-n}. \tag{7.26}$$

(C) The exponent n is assumed to be a function of the ratio of the permeability threshold values of the real permeability map and the permeability threshold value of the uncorrelated random field with the same permeability probability density function as the real one. In order to ensure that the value for models with spatially uncorrelated distributions is recovered ($n = 0$ and $K_{\text{est}} \approx K_{\text{th}}$), we use the following functional form:

$$n = 1 - \frac{K_{\text{th-rand}}}{K_{\text{th}}}. \tag{7.27}$$

4. *A characteristic shape factor multiply by the threshold value*

To estimate the effective permeability in this approach, we use an appropriately defined characteristic shape factor (SF) multiplied by the threshold value (K_{th}):

$$K_{\text{est}} = \text{SF} \cdot K_{\text{th}}. \tag{7.28}$$

There are different ways to find a suitable shape factor value (for instance, a shape factor can be developed by using the approximate streamline method proposed by Begg and King (1985)). We use the shape factor defined as the strength of the connected cluster (the product of p and P), raised to the power of its fractal dimension all divided by the strength of the connected cluster to the power of its fractal dimension for the uncorrelated random field,

$$\mathrm{SF} = \frac{(p \cdot P)_{\mathrm{real}}^{D}}{(p \cdot P)_{\mathrm{rand}}^{1.89}}. \tag{7.29}$$

The exponent D is the fractal dimension of the connected cluster in the real permeability map and 1.89 is the fractal dimension of the uncorrelated percolation cluster.

We now compare these estimation methods on layer 14 from the SPE 10th comparative model with the result of direct numerical simulation, which was $K_{\mathrm{eff}} = 18.5\,\mathrm{md}$. For layer 14, we find that $K_{\mathrm{th}} = 6.89\,\mathrm{md}$. The statistics of the two sub-networks: (1) high permeability values ($K \geq K_{\mathrm{th}}$) and (2) low permeability values ($K < K_{\mathrm{th}}$) as well as the whole permeability distribution are summarised in Table 7.3.

The method of Hunt and Idris (Eq. (7.13)) gives $K_{\mathrm{est}} = \frac{647.2(2.58)}{20000} = 1.16\,\mathrm{md}$. The results for the other methods described here can be summarised as follows:

Equation 7.20 ($m = 1$):

$$K_{\mathrm{est}} = 0.798(6.89) + 0.202(52.77) = 16.1\,\mathrm{md}.$$

Equation 7.20 ($m = 1.13$):

$$K_{\mathrm{est}} = 0.798(6.89)^{1.13} + 0.202(52.77)^{1.13} = 17.2\,\mathrm{md}.$$

Table 7.3. The summary statistics of the two sub-networks with high permeability values ($K \geq K_{\mathrm{th}}$) and low permeability values ($K < K_{\mathrm{th}}$) as well as the whole permeability distribution of layer 14 in 10th SPE model.

Whole network			**Sub-network ($K \geq K_{\mathrm{th}}$)**			**Sub-network ($K < K_{\mathrm{th}}$)**	
K_g	K_{ar}	K_{max}	p	K_{h+}	K_{g+}	K_{h-}	K_{g-}
52.77	647.2	20,000	0.798	35.9	113.4	1.81	2.58

Equation (7.24), using Eq. (7.25) to estimate n:

$$6.89 = (0.798(113.4)^n + 0.20(2.58)^n)^{1/n} \Rightarrow n = -1.616,$$

$$K_{\text{est}} = 6.89(0.798)^{-1.616} = 9.92\,\text{md}.$$

Equation (7.24), using Eq. (7.26) to estimate n:

$$6.89 = (113.4)^n(2.58)^{1-n} \Rightarrow n = 0.26,$$

$$K_{\text{est}} = 6.89(0.798)^{0.26} = 6.50\,\text{md}.$$

The comparison of these estimates with the calculation from the pressure solver (18.5 md) shows that the estimation from Eq. (7.20) is the most appropriate.

Furthermore, these methods gave reasonable estimates (Masihi *et al.*, 2016) for the effective permeability of the other layers of the 10th SPE model. A cross plot of the estimated effective permeability value (when we used the weighted power average between the threshold value and geometric value with $m = 1.89/D$) against the effective permeability calculated with a simulator for the top and lower layers of the 10th SPE model is shown in Figure 7.23.

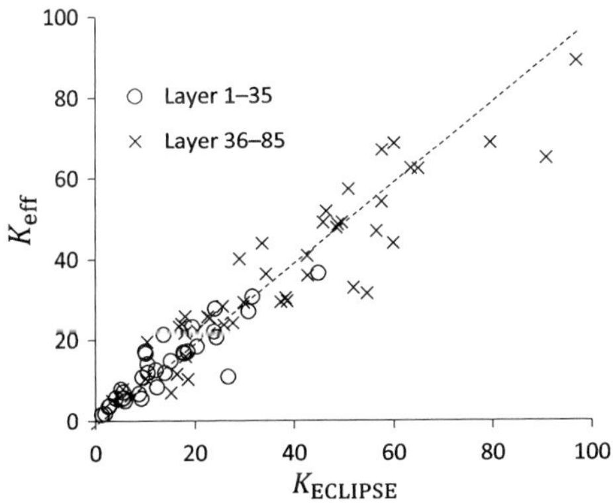

Figure 7.23. The cross plot of the estimated effective permeability values (when we used the weighted power average between the threshold value and geometric value with $m = 1.89/D$) against the numerical effective permeability values calculated using the commercial simulator ECLIPSE for the top and lower layers of 10th SPE model.

Table 7.4. Summary of the percentage of normalised errors (and their variability in various layers) obtained on the effective permeability estimations using four methods presented in this section compared with the effective permeability values calculated by a commercial simulator.

	Methods		% Error layers 1–35	% Error layers 36–85
1	The threshold value (K_{th})	Eq. (7.19)	121(±99)%	48(±17.5)%
2A	A weighted power average between the threshold value and geometric value of permeability distribution	Eqs. (7.20) and (7.21)	1.5(±29)%	4.5(±26)%
2B		Eqs. (7.20) and (7.22)	6.2(±29.7)%	1.7(±27.9)%
2C		Eqs. (7.20) and (7.23)	1(±29)%	1.3(±28)%
3A	A power law function of occupancy probability scaled by the threshold value	Eqs. (7.24) and (7.25)	29(±27)%	31(±41)%
3B		Eqs. (7.24) and (7.26)	52.5(±17)%	6.5(±32)%
3C		Eqs. (7.24) and (7.27)	22(±26)%	12(±23)%
4	A characteristic shape factor multiplied by threshold value	Eqs. (7.28) and (7.29)	6.5(±25)%	10(±32)%

A summary of the relative errors for the different estimates compared with simulator values are shown in Table 7.4. It should be emphasised that the estimates using these approaches gave much better estimates based on the *critical path* approximation (Eq. (7.13)).

With these methods, we can have reliable estimates of effective permeability for very heterogeneous models very quickly. As concluded by Masihi *et al.* (2016), a power averaging method by using a moment from the permeability distribution (such as the geometric mean) and the permeability threshold can give good estimates for the effective permeability.

7.3 Problems

7.1. Consider a cross-sectional reservoir model with well spacing of $L = 1{,}000\,\mathrm{m}$ and thickness of 200 m made by 130 overlapping square sands of side length $a = 100\,\mathrm{m}$ and width $b = 20\,\mathrm{m}$ as shown below. The reservoir net-to-gross ratio is $p = 70\%$. Sands have permeability $K_{\mathrm{ss}} = 10^{-13}\,\mathrm{m}^2$ ($\sim$100 md) and background permeability is zero. Use the formulation developed in this chapter for predicting the connected sand fraction between two wells, the backbone fraction, the dangling-ends fraction and the effective permeability. Moreover, the uncertainty associated with these percolation quantities.

7.2. Use the data collapse shown in Chapters 4 and 6 for the connected sand fraction results of overlapping sandbodies models in 3D with the same methodology described in this chapter for developing the mathematical formulism to make prediction of mean and standard deviation of connected sand fraction applicable to critical region and away from the critical region.

7.3. Use the data collapse shown in Chapters 4 and 6 for the effective permeability results of overlapping sandbodies models in 3D with the same methodology described in this chapter for developing the mathematical formulism to make prediction of mean and standard deviation of effective permeability applicable to critical region and away from the critical region.

7.4. Use the data collapse shown in Chapter 5 for the connected fraction results of randomly oriented fracture network model in 2D with the same methodology described in this chapter for developing the mathematical formulism to make prediction of mean and standard deviation of connected fraction applicable to critical region and away from the critical region.

7.5. For a simple binary sand/shale medium such as a sample with sand permeability of $K_{ss} = 1$ that contains a small fraction of shale of negligible permeability ($K_{sh} = 0$), use the effective medium methods (Eq. (7.17) or (7.18)) to estimate the effective permeability at shale fractions greater than 0.8 and plot the results against sand fraction and conclude.

7.6. Consider the threshold point at which for the first time, the connected cluster of cells with high permeability values appearing along the flow direction has been examined and $K_{th} = 106.84$ md for the permeability map of layer 57 in 10th SPE model shown in Figure 7.21. The statistics of the two sub-networks: (1) high permeability values ($K \geq K_{th}$) and (2) low permeability values ($K < K_{th}$) as well as the whole permeability distribution is summarised in the following table. Use the methods given by Eqs. (7.13), (7.20) and (7.24) to estimate the effective permeability of layer 57 and compare your estimation with the pressure solver result which is $K_{eff} = 63.4$ md.

Whole network			**Sub-network ($K \geq K_{th}$)**			**Sub-network ($K < K_{th}$)**	
K_g	K_{ar}	K_{max}	p	K_{h+}	K_{g+}	K_{h-}	K_{g-}
19.25	372.4	20,000	0.468	299	429.5	0.111	1.253

References

Ababou, R., Identification of effective conductivity tensor in randomly heterogeneous and stratified aquifers. Fifth Canadian-American Conference on Hydrogeology, September 18–20, Calgary, Alberta, Canada, 1990.

Ambegaokar, V.N., Halperin, B.I. and Langer, J.S., Hopping conductivity in disordered systems. *Physical Review B*, 4: 2612–2621, 1971.

Begg, S.H and King, P.R., Modelling the effects of shales on reservoir performance: calculation of effective vertical permeability. Paper presented at the SPE 1985 Reservoir Simulation Symposium, SPE 13529, 1985.

Berkowitz, B. and Balberg, I., Percolation theory and its application to ground water hydrology. *Water Resources Research*, 29(4): 775 794, 1993.

Christie, M.A. and Blunt, M.J., Tenth SPE comparative solution project: a comparison of up scaling techniques. *SPE Reservoir Evaluation and Engineering*, 4: 308, 317, 2001.

Durlofsky, L., Numerical calculations of equivalent gridlock permeability tensors for heterogeneous porous media. *Water Resources Research*, 27(5): 699, 1991.

Hoshen, J. and Kopelman, R., Percolation and cluster distribution. I. Cluster multiple labeling technique and critical concentration algorithm. *Physical Review B*, 14: 3438, 1976.

Hunt, A.G. and Ewing, R., *Percolation Theory for Flow in Porous Media*. Lecture Notes in Physics, vol. 771. Springer, Berlin, 2005.

Hunt, A.G. and Idriss, B., Percolation-based effective conductivity calculations for bimodal distributions of local conductance. *Philosophical Magazine*, 89(22–24): 1–21 and 1989–2007, 2009.

Katz, A.J. and Thompson, A.H., Quantitative prediction of permeability in porous rock. *Physical Review B*, 34: 8179, 1986.

Koltermann, C.E. and Gorelick, S.M., Heterogeneity in sedimentary deposits: a review of structure-imitating, process-imitating, and descriptive approaches. *Water Resources Research*, 32(9): 2617–2658, 1996.

Masihi, M., King, P.R. and Nurafza, P., The effect of anisotropy on finite size scaling in percolation theory. *Physical Review E*, 74: 042102, 2006.

Masihi, M., Gago, P.A. and King, P.R., Estimation of the effective permeability of heterogeneous porous media by using percolation concepts. *Transport in Porous Media*, 114(1): 169–199, doi: 10.1007/s11242-016-0732-9, 2016.

McCarthy, J.F., Analytical models of the effective permeability of sand-shale reservoirs. *Geophysical Journal International*, 105: 513–527, 1991.

Moreno, L. and Tsang, C.F., Flow channelling in strongly heterogeneous porous media: a numerical study. *Water Resources Research*, 30: 1421, 1994.

Pickup, G.E. and Sorbie, K.S., Scale up of two phase flow in porous media using phase permeability. SPE 28586, SPE Annual Technical Conference and Exhibition, New Orleans, 1994.

Renard, P.H. and de Marsily, G., Calculating equivalent permeability: a review. *Advances in Water Resources*, 20(5–6): 253–278, 1997.

Sahimi, M., Flow phenomena in rocks: from continuum models to fractals, percolation, cellular automata, and simulated annealing. *Review of Modern Physics*, 65(4): 1393–1534, 1993.

Scheibe, T. and Yabusaki, S., Scaling of flow and transport behavior in heterogeneous groundwater systems. *Advances in Water Resources*, 22: 223–238, 1998.

Shah, C.B. and Yortsos, Y.C., The permeability of strongly disordered systems. *Physics of Fluids*, 8: 280–282, 1996.

Chapter 8

Dynamic reservoir prediction using percolation

As has been said before, the presence of geological heterogeneities has a significant impact on flow and hydrocarbon recovery (Bear, 1972). The consequence of the uncertainty about the exact location of the heterogeneities is a large uncertainty in the prediction of reservoir performance. The most common method of oil recovery is by immiscible displacement during secondary recovery. Usually water is injected which maintains the reservoir pressure (the principal driving force for oil production) and replaces the oil in the pore space. Ultimately, the injected fluid breaks through at the production wells reducing the amount of oil that can be produced. For economic purposes, it is important to know when the injected fluid will break through (the breakthrough time) and what the rate of decline in oil production will be (the post-breakthrough behaviour). These are strongly dependent on the reservoir structure.

The conventional approach to the prediction of the breakthrough and post-breakthrough behaviour is to build detailed numerical models of the reservoir geology and then perform flow simulations (Mattex and Dalton, 1990; Ertekin *et al.*, 2001). This process is repeated for the many possible realisations of the subsurface to give an estimate of the uncertainty in production. These models are used to aid investment decisions and to choose between various different development scenarios. If each of these potential development plans is fully risked by running multiple realisations for each case, the computational effort required would be prohibitive. Given these practical limitations, some techniques such as renormalisation (King *et al.*, 1993) or fast simulation (Thiele *et al.*, 1995) can be used to estimate the uncertainty in a reasonable time. Percolation theory is an alternative, more phenomenological approach that is considerably quicker (King *et al.*, 2002b; López *et al.*, 2003; Andrade *et al.*, 2011; Oliveira *et al.*, 2010; Wettstein *et al.*, 2012).

In this chapter, we use percolation concepts to estimate the uncertainty in dynamic reservoir performance. In particular, we shall concentrate on breakthrough and post-breakthrough scaling in oil production. The models used will be validated against real reservoir data.

8.1 Breakthrough-time prediction

The breakthrough time is the time taken for the injected fluid to reach a production well. It is an important indicator of reservoir performance. In this section, we show how to use percolation theory concepts to determine the probability distribution for the breakthrough time t_{br} between an injector and a producer in a given reservoir. Preliminary analyses were given by Dokholyan *et al.* (1998), Lee *et al.* (1999) and King *et al.* (1999a, 1999b) and then extended by Andrade *et al.* (2000) to 3D and to account for finite-size effects. King *et al.* (2002a, 2002b) and Sadeghnejad *et al.* (2012) validated the proposed scaling laws by comparison with simulations on real reservoir data sets.

To develop the scaling laws, we start by looking at invasion by an idealised simple passive tracer. This is effectively single-phase flow where tracer particles are simply convected along the streamlines. There is no molecular diffusion. Then the "breakthrough time" is just the fastest transit time of any of the particles. In this section, we explore the approach in some detail to determine the appropriate distribution of the breakthrough time in reservoir engineering terminology as suggested by King *et al.* (2001, 2002a, 2002b).

8.1.1 *Scaling for the average breakthrough time*

We first consider the standard simple reservoir model described in Chapter 3. This is composed of only two rock types: permeable sandbodies distributed randomly and impermeable background shale. The volume fraction of permeable sands or net-to-gross ratio is the equivalent of the occupancy probability p of lattice percolation. There is a single injector–producer well pair with spacing r. These are considered to be connected if there is a flow path through the permeable sands from the injector to the producer.

We assume that the injected fluid has the same viscosity and density as the displaced fluid. We shall discuss the effect of viscosity later. We assume straight-line relative permeabilities (in other words, the phase mobility is just proportional to its volume fraction, known as the saturation or concentration). This means that as the injected fluid displaces the oil, the pressure field is unchanging with time. This displacement is essentially the same as the transport of a passive tracer; the injected fluid is not absorbed by the surroundings and moves only by convection ignoring molecular diffusion. The pressure field is determined by solving the single-phase flow equations. That is, the local flow rate is given by Darcy's law $v = -K\nabla P$ (Dullien, 1992) which coupled with the conservation condition for an incompressible fluid ($\nabla \cdot v = 0$) gives the governing equation for pressure ($\nabla \cdot K\nabla P = 0$) subject to appropriate boundary conditions.

The injected flow follows the streamlines (normals to the isobars) of this flow. The wells operate under fixed total flux conditions (the condition of a fixed pressure difference between two wells will be described later). The Euclidean distance between the injector and a producer is r. Each different realisation of the reservoir gives a different set of streamlines from the injector to the producer. Figure 8.1 shows schematically the streamlines within the percolating cluster between the two wells in 2D realisation.

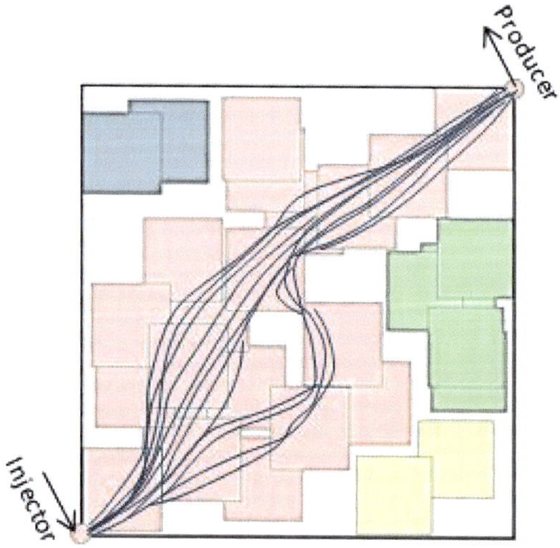

Figure 8.1. Illustration of the streamlines within the percolating cluster of an overlapping sandbody in 2D connecting an injector–producer well pair.

Injected fluid is passively convected along these streamlines shown in Figure 8.1. Other displacement mechanisms along the streamlines could be imagined and handled in a similar way to those described here. The time when the first tracer reaches the producer for a given realisation is called the breakthrough time. This is related to the problem of finding the probability distribution for the shortest path (also called the minimal or chemical path) between two points on a percolating cluster as described by Dokholyan *et al.* (1998). This is based on the hypothesis that transport properties in percolation clusters may be related to some geometric property of the cluster such as shortest path or optimal path (López *et al.*, 2005). Other dynamic properties such as conductivity and diffusion may also be expressed in terms of the shortest path.

We first consider the trivial case of a nearly homogeneous medium (when the occupation fraction of sands $p \approx 1$). In a completely homogeneous medium, the shortest streamline between two points would be a straight line and the longer ones would be 1D curves. If there is a small amount of disorder, the shortest streamline would no longer be straight but be perturbed by random fluctuations caused by the heterogeneities. This is very similar to a random walk which has a Gaussian distribution for the end-to-end distance. So taken over an ensemble of realisations, the probability distribution path distance along the shortest streamline would be Gaussian (Havlin and ben-Avraham, 1987). However, in a more disordered medium, the heterogeneities increase the fluctuations in the streamlines (as shown in Figure 8.1) and they are more similar to self-avoiding random walks. This argument is backed up by numerical and analytical studies (Bear, 1972; López *et al.*, 2003) which find that the average breakthrough time in mildly heterogeneous systems follows the scaling

$$t_{\mathrm{br}} \propto r^2, \tag{8.1}$$

which is what you get for a diffusive process (see Section 9.5 for more details). It should be noted that correlations in the permeability lead to a lower value of the exponent (1.8 was found by Oliveira *et al.* (2012)).

For highly heterogeneous systems, the situation is more complicated. Close to the percolation threshold, the flow is very localised (Figure 8.2 shows the magnitude of velocity in a percolating cluster.

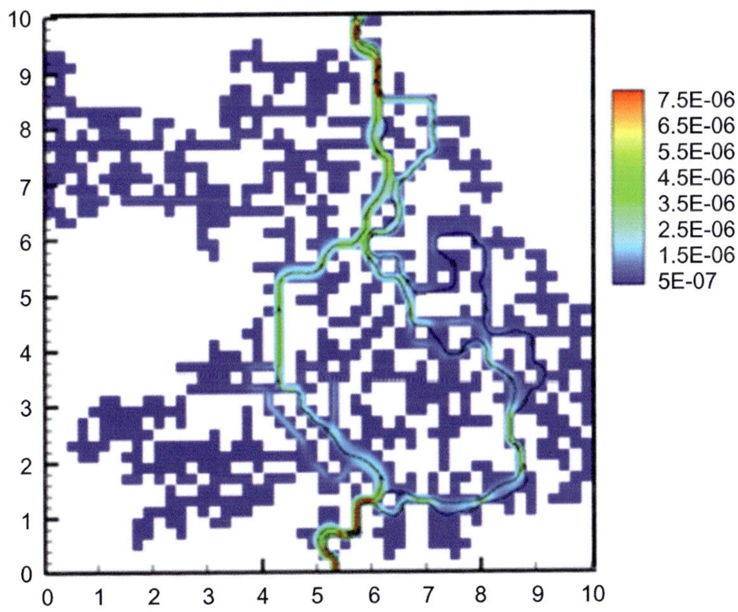

Figure 8.2. Velocity magnitude on a percolating cluster in 2D.

The scaling for the shortest path (Havlin *et al.*, 1984) between two points a distance r apart on the backbone follows the power law $l_{\text{min}} \propto r^{d_{\text{min}}}$ (Eq. (2.48)) where the exponent d_{min} is 1.13 and 1.37 in 2D and 3D respectively (Havlin *et al.*, 1984). If we assume that all velocities on the streamline are the same, then the shortest transit time between two points separated by r is expected to be proportional to the shortest path length. As shown by Andrade *et al.* (2000), the shortest transit time (the breakthrough time) on the backbone is strongly correlated with the shortest path length. The relation is $t_{\text{br}} \propto l_{\text{min}}^{1.17}$ in 2D. Hence, the shortest transit time has a power law scaling with the well separation (López *et al.*, 2003),

$$t_{\text{br}} \propto r^{d_t}, \tag{8.2}$$

with exponent $d_t = 1.33$ in 2D. This is larger than the fractal dimension of the shortest path but smaller than the fractal dimension of the backbone d_B ($d_{\text{min}} < d_t < d_B$, the value of d_B is given in Table 2.3). The exponent d_t is the fractal dimension of the sub-network traversed by the tracer particles at the time of breakthrough.

8.1.2 *Probability distribution for the breakthrough time*

We have assumed that travelling time scales in the same way as the path length. So the following form is proposed for the break-through time which has been validated numerically. The probability distribution for the breakthrough time (t_{br}) is conditional on the well spacing (r), the reservoir size (L) and net-to-gross ratio (p) dimensions (Lee *et al.*, 1999; Andrade *et al.*, 2000; King *et al.*, 2001, 2002a, 2002b),

$$P(t_{\text{br}}|r, L, p) \propto \frac{1}{r^{d_t}} \left(\frac{t_{\text{br}}}{r^{d_t}}\right)^{-g} f_1\left(\frac{t_{\text{br}}}{r^{d_t}}\right) f_2\left(\frac{t_{\text{br}}}{L^{d_t}}\right) f_3\left(\frac{t_{\text{br}}}{\xi^{d_t}}\right). \tag{8.3}$$

The proposed functional forms for the functions f_1, f_2 and f_3 are

$$f_1(x) = e^{-ax^{-\varphi}}, \quad f_2(x) = e^{-bx^{\psi}} \quad \text{and} \quad f_3(x) = e^{-cx},$$

Table 8.1. Summary of coefficients and exponents used in the scaling laws for the breakthrough time (Andrade *et al.*, 2000; King *et al.*, 2001, 2002a, 2002b).

Term	**2D**	**3D**
a	1.1	2.5
b	5	2.3
c	1.6 (if $p < p_c$)	2.9 (if $p < p_c$)
	2.6 (if $p > p_c$)	—
d	1.33 ± 0.05	1.45 ± 0.1
g	1.9 ± 0.1	2.1 ± 0.1
φ	3	1.6
ψ	3	2

and $\xi = |p - p_c|^{-\nu}$ is the usual percolation correlation length. In this scaling, d_t is the fractal dimension of the path with the shortest transit time, g is the exponent for the geometric spacing effect and a, b, c are coefficients that need to be determined numerically. The exponents $\varphi = 1/(d-1)$ and ψ have been determined numerically and are summarised in Table 8.1.

The same scaling form is expected to apply (but with new exponents and coefficients) to more complex systems such as when there is a long-range spatial correlation in the medium (Araújo *et al.*, 2002). A similar scaling to Eq. (8.3) can also be used for the condition of a fixed pressure difference between two wells with the new scale time $t_p \equiv t/r^{d_t+\mu}$ instead of t/r^{d_t} with parameters obtained numerically ($g = 1.5$) where μ is the characteristic exponent of resistivity dependence on distance with numerical value of 0.98 (Grassberger, 1989; Andrade *et al.*, 2000; López *et al.*, 2003).

There is a rationale behind the functional form in Eq. (8.3) and a physical significance for the functions f_1, f_2 and f_3 (King *et al.*, 2001). In a disordered medium (as postulated by de Gennes (1979) and described by King *et al.*, (2002a, 2002b)), the probability distribution for the end-to-end distance between two points is not Gaussian (as it would be in a mildly heterogeneous medium) but it is a "stretched Gaussian" distribution like function f_1 in Eq. (8.3).

In reality, all systems are finite. Streamlines longer than the system size are truncated by finite boundaries. The upper cut-off is accounted for by the function f_2 in Eq. (8.3). Away from the percolation threshold, the clusters have a finite size with a typical size (given by the correlation length ξ). This also truncates the longer streamlines through the function f_3 in Eq. (8.3). Dokholyan *et al.* (1998) assumed that the combination of these effects would be multiplicative. It is then further assumed that the functional form for the scaling of the breakthrough time should be of the same form as that for the shortest path length, leading to Eq. (8.3). The validity of this assumption and the parameters need to determined by Monte Carlo simulations.

The variables in this scaling relationship are dimensionless:

$$t' = \frac{t}{t_s}, \quad r' = \frac{r}{r_s}, \quad L' = \frac{L}{r_s}, \tag{8.4}$$

where t, r and L are respectively the time, well spacing and reservoir length and t', r' and L' their corresponding dimensionless values. They are non-dimensionalised by the typical sandbody length r_s and the corresponding typical time t_s for the flow to transit through one sandbody. This transit time can be estimated by first assuming that the sandbodies are placed end to end (just touching each other) in the direction of the flow. Then the pressure drop across a single sandbody (Δp_s) is related to the total pressure drop $\Delta p = p_{\text{inj}} - p_{\text{pro}}$ between two wells through $\Delta p_s = \frac{p_{\text{inj}} - p_{\text{pro}}}{r/r_s}$.

We then consider the pressure drop for incompressible fluid flow through a homogeneous rock with radial flow (Craft and Hawkins, 1991; Ahmed, 2001),

$$\Delta p = \frac{\mu q}{2\pi K h} \ln\left(\frac{r}{r_w}\right), \tag{8.5a}$$

where variables are as follows: volumetric flow rate q, fluid viscosity μ, wellbore radius r_w, pressure drop Δp, sandbody length r, thickness h and permeability K.

Solving this for the flow rate in a single sandbody gives

$$q = \frac{2\pi K h \Delta p_s}{\mu \ln\left(\frac{r_s}{r_w}\right)}, \tag{8.5b}$$

where we use the pressure drop (Δp_s) across a single body from above. The transit time for a single body would then be $t_s = \frac{r_s^2 h}{q}$. Substituting the flow rate from Eq. (8.5b) gives

$$t_s = \frac{\mu r_s^2 \ln\left(\frac{r_s}{r_w}\right)}{2\pi K(p_{\text{inj}} - p_{\text{pro}})/(r/r_s)}. \tag{8.6a}$$

In reality, this is a bit simplistic as it ignores the overlaps between the bodies. Numerical simulations indicate that the following expression is a better fit (King *et al.*, 2001),

$$t_s = \frac{12\mu r_s^2 \ln\left(\frac{r_s}{r_w}\right)}{15K(p_{\text{inj}} - p_{\text{pro}})/(r/r_s)}, \tag{8.6b}$$

where r is the well spacing.

The dimensionless lengths in Eq. (8.4) can be defined along each direction. For example, $r_x = \frac{\text{Well spacing}}{\text{Sand size}}$ and $L_x = \frac{\text{Reservoir size}}{\text{Sand size}}$ in the x direction. Then the relevant length scale used for function f_2 in Eq. (8.3) is the minimum length (L_m) from L_x, L_y and L_z. Similar minimum-length arguments should be used for the well spacing (r_m) in the scaling (Eq. (8.3)). The validity of using this minimum length scales has been shown numerically by Andrade *et al.* (2000).

So far, we have considered passive tracer transport. For other flow regimes, as a first approximation, we use streamlines found above. We could then use Buckley–Leverett displacement along the fastest streamline (Buckley and Leverett, 1942). This approach ignores the coupling between viscous displacement and the flow pattern. The shock front speed from the standard Welge construction gives an estimate of the breakthrough time (King *et al.*, 2002a, 2002b) We replace the single-phase viscosity μ in Eq. (8.6) by an effective viscosity estimated from the Buckley–Leverett approximation.

Now with these new dimensionless variables, the probability function for the breakthrough time along, e.g., the x direction is

$$P_x(t|r, L, p) = \frac{1}{r_m^{d_t}} \left(\frac{t_x}{r_m^{d_t}}\right)^{-g} e^{-a\left(\frac{t_x}{r_m^{d_t}}\right)^{-\varphi}} e^{-b\left(\frac{t_x}{L_m^{d_t}}\right)^{\psi}} e^{-c\left(\frac{t_x}{\xi^{d_t}}\right)}, \tag{8.7}$$

where t_x is the dimensionless breakthrough time in the x direction. Similarly we can determine the distribution function of breakthrough time in the y direction. This scaling relation (Eq. (8.7)) can be encoded in a spreadsheet to predict the probability distribution for the breakthrough time along the x and y directions in fractions of a second. This should be contrasted with the hours required for conventional flow simulations.

As an example of how to use these formulae, consider a geometrically simple reservoir. It is rectangular with dimensions 9,000, 2,000 and 150 m respectively in the x, y and z directions. It has a permeability of 5 d, and the reservoir net-to-gross ratio is 53%. The sandbody sizes are 7,000 300 and 15 m in the x, y and z directions respectively. Their long axes are aligned with the long axis of the reservoir. The fluids are incompressible both with the same viscosity of 1.2×10^{-3} Pa s, and the flow is essentially 2D in this thin reservoir. The injector bottom-hole pressure is 2.94×10^7 Pa and production pressure is 2.23×10^7 Pa. The wellbore radius is 0.06 m, and well spacing is 1,200 m. The aim is to determine the probability distribution for the breakthrough time between two wells aligned parallel to the shortest axis.

First we use Eq. (8.6) to calculate the typical time for the passage of flow along the width of a sandbody of size $r_s = 300$ m which gives $t_s = 0.666$ y. The minimum reservoir is length scale and well spacing are $L_m = 1.285$ and $r_m = 0.171$. Then using Eq. (8.7), the distribution function for the breakthrough time between is shown in Figure 8.3. This is very similar to the second case study (Section 8.1.3.2).

The breakthrough time scaling (Eq. (8.3)) can also be used to find the interwell connectivity (Li *et al.*, 2009). This is defined as the discrete random variable R_{ij} which is the probability that wells i and j are on the same flow path through the good sands. This is similar to the spanning probability described in Chapter 2. A finite breakthrough time indicates that two wells are in hydraulic communication. So, the integral of the product of the sand fraction and the breakthrough probability distribution (which gives $p \cdot P$ $(t_{\text{br}}|r, L, p)$) gives the probability that two wells being connected

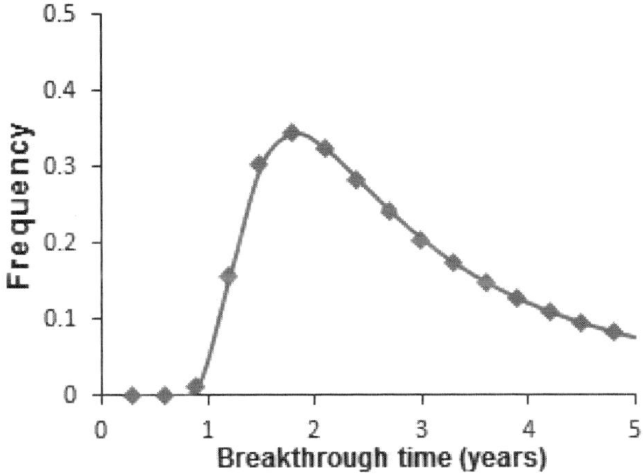

Figure 8.3. Probability distribution for the breakthrough time between two wells placed along the width of the reservoir from the scaling law (Eq. (8.7)).

as (Li *et al.*, 2009)

$$\mathrm{P}(R_{ij} = 1|r, L, p)$$
$$= p \int_0^\infty \frac{1}{r^d} \left(\frac{t_{\mathrm{br}}}{r^d}\right)^{-g} f_1\left(\frac{t_{\mathrm{br}}}{r^d}\right) f_2\left(\frac{t_{\mathrm{br}}}{L^d}\right) f_3\left(\frac{t_{\mathrm{br}}}{\xi^d}\right) dt_{\mathrm{br}}. \tag{8.8}$$

A case study to validate this scaling will be described later.

8.1.3 *Case studies*

In order to check the validity of the breakthrough scaling, we use some real field cases. For a given reservoir model, we sample different realisations of the locations of the wells. Alternatively, for the same well locations, we generate different reservoir models with the same underlying statistics. For each realisation, we run conventional flow simulation to get the statistics of the breakthrough times. We compare the histogram of breakthrough times for the given reservoir size, the net-to-gross ratio and well spacing with the predictions from the scaling (Eq. (8.7)). In this section, we consider four case studies.

8.1.3.1 *Case study 1*

This is a turbidite reservoir model with dimensions of $700 \times 700 \times 170\,\mathrm{m}^3$ from a North Sea field with an approximately bimodal permeability distribution in the range of 100–900 md. The two modes are at 200 and 300 md (King *et al.*, 1999a, 1999b). The sands were deposited by submarine turbidity currents in a deepwater environment. The resulting sands consist of channels from the avalanche scour and lobe-shaped sandbodies from the resulting avalanche plumes. A cut-off in permeability at 273 md is applied to separate the sand into high and low permeabilities. The cut-off was chosen as it corresponds to the threshold value for the system at which the incipient spanning cluster appears. Then simulations of the flow were carried out to determine the distribution of breakthrough times for well pairs at different locations and a variety of separations (solid lines in Figure 8.4) which shows an agreement with the breakthrough-time predictions by using the scaling law.

8.1.3.2 *Case study 2*

Another deepwater turbidite reservoir was described by King *et al.* (2001, 2002a, 2002b). In this case, the reservoir was 10,000 m long,

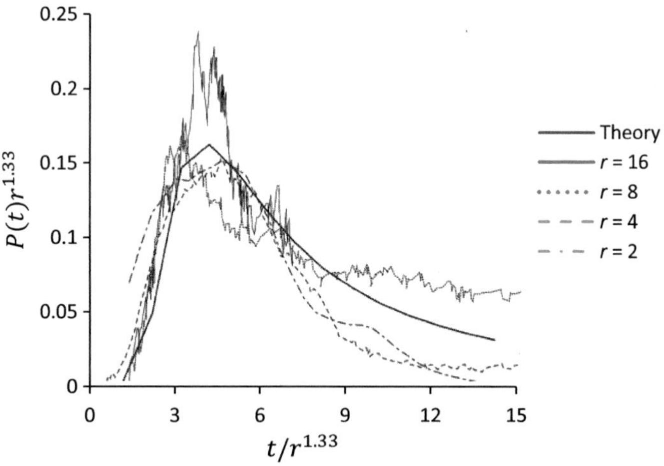

Figure 8.4. Comparison of simulation results of the breakthrough time at four different well spacing $r = 2$, 4, 8 and 16 with the prediction from the scaling laws (Eq. (8.7)).

1,500 m wide and 150 m thick. The turbidite channels (that make most of the net pay) are 8,000 m long, 200 m wide and 15 m thick. Their long axes are aligned with the long axis of the reservoir. The reservoir net-to-gross ratio is $p = 50\%$. Well spacing is $r = 1{,}500$ m, and two cases were considered: one with the wells aligned with the longer axis of the reservoir and the other where they were perpendicular. Injector and producer wells have pressures of 3.4×10^7 Pa and 2.6×10^7 Pa respectively. The sandbody permeability is 4,000 md, and the fluid viscosity is 10^{-3} Pa s. With three existing injectors and three producers in a linear pattern in the field, there were only nine samples of the breakthrough time observed from conventional flow simulations. The statistics for breakthrough times of various well pairs in the model have been collected. Then from the scaling law (Eq. (8.7)), the probability distribution of the breakthrough time is shown by the solid curve in Figure 8.5. This should be compared with the results obtained from the conventional flow simulations (the histogram). As can be seen, the agreement between the prediction from the scaling law and the numerical flow simulation results is good.

8.1.3.3 *Case study 3*

Sadeghnejad *et al.* (2012) used the data of an offshore oil field described in Section 6.4. The reservoir is approximately

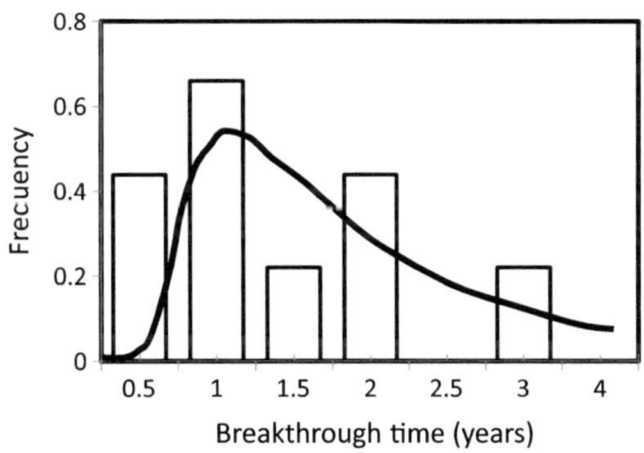

Figure 8.5. Comparison of the prediction of breakthrough time using scaling laws (Eq. (8.7)) (solid curve) with results from the numerical flow simulations.

Table 8.2. Reservoir properties and operation data used in case study 3.

Parameter	Value
Injector pressure (Pa)	2.43×10^7
Producer pressure (Pa)	2.38×10^7
Sand sizes (m)	$1{,}200 \times 1{,}300 \times 15$
Sand permeability (md)	8,000
Well diameter (m)	0.18
Fluid viscosity (Pa s)	3.6×10^{-3}
Oil density (kg/m^3)	935.5
Water density (kg/m^3)	1,140
Gas density (kg/m^3)	1.297
Solution gas oil ratio (m^3/m^3)	89
Bubble point pressure (Pa)	1.85×10^7

$5{,}200 \times 15{,}100 \times 110$ m, and the model contains $50 \times 100 \times 20$ simulation grid blocks. The average net-to-gross ratio of the reservoir is $p = 54.6\%$. Three different well spacings of 500, 750 and 1,000 m aligned either along or perpendicular to the long axis of the reservoir have been considered. Constant rate and constant bottom-hole pressure were used for boundary conditions in the injection and the production wells. A summary of the reservoir properties and operating conditions is shown in Table 8.2. To compare the prediction from the scaling law with the breakthrough times computed from the conventional flow simulations, different well configurations in different locations of the field were considered randomly.

During the simulation, the pressure of the reservoir is maintained above the bubble point pressure to avoid gas production and to ensure that there was only two-phase flow. Nearly 50 realisations were considered for each well spacing r, and a total of 300 flow simulations were carried out. The typical simulation outputs for the case of a well spacing of 500 m are shown in Figure 8.6.

Then the probability distribution of the breakthrough times was determined using the scaling law (Eq. (8.7)) for each well spacing and the results were compared with the histogram of the breakthrough time computed from the reservoir flow simulations. The comparison of the results for the breakthrough times predicted (solid curves) and

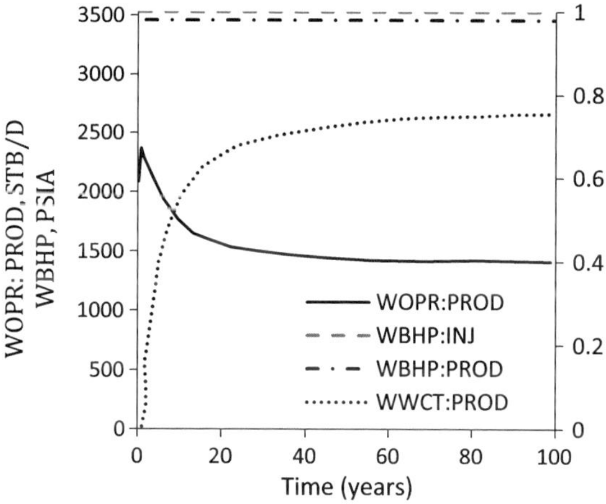

Figure 8.6. Typical oil production, water cut and bottom-hole pressure in production and injection well for a realisation with well spacing of $r = 500\,\text{m}$ obtained from a conventional flow simulator. WOPR: well oil production rate, WBHP: well bottom hole pressure, WWCT: well water cut, PROD: production, INJ: injection, PSIA: pounds per square inch absolute and STB/D: stock tank barrel per day.

the simulation results (histogram) for well spacings of 500 and 750 m are shown in Figure 8.7.

8.1.3.4 *Case study 4*

As explained in the derivation of Eq. (8.8), the integral of the breakthrough-time distribution function can be used to estimate the interwell connectivity. This can be used to design an appropriate well spacing for an infill drilling programme. Li *et al.* (2009) used Eq. (8.8) to evaluate the probability that two wells are connected in the interval D1 of the Monument Butte field in Utah. This is a stratigraphically thin interval with average net thickness of 3.66 m. A square region of side 2,500 m in the northwest portion of the field was used. The well spacing was approximately 400 m. Based on the available correlations and maps, the best estimate for the sandbody size was 780 m. They were separated by thick shale layers and the resulting areal net-to-gross ratio was $p = 0.6208$. Note that this is below the infinite threshold value of $p = 0.668$.

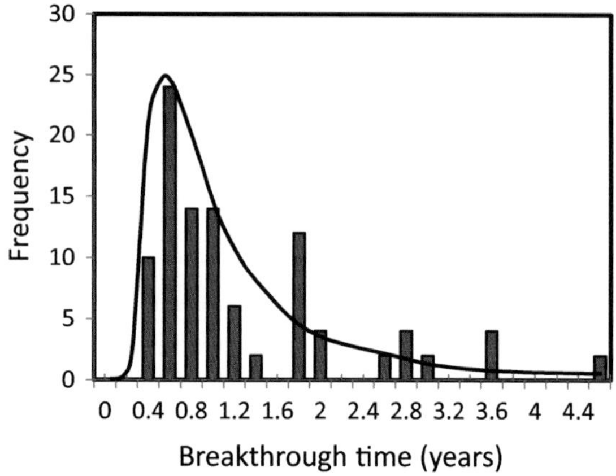

(a)

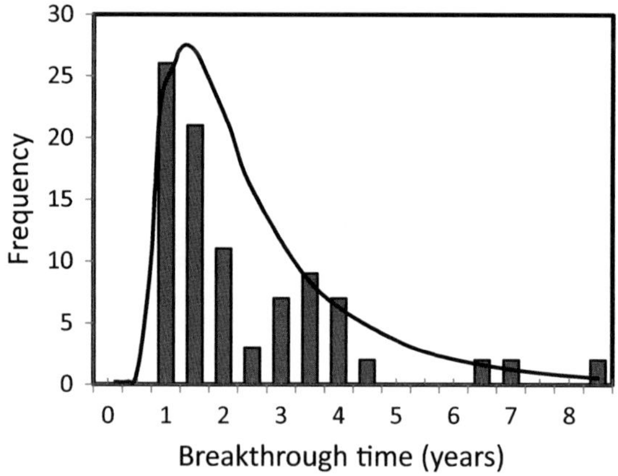

(b)

Figure 8.7. Comparison of the prediction of the breakthrough time using scaling laws (Eq. (8.7)) (solid curve) with the breakthrough results determined from the numerical flow simulations (histogram) for well spacings of (a) 500 m and (b) 750 m.

A Monte Carlo model with square bodies randomly placed in a square system was used to find connected paths between two points spaced a distance r apart. Using this, the interwell connectivity analysis (Eq. (8.8)) was compared with a history-matched flow

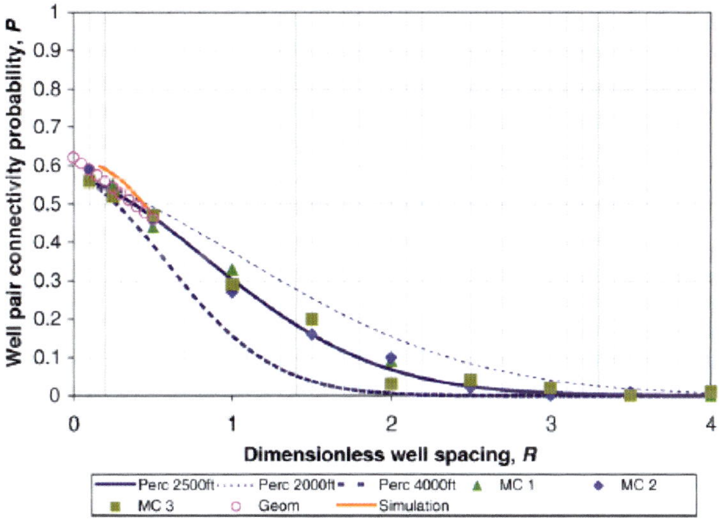

Figure 8.8. Interwell connection probability in the D1 interval of the Monument Butte field in Utah. The point shows Monte Carlo results (Li *et al.*, 2009).

simulation. The results shown in Figure 8.8 show that the predictions by this method agree well with simulation results analysis while this method is very fast compared with the very time-consuming simulation method.

8.2 Post-breakthrough behaviour

So far, we have focussed on breakthrough time based on the shortest path (or, more accurately, the path with lowest transit time). We now examine the post-breakthrough production decay. This is effectively the distribution of the longer paths longer than the shortest. We present a preliminary analysis of this problem (King *et al.*, 1999b, 2002b; López *et al.*, 2003).

8.2.1 *Scaling for the average post-breakthrough production decay*

Consider a percolating system close to the percolation threshold. We take a pair of wells (an injector and a producer operating under constant rate) with spacing r that is large compared with the sandbody size (effectively $L \to \infty$). At time $t = 0$, we start injection (in well A) at constant rate and produce from well B. We assume both fluids are incompressible and have equal viscosities. Then the scaling for average production decline (averaged over a large number of realisations) after the breakthrough can be analysed in a similar way to that discussed in Section 8.2 (King *et al.*, 1999b; López *et al.*, 2003). Moreover, we will see how knowledge of the breakthrough time improves the forecast of future reservoir production (Røislien, 2004).

In the simple 2D conceptual model described in Section 8.2, the probability distribution of the travelling times $P(t, r)$ can be calculated by counting the number of particles that travel from the injector A to producer B in a time between t and $t + dt$. This must be averaged over realisations using Monte Carlo simulation. Physically, this is the rate of increase of water production in the production well at time t. It is zero for $t < t_{\mathrm{br}}$, positive for $t > t_{\mathrm{br}}$ and approaches zero again when $t \to \infty$. Assuming the streamlines do not change with time, the production curve $C(t, r)$ is the fraction of oil produced at time t. It must be unity ($C = 1$) initially. As production continues, it can be calculated from

$$C(t, r) \equiv 1 - \int_0^t P(t, r)dt, \tag{8.9a}$$

or

$$P(t, r) = -\frac{dC(t, r)}{dt}. \tag{8.9b}$$

Before analysing the production decline in the model near the threshold, first consider an analytical solution of the production for a homogeneous system (López *et al.*, 2003). We solve Laplace's equation $\nabla^2 P = 0$ for the pressure P in an infinite domain with two wells. The effects of wellbore pressure and diameter and the condition choice of constant rate boundary do not alter the essential results. We will also discuss the effects of different viscosities of two fluids later. The single-phase pressure field can be calculated by using bipolar cylindrical coordinates (Lockwood, 1967; King *et al.*,

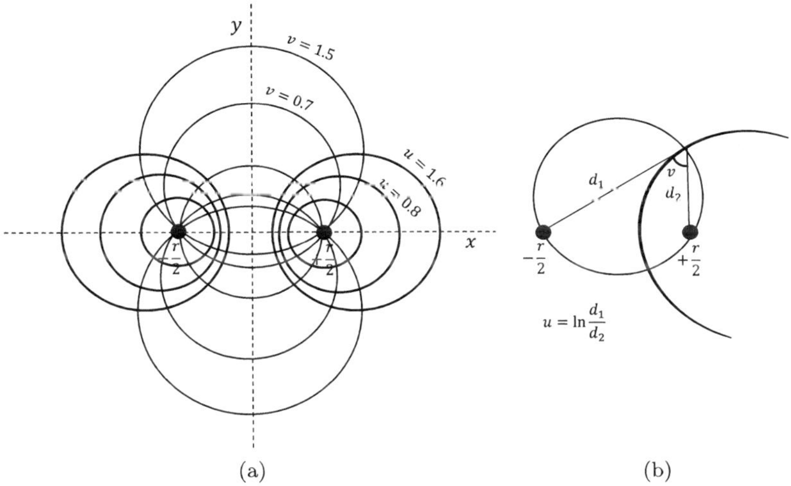

Figure 8.9. Illustration of bipolar coordinates showing (a) isobars (thick lines) and streamlines (thin lines) and (b) geometrical interpretation of variables u and v.

1999a, 1999b),

$$x = \frac{r}{2}\frac{\sinh u}{\cosh u - \cos v}, \quad y = \frac{r}{2}\frac{\sin v}{\cosh u - \cos v}, \tag{8.10}$$

where pressure is associated with coordinate u and the stream function with the coordinate v (see Figure 8.9).

In this coordinate system, the positions of the injector A and the producer B are $\left(-\frac{r}{2}, 0\right)$ and $\left(+\frac{r}{2}, 0\right)$ respectively, so the well spacing is r. Curves of constant u are isobars, and curves of constant v are streamlines, and both curves are arcs of coaxial circles. Then the length of streamline v is $(\pi - v)/r \sin v$ and transit time along the streamline v is $t(v) = \frac{r^2}{2K}\frac{1+(\pi-v)\cot v}{\sin^2 v}$, where K is the permeability. Then the production decline is (King *et al.*, 1999a, 1999b, 2002b)

$$C(t) \propto \left(\frac{t}{r^2}\right)^{-1/3}. \tag{8.11}$$

We now turn to finite, weakly disordered systems where the average travelling time $P(t, r, L)$ can be found from Monte Carlo

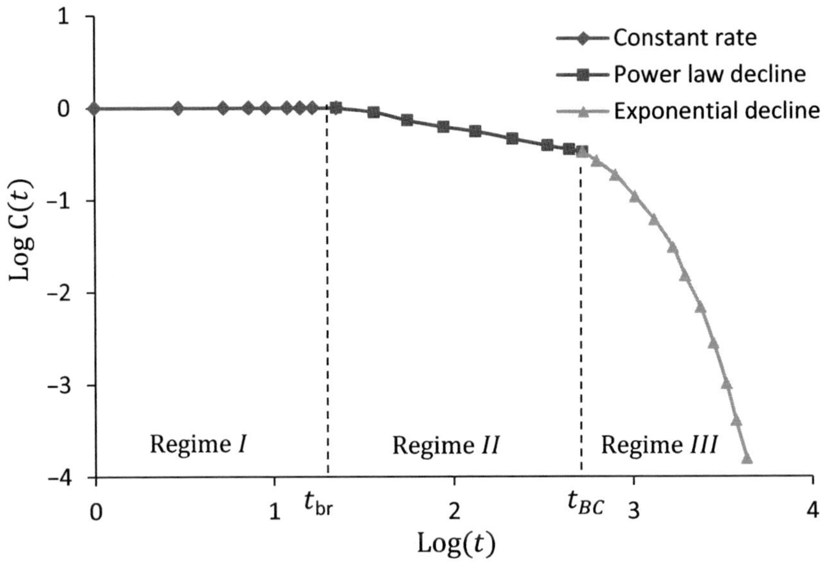

Figure 8.10. Illustration of three time intervals to describe the average production.

simulations. We integrate this to find the average production decline through Eq. (8.9). Three main time intervals can be classified when plotted against dimensionless time $t' = t/t_{\mathrm{br}}$ (López *et al.*, 2003; Oliveira *et al.*, 2012) as shown schematically in Figure 8.10.

Regime I: This is the time interval before injected water appears in the production well. In this regime, only oil is produced:

$$t < t_{\mathrm{br}} \quad \text{and} \quad C(t) = 1. \tag{8.12}$$

Regime II: This is the interval after breakthrough time t_{br} and before the effects of the finite boundary become significant at time t_{BC}. In this regime, a power law decay in oil production is observed which is the same as that for a homogeneous system:

$$t_{\mathrm{br}} < t < t_{BC} \quad \text{and} \quad C(t) \propto t^{-1/3}. \tag{8.13}$$

Regime III: This is when the effects of the boundary become important. In this regime, an exponential decay in oil production is observed:

$$t > t_{BC} \quad \text{and} \quad C(t) \propto e^{-\rho t}. \tag{8.14}$$

The exponential decay was proposed by López *et al.* (2003) from Monte Carlo simulation. However, it should be noted that Oliveira *et al.* (2012) found power law decay for regime III ($C(t) \propto t^{-5/2}$). The reason for this discrepancy is probably due to the use of relative permeability curves by Oliveira rather than the passive tracer used by López *et al.* However, both methods result in the same power law scaling (Eq. (8.13)) for regime II.

Finally we examine the response for a model close to the percolation threshold. We expect the production decline behaviour to be more complex as it is affected by both the finite boundaries and the finite cluster sizes. As before, there are several regimes. In regime I, only oil will be produced and Eq. (8.12) is valid. After the breakthrough time, we expect the production decay to be a power law with a different exponent from that observed in Eq. (8.11) (King *et al.*, 1999b, 2002b),

$$t > t_{\mathrm{br}} \quad \text{and} \quad C(t) \propto \left(\frac{t}{r^{d_B}}\right)^{-\beta}, \tag{8.15}$$

where d_B is the fractal dimension of the backbone of the percolating cluster (given in Table 2.3). From the numerical simulations, the exponent β was found to be 0.63 ± 0.05.

8.2.2 *Probability distribution for the production decay*

By analysing the results from Monte Carlo simulations, López *et al.* (2003) suggested the following scaling for the travelling time distribution in 2D for the case of constant well pressure which covers all observed time regimes,

$$P(t, r, L) \propto \left(\frac{t}{r^{d_B}}\right)^{-g_{p1}} F_p\left(\frac{t}{r^{d_B+\mu}}\right) G\left(\frac{t}{L^{d_p} \cdot r^{d_B+\mu-d_p}}\right), \tag{8.16}$$

where functions $F_p(x) = e^{-bx^{-\phi}}$ and $G(x) = \begin{cases} \text{const}, & x \ll 1, \\ x^{-g_{p2}+g_{p1}}, & x \gg 1 \end{cases}$.
The exponents and coefficients determined numerically in 2D are

$\phi = -1.42$, $g_{p1} = 1.42$, $g_{p2} = 2.43$, $d_p = 2.78$ and μ is the conductivity exponent.

The scaling laws proposed for the breakthrough time and the post-breakthrough production decline for constant injection rate and constant pressure follow a similar form. However, it is necessary to use the time scale $t_c \equiv t/r^{d_B}$ for constant injection rate and $t_p \equiv t/r^{d_B+\mu}$ for constant injection pressure (Andrade *et al.*, 2000; López *et al.*, 2003).

Oliveira *et al.* (2012) investigated the influence of the viscosity ratio (μ_o/μ_w) on the production decline curves through simulation. They observed that despite the shorter breakthrough time at a higher viscosity ratio, the water front advances more slowly towards the reservoir boundary displacing less oil. For example, in a homogeneous medium, the production decay at the crossover time where $t_{BC} \propto (\mu_o/\mu_w)$ scales as

$$C(t) \propto (\mu_o/\mu_w)^{-0.85}. \tag{8.17}$$

It should be emphasised that the analysis discussed so far was in 2D where wells are represented by points and in 3D where vertical well pairs were assumed; however, for deviated wells in 3D, some modifications in the scaling of the breakthrough time and post-breakthrough production decline may be needed to account for the change in the backbone mass (Silva *et al.*, 2003).

8.2.3 *Uncertainty in post-breakthrough oil production decline*

We have described the form of the scaling for the average oil production curve $C(t, r, L)$ after breakthrough for both the homogeneous and percolation cases. In this section, we focus on the 2D percolation problem near the threshold. We use Monte Carlo simulation to estimate the uncertainty in the post-breakthrough oil production curves. Figure 8.11 shows 141 production curves against dimensionless time, i.e., $\frac{t}{r^{d_B}}$ for a model of size $L = 514$, well spacing $r = 46$ (Røislien, 2004). We then extract a subset of the production curves with breakthrough time $t_{\rm br} \in [t \pm dt]$. For example,

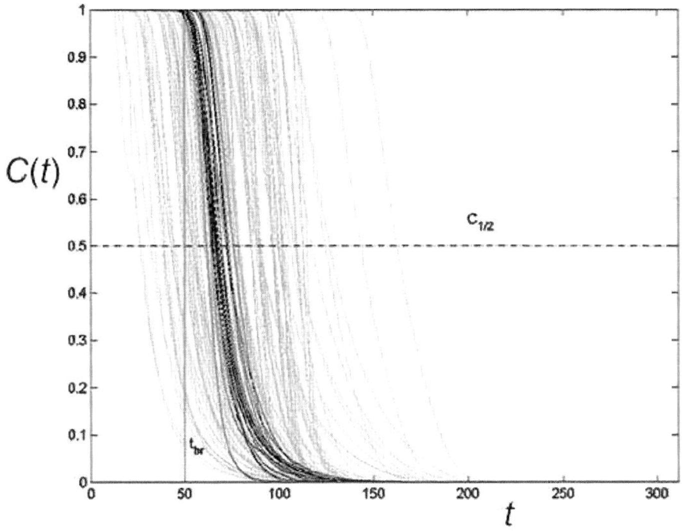

Figure 8.11. Illustration of oil production curves of an inhomogeneous percolation medium of size $L = 514$ and well spacing $r = 46$. There are 141 production curves; a subset of 18 curves with the breakthrough time in the range $t_{\text{br}} \in [50 \pm 3]$ are in bold (Røislien, 2004).

18 curves conditioned on the breakthrough time $t_{\text{br}} = 50$ are shown in Figure 8.11.

We then analyse the statistics of these curves: (1) all unconditional production data and (2) the subset of production data conditional on the given breakthrough time t_{br}. In particular, we calculate the empirical mean $\mu_c(t)$ and variance $\sigma_c^2(t)$ (representing the uncertainty in prediction of oil production). These are shown in Figure 8.12 for the unconditional and conditional production data. As can be seen, the empirical means $\mu_c(t)$ follow a similar trend as observed for the production data (Figure 8.11) and the variances $\sigma_c^2(t)$ appear to be Gaussian. The variance (which is the uncertainty in production prediction) is reduced considerably if the breakthrough time is known.

We can repeat the analysis at other breakthrough times (e.g., $t_{\text{br}} = 40, 50, \ldots, 90$) to determine the mean and variance in these production curves as shown in Figure 8.13.

Røislien (2004) found that it is possible to rescale the production curves to get a data collapse onto one single curve. The relevant

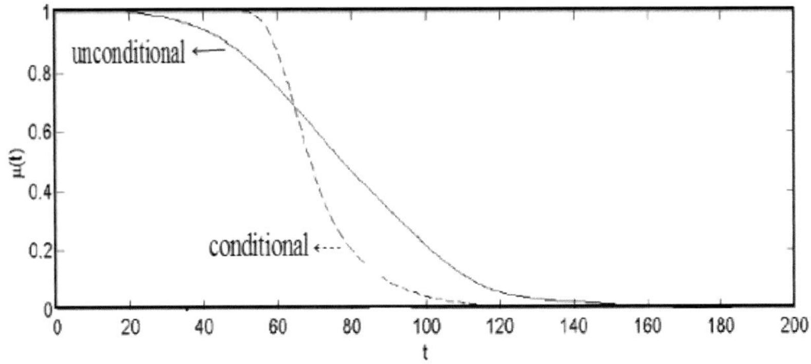

(a)

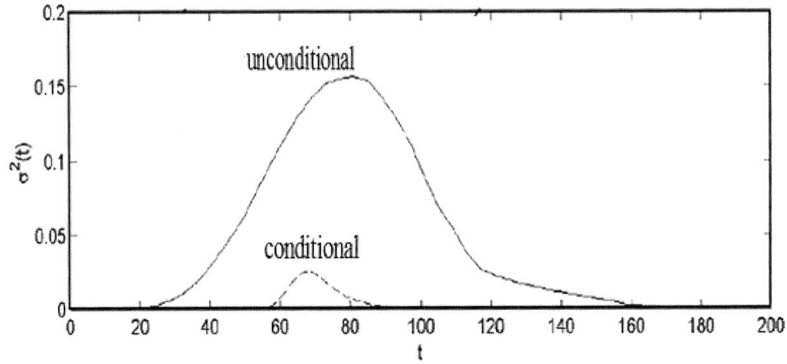

(b)

Figure 8.12. Illustration of calculated (a) empirical mean $\mu_c(t)$ and (b) variance $\sigma_c^2(t)$ for the unconditional and conditional production data given in Figure 8.11 (Røislien, 2004).

scaling parameter is the breakthrough time (Figure 8.14). If we know the breakthrough time we can make an appropriate estimate of the mean and variance in the future production. This should be considered a hypothesis as only one system has been studied.

The conditional production data curves (Figure 8.13) indicate that the means and variances are very similar for different breakthrough times. This indicates that there is a strong correlation between the breakthrough time t_{br} and any characteristic post-breakthrough time, e.g., $t_{1/2}$ which is the time required for production to fall by 50% (or water cut to increase to 50%). It is interesting to note that this time ($t_{1/2}$) roughly corresponds to the maximum

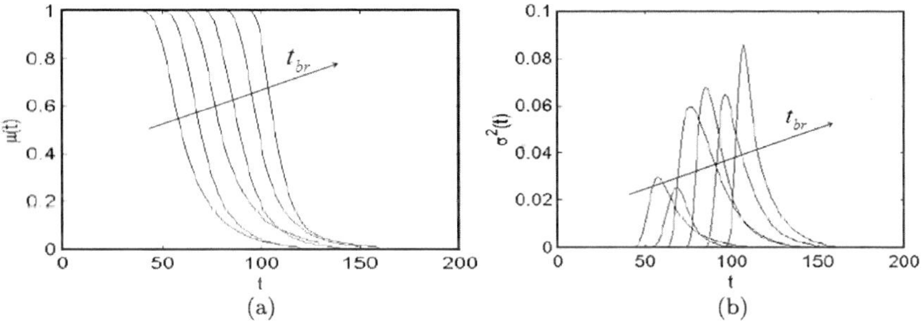

Figure 8.13. (a) Mean $\mu_c(t)$ and (b) variance $\sigma_c^2(t)$ in the production curves conditioned to $t_{\text{br}} = 40, 50, \ldots, 90$ (Røislien, 2004).

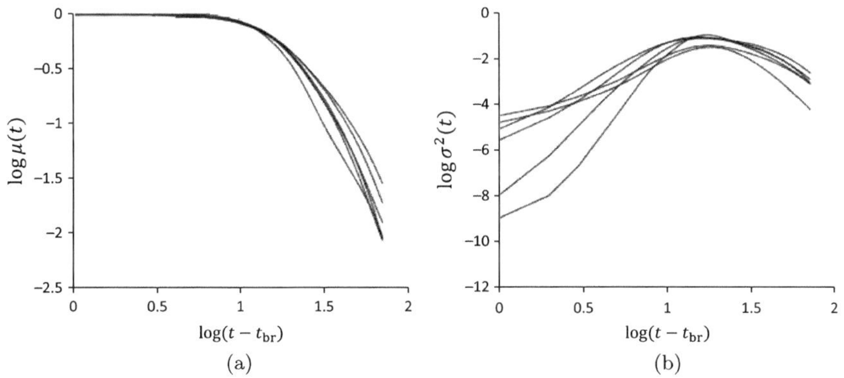

Figure 8.14. (a) Rescaled mean $\mu_c(t)$ and (b) variance $\sigma_c^2(t)$ in the production curves conditioned to $t_{\text{br}} = 40, 50, \ldots, 90$ showing data collapse (Røislien, 2004).

of the variance curve. The log–log plot of this time $t_{1/2}$ against the breakthrough time t_{br} is shown in Figure 8.15. This shows that the time $t_{1/2}$ taken for production to fall to $C_{1/2} = 50\%$ scales as a power of the breakthrough time, with a power of $b = 0.76$:

$$t_{1/2} \propto t_{\text{br}}^b. \tag{8.18}$$

Hence, for a system close to the threshold, the knowledge of the breakthrough time gives a good estimate for (a) the mean and variance in future production and (b) the time for production to fall by 50%. Note that this correlation for long breakthrough times must have corrections to avoid $t_{1/2} < t_{\text{br}}$ which clearly cannot be correct.

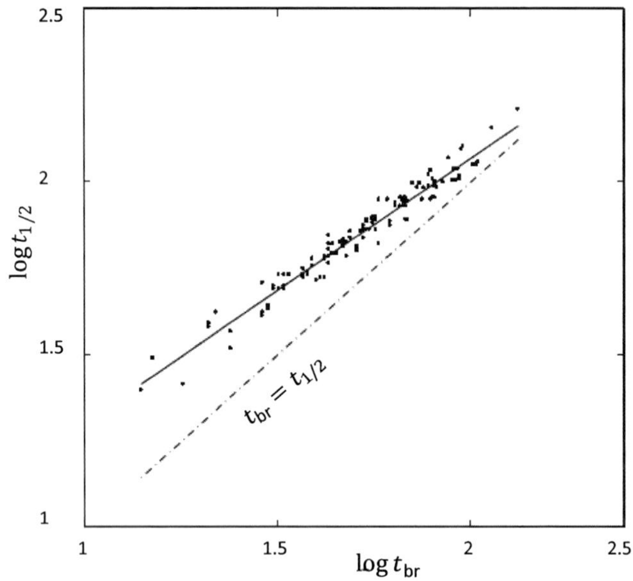

Figure 8.15. Correlation between the breakthrough time t_{br} and the time $t_{1/2}$ taken for production to fall to 50% (Røislien, 2004).

8.2.4 *Case studies*

In this section, we use reservoir simulation models to check the validity of the post-breakthrough scaling relations discussed in the previous section.

8.2.4.1 *Case study 1*

In this section, we examine the post-breakthrough production data of the reservoir B, the offshore oil field with details described in Sections 6.4 and 8.1 (Sadeghnejad *et al.*, 2012). For a given reservoir model, we sample different realisations of the locations of the wells. For each realisation, we can run conventional flow simulations to obtain statistics of the post-breakthrough production and water cut. Figure 8.16 shows the post-breakthrough production decline curves for a well spacing of $r = 500\,\text{m}$.

To investigate the power law scaling between the breakthrough time t_{br} and $t_{1/2}$, we extracted and collected these times from the

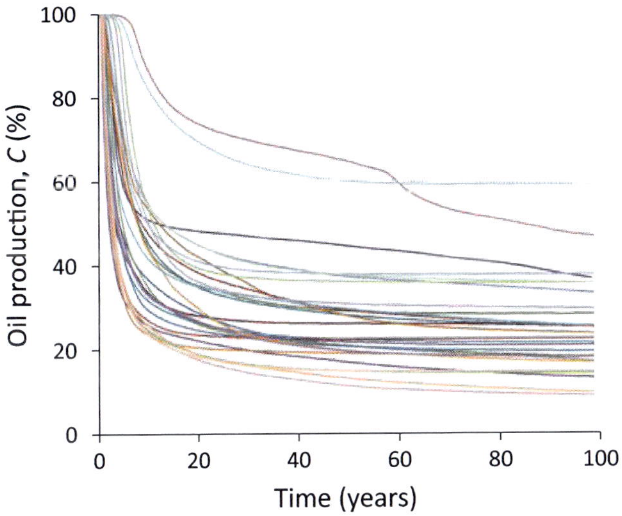

Figure 8.16. Post-breakthrough production decline curves for well spacing $r = 500\,\text{m}$ of the reservoir B.

simulations shown in Figure 8.16. Figure 8.17 shows the log–log plot of time $t_{1/2}$ against t_{br}. The difference between the power found from these results ($b = 0.89$ from Figure 8.17) and the simple percolation model ($b = 0.76$ from Figure 8.15) is probably due to the more complex nature of the real reservoir compared with the simple percolation model. However, the agreement is still reasonably good for practical purposes.

8.2.4.2 *Case study 2*

In this section, we use the post-breakthrough production data determined from different layers of 10th SPE model (Christie and Blunt, 2001). The characteristics of the top 35 layers are different from those of the lower 50 layers (see Section 7.2 for more detail). In particular, Figure 8.18 shows the log–log plot of time to reach 50% water cut in the production well versus the breakthrough time t_{br} for all the top and lower layers and at well pair distance of 220 along the

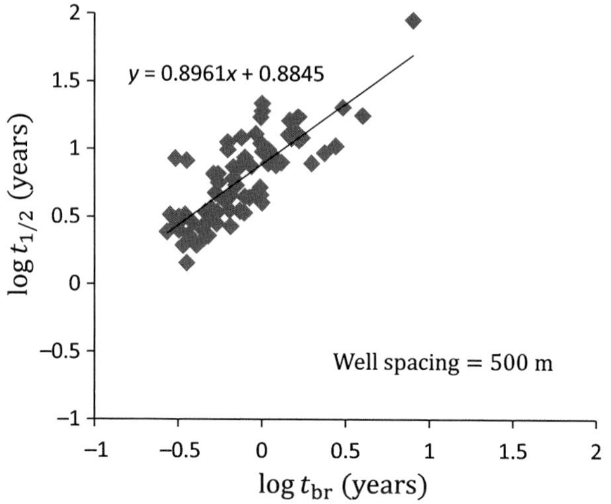

Figure 8.17. Log–log plot of time of reaching to 50% water cut in the production well versus breakthrough time for well pair distance of 500 m in the oil reservoir B.

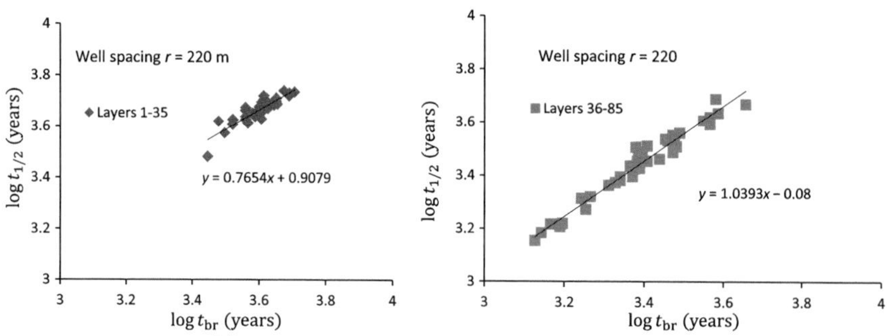

Figure 8.18. Log–log plot of time of reaching to 50% water cut in the production well versus breakthrough time for (a) all the top layers and (b) all the lower layers. Well spacing $r = 220$ m for both (a) and (b).

longer length of the model ($L_y = 220$). As can be seen in Figure 8.18, the exponent of the power law scaling (Eq. (8.18)) is slightly different for the top and lower layers indicating they have different structures. This requires further investigation.

8.3 Problems

8.1. Use the scaling of the breakthrough in homogeneous and inhomogeneous mediums in 2D to show how the breakthrough time will be changed if the well spacing increases by one order of magnitude ($r_2 = 10r_1$).

8.2. Consider a rectangular reservoir of dimensions 7,000, 2,500 and 150 m respectively in the x, y and z directions. This thin reservoir has a permeability of 7 d and a reservoir net-to-gross ratio of 58%. The sandbody sizes are 6,000, 300 and 15 m in the x, y and z directions respectively and have their long axis aligned with the long axis of the reservoir. The fluids are incompressible with viscosity of 1.7 cP and the flow is essentially 2D. Assume injection and production wells are either aligned along the x or y direction with injector pressure of 300 atm and producer pressure of 210 atm, wellbore radius of 0.05 m and well spacing of 1,100 m. Use the scaling of breakthrough probability distribution to determine the distribution breakthrough time between two wells in two directions.

8.3. Use the proposed scaling for the dependency of the breakthrough time to the viscosity ratio (μ_o/μ_w) to find how the breakthrough time between a well pair will be changed if a heavy oil with viscosity $\mu_o = 10{,}000$ cP is used instead of a light oil with viscosity $\mu_o = 2$ cP.

8.4. Use the scaling law between the breakthrough time t_{br} and the post-breakthrough time $t_{1/2}$ (the time taken for production to fall by 50%) to find how the time to reach 50% water cut in the production well changes if the breakthrough time is doubled.

References

Ahmed, T.H., *Reservoir Engineering Handbook, Second Edition*. Gulf Professional, London, 2001.

Andrade, J.S., Buldyrev, V., Dokholyan, V., Sergey, J.R., Havlin, S., King, P.R., Lee, Y., Paul, G. and Stanley, H.E., Flow between two sites on a percolation cluster. *Physical Review E*, 62: 8270–8281, 2000.

Andrade, J.S., Herrmann, H.J., Moreira, A.A. and Oliveira, C.L.N., Transport on exploding percolation clusters. *Physical Review E*, 83: 031133, 2011.

Araújo, A.D., Moreira, A.A., Makse, H.A., Stanley, H.E. and Andrade, J.S. Jr., Traveling length and minimal traveling time for flow through percolation networks with long-range spatial correlations. *Physical Review E*, 66: 046304, 2002.

Bear, J., *Dynamics of Fluids in Porous Media.* American Elsevier Publishing Company, New York, 1972.

Broadbent, S.R. and Hammersley, J.M., Percolation processes: I. Crystals and mazes. *Proceedings of the Cambridge Philosophical Society*, 53: 629–641, 1957.

Buckley, S.E. and Leverett, M.C., Mechanism of fluid displacement in sands. *Transactions of the AIME*, 146: 107–116, 1942.

Bunde, A., Ingram, M.D., Maass, P. and Ngai, K.L., Diffusion with memory: a model for mixed alkali effects in vitreous ionic conductors. *Journal of Physics A: Mathematical and General*, 24: L881–L886,1991.

Chandler, R., Koplik, J., Lerman, K. and Willemsen, J.F., Capillary displacement and percolation in porous media. *Journal of Fluid Mechanics*, 119: 249–267, 1982.

Chayes, J.T., Chayes, L. and Newman, C.M., The stochastic geometry of invasion percolation. *Communications in Mathematical Physics*, 101: 383–407, 1985.

Chen, J.-D. and Wilkinson, D., Pore-scale viscous fingering in porous media. *Physical Review Letters*, 55: 1892–1895, 1985.

Christie, M.A. and Blunt, M.J., Tenth SPE comparative solution project: a comparison of up scaling techniques. *SPE Reservoir Evaluation & Engineering*, 4: 308, 317, 2001.

Clement, E., Baudet, C., Guyon, E. and Hulin, J.P., Invasion front structure in a 3-D model porous medium under a hydrostatic pressure gradient. *Journal of Physics D: Applied Physics*, 20: 60S, 1987.

Combessis, A., Charvin, N., Allais, A., Fournier, J. and Flandin, L., Understanding dynamic percolation mechanisms in carbonaceous polymer nanocomposites through impedance spectroscopy: experiments and modeling. *Journal of Applied Physics*, 116: 034103, 2014.

Craft, B.C. and Hawkins, M.F., *Applied Petroleum Reservoir Engineering, Second Edition.* Prentice Hall, Upper Saddle River, NJ, 1991.

de Gennes, P.G., *Scaling Concepts in Polymer Physics.* Cornell University Press, Ithaca, 1979.

Deng, H., Lin, L., Ji, M., Zhang, S., Yang, M. and Fu, Q., Progress on the morphological control of conductive network in conductive polymer composites and the use as electroactive multifunctional materials. *Progress in Polymer Science*, 39(4): 627–655, 2014.

Dias, M.M. and Wilkinson, D., Percolation with trapping. *Journal of Physics A: Mathematical and General*, 19(15): 3131–3146, 1986.

Dokholyan, N.V., Lee, Y., Buldyrev, S.V., Havlin, S.V., King, P.R. and Stanley, H.E., Scaling of the distribution of shortest paths in percolation. *Journal of Statistical Physics*, 93: 603–613, 1998.

Druger, S., Nitzan, A. and Ratner, M.A., Dynamic bond percolation theory: a microscopic model for diffusion in dynamically disordered systems. I. Definition and one-dimensional case. *The Journal of Chemical Physics*, 79: 3133–3142, 1983.

Dullien, F.A.L., *Porous Media: Fluid Transport and Pore Structure, Second Edition*. Academic Press, London, 1992.

Ertekin, T., Abou-Kassem, J. and King, G., *Basic Applied Reservoir Simulation*. SPE Textbook Series. Society of Petroleum Engineers, Richardson, TX, 2001.

Grassberger, P., Directed percolation in 2+1 dimensions. *Journal of Physics A: Mathematical and General*, 22: 3673–3679, 1989.

Havlin, S., Nossal, R., Trus, B. and Weiss, G.H., Universal substructures of percolation clusters: the skeleton. *Journal of Physics A: Mathematical and General*, 17(18): L957–L960, 1984.

Havlin, S. and ben-Avraham, D., Diffusion in disordered media. *Advances in Physics*, 36: 695–798, 1987.

King, P.R., Muggeridge, A.H. and Price, W.G., Renormalization calculations of immiscible flow. *Transport in Porous Media*, 12: 237–260, 1993.

King, P.R., Buldyrev, S.V., Dokholyan, N.V., Havlin, S., Lee, Y., Paul, G. and Stanley, H.E., Applications of statistical physics to the oil industry: predicting oil recovery using percolation theory. *Physica A: Statistical Mechanics and its Applications*, 274: 60–66, 1999a.

King, P.R., Andrade, J.S., Buldyrev, S.V., Dokholyan, N., Lee, Y., Havlin, S. and Stanley, H.E., Predicting oil recovery using percolation. *Physica A: Statistical Mechanics and its Applications*, 266: 107–114, 1999b.

King, P.R., Buldyrev, S.V., Dokholyan, N.V., Havlin, S., Lee, Y., Paul, G., Stanley, H.E. and Vandesteeg, N., Predicting oil recovery using percolation theory. *Petroleum Geoscience*, 7: S101–S107, 2001.

King, P.R., Buldyrev, S.V., Dokholyan, N.V., Havlin, S., López, E., Paul, G. and Stanley, H.E., Uncertainty in oil production predicted by percolation theory. *Physica A: Statistical Mechanics and its Applications*, 306(1–4): 376–380, 2002a.

King, P.R., Buldyrev, S.V., Dokholyan, N.V., Havlin, S., López, E., Paul, G. and Stanley, H.E., Using percolation theory to predict oil field performance. *Physica A: Statistical Mechanics and its Applications*, 314: 103–108, 2002b.

Lee, Y., Andrade, J.S. Jr., Buldyrev, S.V., Dokholyan, N., Havlin, S., King, P.R., Paul, G. and Stanley, H.E., Traveling time and traveling length in critical percolation clusters. *Physical Review E*, 60(3): 3425, 1999.

Li, W., Jensen, J.L., Ayers, W.B., Hubbard, S.M. and Heidari, M.R., Comparison of interwell connectivity predictions using percolation, geometrical and Monte Carlo models. *Journal of Petroleum Science and Engineering*, 68: 180–186, 2009.

Lockwood, E.H., *A Book of Curves*. Cambridge University Press, Cambridge, UK, pp. 186–190, 1967.

López, E., Buldyrev, S.V., Dokholyan, N.V., Goldmakher, L., Havlin, S., King, P.R. and Stanley, H.E., Postbreakthrough behavior in flow through porous media. *Physical Review E*, 67: 056314, 2003.

López, E., Buldyrev, S.V., Braunstein, L.A., Havlin, S. and Stanley, H.E., Possible connection between the optimal path and flow in percolation clusters. *Physical Review E*, 72: 056131, 2005.

Mattax, C.C. and Dalton, R.L., *Reservoir Simulation.* Society of Petroleum Engineers, Richardson, TX, 1990.

Oliveira, C.L.N., Wittel, F.K., Andrade, J.S. and Herrmann, H.J., Invasion percolation with a hardening interface under gravity. *International Journal of Modern Physics C*, 21: 903, 2010.

Oliveira, C.L.N., Araújo, A.D., Lucena, L.S., Almeida, M.P. and Andrade, J.S. Jr., Post-breakthrough scaling in reservoir field simulation. *Physica A: Statistical Mechanics and its Applications*, 391: 3219–3226, 2012.

Røislien, J., Random Field Models and Near Well Reservoir Characterization. PhD thesis, Department of Petroleum Engineering and Applied Geophysics, Norwegian University of Science and Technology (NTNU), 2004.

Sadeghnejad, S., Masihi, M., Shojaei, A., Pishvaei, M. and King, P.R., Field scale characterization of geological formations using percolation theory. *Transport in Porous Media*, 92(2): 357–372, 2012.

Silva, L.R., Paul, G., Havlin, S., Baker, D.R. and Stanley, H.E., Scaling of cluster mass between two lines in 3d percolation. *Physica A: Statistical Mechanics and its Applications*, 318: 307–318, 2003.

Thiele, M.R., Blunt, M.J. and Orr, F.M. Jr., Modeling flow in heterogeneous media using streamtubes. I. Miscible and immiscible displacements. *In Situ*, 19: 299–339, 1995.

Wettstein, S.J., Wittel, F.K., Araújo, N.A.M., Lanyon, B. and Herrmann, H.J., From invasion percolation to flow in rock fracture networks. *Physica A: Statistical Mechanics and its Applications*, 391: 264–282, 2012.

Chapter 9

Percolation and pore-scale applications

So far, in this book, we have described how percolation theory can be applied to large-scale reservoir modelling. Percolation theory can also be applied at the pore scale (Sahimi, 1994; Selyakov and Kadet, 1996; Berkowitz and Ewing, 1998; Hunt and Ewing, 2005; Blunt, 2016). The study of immiscible flow in porous media at the microscale has been an important subject for reservoir engineering and hydrology communities for some time. Examples include water displacing oil during secondary recovery process or oil displacing water during migration and non-aqueous phase liquid migration in ground water (Dullien, 1992; Fetter, 1992; Vafai, 2005; Blunt, 2016).

This is usually done to get insights into the physics of flow and transport in porous media and how it changes under varying conditions, such as the addition of surfactants to alter interfacial tensions. It is also an important way of estimating macroscopic flow properties (such as relative permeability and capillary pressure) from samples of rock taken from a reservoir. This chapter summarises some of the main applications of percolation theory to the pore-scale flow. In particular, we focus on the use of this approach to determine the average dynamical properties. It should be emphasised that this is a brief summary of the main issues; the details can be found in the literature given above.

Fluid displacements in porous media are controlled simultaneously by several forces. At larger scales, viscous and gravity forces usually dominate (Bear, 1972; Wilkinson, 1984). However, at the pore scale, the dominant force is the interfacial tension between the phases. This is particularly true at the very low flow rates (tens of centimetres per day) typically found in oil reservoir and groundwater applications. The topology of the pore space, surface roughness and wettability of the rocks to the fluids are other important factors (Sahimi, 1994). The type of the displacement depends on two key dimensionless numbers:

(1) The capillary number Ca which is the ratio of the viscous pressure drop at the pore scale to the capillary pressure (Wilkinson, 1984; Meakin *et al.*, 2000; Méheust *et al.*, 2002; Løvoll *et al.*, 2005):

$$Ca = \frac{\Delta p_{\text{visc}}}{\Delta p_{\text{cap}}} = \frac{u\mu l/K}{\sigma \cos\theta/l}, \tag{9.1}$$

where u is the Darcy velocity, K is the permeability of the porous medium, μ is the fluid viscosity, l is a characteristic length (for example, the length of the tubes/throats or a representative radius of curvature of the interface), σ is the interfacial tension between two fluids and θ is the contact angle.

(2) The Bond number B which is the ratio of the hydrostatic pressure drop over a pore due to buoyancy (gravity) to the capillary pressure (Meakin *et al.*, 2000; Løvoll *et al.*, 2005):

$$B = \frac{\Delta p_{\text{grav}}}{\Delta p_{\text{cap}}} = \frac{\Delta\rho g l}{\sigma \cos\theta/l}, \tag{9.2}$$

where $\Delta\rho$ is the density difference between the two fluids and g is the acceleration due to gravity.

The detailed description of the multiphase displacements in porous media is very complex due to the variety of physical phenomena involved. For example, the flow behaviour of two immiscible fluids depends on the wetting properties of two fluids, their viscosity ratio, their respective densities, initial distribution of two phases and the displacement rate.

There are two main fluid displacement processes for immiscible displacement in porous media. Drainage is when the wetting phase (e.g., water) is displaced by the non-wetting-phase (e.g., oil). Imbibition is the opposite when the wetting phase displaces the non-wetting phase. Examples of drainage include migration of oil and gas from a source rock (which is vertically dominated because of buoyancy effects) to the current hydrocarbon accumulation, mercury injection, capillary pressure measurements and injection of CO_2 into a saline aquifer. Examples of imbibition include water flooding to improve oil recovery from reservoirs or the filtration of water into soil.

9.1 Modelling pore-scale fluid displacements

Before we look at pore-scale flow, we briefly review the conventional approach to modelling multiphase flow at the larger scale. This is not only the scale at which we average over many pores but also the scale below which the geological heterogeneities become apparent. This length scale is known as the representative elementary volume (REV) and is typically of the order of centimetres. This is a continuum representation which for single-phase flow is Darcy's law as discussed in Chapter 6. Although Darcy's law was first introduced for single-phase flow, it has been extended for multiphase flow by introducing relative permeabilities (Muskat, 1949). There are alternative approaches in the literature (e.g., Hassanizadeh and Gray (1979), Hilfer (2002) and Gielen *et al.* (2005)) but they are not commonly used in industry. The extended Darcy's law and the continuity equation for flow of an incompressible fluid becomes (Muskat, 1949; Bear, 1972; Peaceman, 1977)

$$u_i = -\frac{KK_{r_i}}{\mu_i}\nabla P_i, \tag{9.3}$$

$$\phi\frac{\partial S_i}{\partial t} + u_i\nabla S_i = 0, \tag{9.4}$$

where u_i, P_i, S_i, μ_i and K_{r_i} are respectively the velocity, pressure, saturation, viscosity and relative permeability of phase i, and ϕ and K are the porosity and absolute permeability of the porous medium. The relative permeability gives a measure of the increased "resistance" to flow of one fluid phase caused by the presence of the other fluid in the pore space. The relative permeabilities are non-linear functions of the pore occupancy (known as the fluid saturation). In addition, there is a difference between the phase pressures which is known as the capillary pressure. Again this is a function of the saturations ($P_c(S)$) and comes about because of the interfacial tension between the fluids.

These are the equations which are solved at the reservoir scale to make predictions of oil recovery rates, for example. To do this, information about the phase relative permeabilities is required. In principle, these can be obtained by measurement of samples from the reservoir, but in practice, it is costly to obtain a large number of samples. Also it is impractical to measure these properties for varying conditions, such as varying interfacial tension or contact angle during a surfactant flood.

An alternative is to make first principles estimates of these quantities by modelling the physics of the flow at the pore scale. This entails generating a model of the pore space and then solving the flow equations (typically Stokes flow) numerically. The details of how this is done are not relevant here and a thorough review is given by Blunt (2016). In this section, we describe the main concepts to show how percolation theory can be used to help understand pore-scale flow. We briefly review network modelling in the rest of this section.

The starting point is typically a high-resolution image obtained from micro- to nano-CT scanning (Flannery *et al.*, 1987; Spanne *et al.*, 1994; Fredrich, 1999; Ketcham and Carlson, 2001; Wildenschild *et al.*, 2002; Tomutsa and Silin, 2007; Dong and Blunt, 2009; Wildenschild and Sheppard, 2013; Blunt *et al.*, 2013). As an example, the 2D cross sections of 3D greyscale images of four carbonate rock samples (1) Ketton limestone, (2) Mount Gambier limestone, (3) Estaillades limestone and (4) Indiana limestone are shown in Figure 9.1 which illustrates different degrees of heterogeneity at the pore scale. The porosity and permeability of the samples shown in Figure 9.1 as reported by Bijeljic *et al.* (2013) are (1) Ketton limestone (14.9%, 8,476 md), (2) Mount Gambier limestone (55.6%,

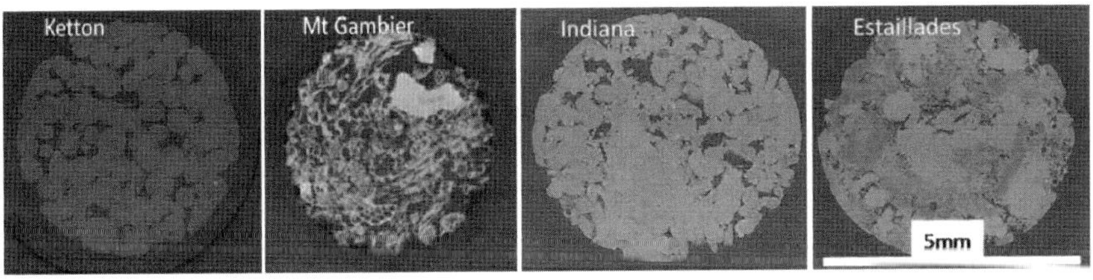

Figure 9.1. Illustration of 2D cross sections of 3D greyscale images of four carbonate rock samples: (1) Ketton limestone, (2) Mount Gambier limestone, (3) Estaillades limestone and (4) Indiana limestone (Bijeljic *et al.*, 2013).

16,607 md), (3) Estaillades limestone (13.3%, 328 md) and (4) Indiana limestone (11%, 292 md).

The pore-space images (after filtering and removing noise) are used to reconstruct an equivalent pore network (Dunsmuir *et al.*, 1991; Bakke and Øren, 1997; Biswal *et al.*, 2007; Guises *et al.*, 2009; Hilfer and Zauner, 2011). The detailed description of various methods for this reconstruction can be found elsewhere (Journel, 1989; Adler *et al.*, 1990; Coker *et al.*, 1996; Hazlett, 1997; Yeong and Torquato, 1998; Deutsch, 2002; Okabe and Blunt, 2004, 2005; Latief *et al.*, 2010; Caers, 2005).

To describe the pore-scale flow, we need a simplified geometrical representation of the pore space which preserves its main characteristics (porosity and permeability). This can be achieved by constructing a network of pores connected by throats with idealised geometries. An example of a sandstone pore-space image of 3 mm size and its equivalent network of pores and throats is shown in Figure 9.2. The extraction of this network is complex and readers are referred to the following references for details: Bryant *et al.* (1993), Thovert *et al.* (1993), Lindquist *et al.* (1996), Baldwin *et al.* (1996), Silin and Patzek (2006), Dong and Blunt (2009), Schluter *et al.* (2014) and Blunt (2016).

These equivalent networks are quite complex and the solution of the pore-scale flow numerically can be time consuming. This is required to make quantitative prediction of the macroscopic flow. The discussion of this is beyond this book; interested readers are referred to the recent book by Blunt (2016). However, to gain

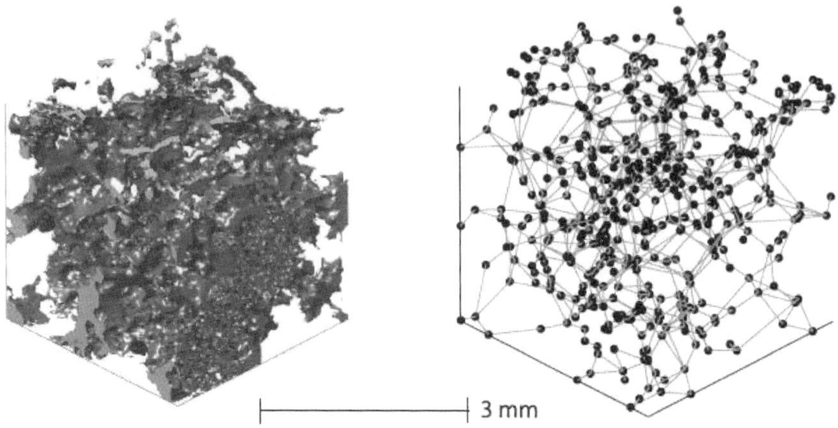

(a) (b)

Figure 9.2. Illustration of (a) the pore space of a 3 mm size Berea sandstone and (b) the equivalent topological network of pores connected by throats (Bakke and Øren, 1997; Valvatne and Blunt, 2004).

an insight into the physics of the flow, it is sufficient to examine simple networks and a rule-based model of the physics of the displacement. This is known as invasion percolation which we now describe.

9.2 Invasion percolation

Invasion percolation is a simple model of displacement based on the movement of interfaces through the medium. It can be used to estimate the physical properties such as capillary pressure and relative permeability curves (Heiba *et al.*, 1992; Mani and Mohanty, 1998). This model has widely been used in studying dynamics of flow displacements at pore scale (e.g., Sahimi (1994), Selyakov and Kadet (1996), Berkowitz and Ewing (1998), Hunt and Ewing (2005) and Blunt (2016)).

We consider the displacement of two immiscible fluids. This is best described for drainage where the non-wetting fluid displaces the wetting one. The interfacial pressure drop across the interface is given by the Young–Laplace equation. This equates to the pressure threshold required to move the interface into a pore throat of radius r,

$$P_c^* = \frac{2\sigma \cos\theta}{r}, \tag{9.5}$$

where σ is the interfacial tension, θ is the contact angle and r is the pore throat radius. Initially all the bonds are empty and then they are filled randomly one at a time in the order of lowest threshold pressure first. This implies that the pores are filled in order of the largest pore throat radius first. Only sites already connected to the injected phase are filled (Wilkinson and Willemsen, 1983; Lenormand and Bories, 1980; Chandler *et al.*, 1982). The reader is referred to Løvoll *et al.* (2005) and Toussaint *et al.* (2005, 2012) for various invasion scenarios.

Invasion percolation uses a sorted list of entry pressures (related to the bond radii) in order to choose the next bond to be filled. This list initially contains the bonds adjacent to the inlet face, and at each simulation step, it is updated by adding the entry pressures of the new bonds connected to the displacing phase which are not already on the list. These are then re-sorted to find the largest pore throat (lowest entry pressure). This defines the next bond to be occupied. Updating the entry pressure list is the most time-consuming part of the invasion-percolation simulation which entails efficient data algorithms (Sheppard *et al.*, 1999; Masson and Pride, 2014; Masson, 2016). At each step, the top-ranked bond is filled. By continuing the simulation of this process, we get a percolating cluster at breakthrough of the non-wetting fluid to the outlet face of the system. Figure 9.3 shows a schematic of invasion-percolation process in 2D where oil comes from the left boundary; the bonds will fill in the order $1, 2, \ldots$ where the thickness illustrates the bond radii.

Figure 9.4 shows the outcome of invasion percolation on a 200×200 lattice where the flow is from the left to the right side of the medium. It is compared with ordinary site percolation. It can be seen that invasion-percolation model just selects the percolation cluster. All the other clusters, which are disconnected from the main cluster, are not occupied.

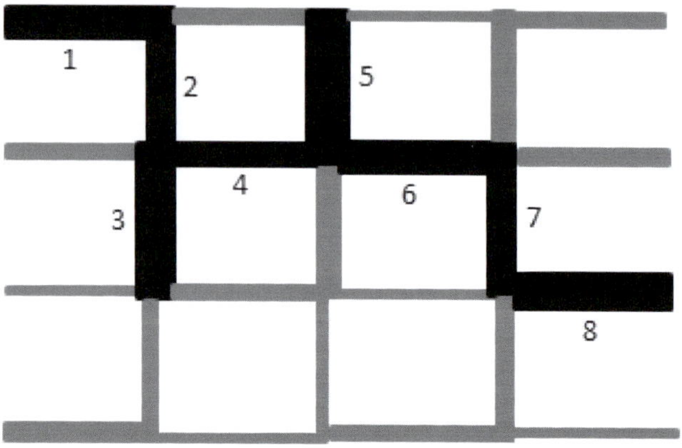

Figure 9.3. Illustration of percolation threshold (or breakthrough) of a network of bonds with varying radii under invasion-percolation process. The numbers represent the filling order of bonds.

(a)

(b)

Figure 9.4. Two realisations of (a) the structure of the invasion-percolation cluster at breakthrough and (b) percolation cluster (in blue) and other small clusters in ordinary (random) percolation.

As invasion percolation has widely been used for describing two-phase displacements, we give a more detailed analysis and scaling properties in Section 9.3.

In the rest of this section, we discuss imbibition which is more difficult. In principal, we could use the same argument of capillary-dominated flow, but now the Young–Laplace law would indicate that the smaller pores are filled first. The capillary pressure in imbibition is lower (compared with drainage) because the displacement processes (such as snap-off) need a lower threshold pressure in imbibition than piston-like advance in drainage. The competition between snap-off and pore filling in the displacement of the non-wetting phase by a wetting phase results in different flow patterns (Blunt, 2016). This is linked with trapping of the non-wetting phase. Trapping can occur by flow of the wetting phase through a connected layer in the thin edges of the pore space. The wetting phase swells from the edges disconnecting the non-wetting phase which is then trapped. If the disorder in the medium is not too high, then the invading fluid moves more uniformly, known as frontal advance. This is particularly true when snap off is reduced by contact angles close to 90° (moderately water-wet rocks). But in highly water-wet media, the wetting film plays a very strong role and cluster growth occurs (Chatzis and Morrow, 1981; Chandler *et al.*, 1982; Homsy, 1987; Lenormand *et al.*, 1988; Cieplak and Robbins, 1988; Blunt and Scher, 1995; Yortsos *et al.*, 1997; Alava *et al.*, 2004). The range of applicability of these four patterns depends on the degree of heterogeneity and the wettability. Figure 9.5 shows the schematic of the various flow regimes.

9.3 Invasion-percolation-type displacement

We have already described the basic idea of invasion-percolation model in Section 9.2. This is a type of dynamic percolation used for studying two-phase immiscible displacement in pore-scale porous media. Hence, in this section, we describe the percolation quantities of interest in invasion percolation and determine their scaling properties.

We shall first concentrate on simple invasion percolation (without trapping effects) on a network of bonds. As in ordinary percolation, the main quantities of interest are the number of the bonds which become filled by the oil and the volume fraction of the filled bonds. This is basically the saturation of the non-wetting phase S_{nw}. Assume

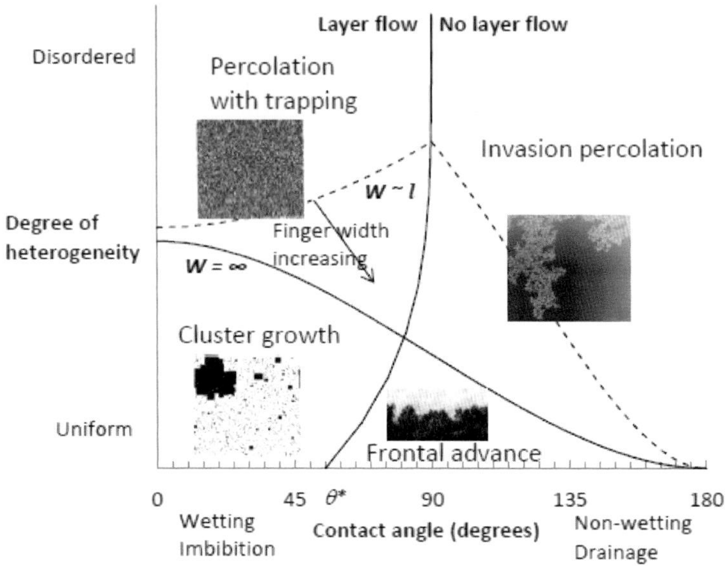

Figure 9.5. Illustration of the range of applicability of four flow patterns described by percolation with trapping, invasion percolation with trapping, frontal advance and cluster growth based on the degree of heterogeneity and the wettability (Blunt, 2016).

that the bond radii have a distribution $f(r)$ with bonds of size between r and $r+dr$ given by

$$f(r)dr. \tag{9.6}$$

Then the fraction of filled bonds p with radius larger than a critical radius r_c, i.e., $r \geq r_c$, can be calculated by

$$p = \frac{\int_{r_c}^{r_{\max}} f(r)dr}{\int_{r_{\min}}^{r_{\max}} f(r)dr}, \tag{9.7}$$

where $r_{\min}$ and $r_{\max}$ are the minimum and maximum bond radii in the network. As in ordinary percolation, there are two length scales involved in the percolation analysis. These are the bond length l and the correlation length ξ, which increases as the displacing fluid advances through the medium and reaches the order of system size $\xi \sim L$ at the threshold (which is the equivalent to breakthrough). Hence, as described in Chapter 2, the relevant scaling variable at the threshold is the ratio of these two lengths $(L/l)^d$. The number of injected bonds M (the cluster mass in percolation terminology) is a fractal property and has power law scaling

$$M \propto \left(\frac{L}{l}\right)^D, \tag{9.8}$$

with a fractal dimension of $D = 91/48 \approx 1.89$ and 2.52 in 2D and 3D respectively (Stauffer and Aharony, 1991) using the fractal dimensions determined for ordinary percolation discussed in Chapter 2 (see Table 2.3). As the total number of bonds in the system scales as $(L/l)^d$, where d is the spatial dimension, then using Eq. (9.8), the saturation of the non-wetting phase (the volume fraction filled by the injected fluid) scales as a power law,

$$S_{\mathrm{nw}} \propto \left(\frac{L}{l}\right)^{D-d}, \tag{9.9}$$

with $D - d = -0.11$ and -0.48 for 2D and 3D systems. For example, for a system with $L = 150$ and $l = 10$, Eq. (9.9) gives the non-wetting-phase saturation at breakthroughs of around 74% and 27% in 2D and 3D. It should be noted that for a very large system size ($L \gg 1$), the non-wetting-phase saturation tends to zero.

As in ordinary percolation (see Chapter 6), we can consider the backbone of the non-wetting phase (Herrmann *et al.*, 1984; Havlin and ben-Avraham, 1987; Grassberger, 1992) through which the main flow takes place. There are also dangling ends through which there is no flow. The number of bonds in the backbone M_b is fractal with a power law,

$$M_b \propto \left(\frac{L}{l}\right)^{D_b}. \tag{9.10}$$

Hence, by using Eq. (9.8), the fraction of filled bonds on the backbone scales as

$$\frac{M_b}{M} \propto \left(\frac{L}{l}\right)^{D_b - D}. \tag{9.11}$$

Moreover, the flow moves mainly in the shortest path (with fractal dimension $D_{\min}$) within the backbone from the inlet to the outlet. This is not a straight-line path on the backbone as discussed before (see Section 8.1.1) but a tortuous path, whose tortuosity τ scales as

$$\tau \propto \left(\frac{L}{l}\right)^{D_{\min}-1}. \tag{9.12}$$

There is also a sub-network of red bonds (these are the bonds which if removed cause the network to become disconnected, with $D - d = -0.11$ and -0.48 for 2D and 3D systems) that follows a power law scaling,

$$M_{\text{red}} \propto \left(\frac{L}{l}\right)^{D_{\text{red}}}, \tag{9.13}$$

where the exponent D_{red} is 3/4 and 1.14 in 2D and 3D respectively (Stauffer and Aharony, 1991). These are also important during imbibition where filling any red bond will disconnect the non-wetting phase. Moreover, while these scaling laws are the same for ordinary percolation and invasion percolation, the scaling exponents or functions may differ showing that the clusters are not strictly similar (Angel *et al.*, 2008). We will later see how these scaling laws can be used to determine the behaviour of capillary pressure and relative permeability functions for the microscale two-phase displacements.

9.3.1 *Trapping effects*

So far, we have considered simple invasion-percolation model without trapping effects which means that the displaced fluid (the wetting phase in drainage) is compressible and so every bond on the interface can be accessed by the injected fluid. Alternatively, we may consider invasion percolation with trapping. In this case, the displaced fluid is incompressible which results in isolated patches of the displaced fluid (the called residual saturation) as they are entirely surrounded by the injected fluid, and the injected fluid is not allowed to enter the trapped patches (Dias and Wilkinson, 1986; Bo *et al.*, 2006; Masson, 2016). This kind of bypass trapping is a large-scale phenomenon due to the connection of multiple bonds (or pores) and is different from trapping in a single pore where water cannot completely displace the oil (Wilkinson, 1984).

To simulate this process, we need to have a new rule that the invading fluid cannot invade the trapped regions of displaced fluid. This requires a global search after each step to check whether trapping has taken occurred. Typically cluster-labelling algorithms (Hoshen and Kopelman, 1976) have been used to identify the trapped clusters formed by the displaced fluid at each time step. However, more efficient algorithms based on the analysis of the neighbours of each newly invaded site to check for trapping are also possible (Babalievski, 1998; Masson, 2016).

As the process continues, more the trapped clusters grow in size. If we pursue the invasion process beyond the breakthrough, a time is reached when the displaced fluid consists of isolated clusters only and the trapped fluid cannot move anymore. At this point, the process stops and the volume fraction of trapped fluid is the residual saturation. This is about 0.34 in 3D which may be compared to the threshold of 0.312 in ordinary percolation (Wilkinson, 1984). In reality, gravity and viscous forces will affect this residual value. There are two key moments in the invasion-percolation process:

(1) breakthrough of injected fluid at the outlet face,
(2) the point where all remaining displaced fluid is trapped.

These are referred to as the two percolation thresholds of invasion percolation (Wilkinson, 1986). Due to the wettability preference of the medium, the first threshold is relevant for drainage and the second for imbibition.

Numerical studies show that a similar power law scaling to Eqs. (9.8) and (9.9) can be applied to the case of invasion with trapping. The fractal dimension of the injected cluster mass is $D = 1.82$ in 2D which is less than the $91/48$ ($\sim$1.896) of ordinary percolation with no significant difference in 3D (Lenormand and Zarcone, 1985). This is because the trapping rule prevents the injected fluid from filling some bonds which it would otherwise have filled giving a lower fractal dimension. However, the trapping rule

Table 9.1. Summary of the fractal dimensions of percolation sub-networks of various invasion-percolation processes, i.e., invasion percolation without trapping (IP), invasion percolation with trapping (TIP) (ben-Avraham and Havlin, 2000; Sheppared *et al.*, 1999; Stauffer and Aharony, 1991; Sahimi, 1994; Grassberger, 1997; Herrmann *et al.*, 1984).

Fractal dimension	Type of process	2D	3D
D_c	IP	1.8949	2.528
	Site TIP	1.825	2.528
	Bond TIP	1.825	2.528
D_B	IP	1.642	1.868
	Site TIP	1.217	1.861
	Bond TIP	1.217	1.458
$D_{\min}$	IP	1.129	1.3697
	Site TIP	1.214	1.3697
	Bond TIP	1.217	1.458
D_{red}	IP	0.75	1.141

has much less effect in 3D as the displacing fluid must form closed surfaces in order to disconnect the displaced fluid. So the volume fraction of the injected fluid follows the power law scaling (Eq. (9.9)) with exponent -0.18 in 2D. It should be emphasised that although the properties of invasion percolation such as the scaling laws are consistent with that of ordinary percolation, the spanning clusters are not precisely the same.

Similar comparisons between the properties of the percolation sub-networks in ordinary and invasion percolation can be considered. For example, the fractal dimension of the minimal path in 2D is about $D_{\min} = 1.22$ which is greater than 1.13 in ordinary percolation. The fractal dimensions of the percolating cluster and the other percolation sub-networks are summarised in Table 9.1.

So far, we have described invasion percolation for slow displacement governed by the heterogeneity of the capillary forces on the interface. However, fluid flow displacement through a porous medium is also affected by viscous pressure drops and gravity (Bear, 1972). The dimensionless numbers to characterise the effects of these forces are the capillary number Ca and Bond number B (Eqs. (9.1) and (9.2)) and the viscosity ratio μ_i/μ_d where i and d stand for the injected and displaced fluid respectively. We now turn to the impact of these other driving mechanisms.

9.3.2 *Invasion percolation in the presence of gravity*

Invasion percolation in the presence of gravity forces is also called invasion percolation with gradient (IPG) (Wilkinson, 1984, 1986; Birovljev *et al.*, 1991; Blunt *et al.*, 1992; Frette *et al.*, 1992; Meakin *et al.*, 2000; Masoum and Masihi, 2010). There are practical cases where the gravity forces, $(\rho_w - \rho_{\mathrm{nw}})g\Delta z$, acting on the density difference between the two fluids may be comparable to the capillary forces, such as in a relatively thick reservoir. The effect of gravity is to modify the locally dominant forces depending on the scale of the study. To illustrate this, we consider a slow flow displacement process (so we can neglect the viscous forces) on the same network as before but in the presence of the gravity. We consider this type of displacement in terms of the dimensionless Bond number B (Eq. (9.2)). Typically the Bond number is around 3×10^{-4} for the pore scale which means that the effect of the gravity is very low.

A simple modification to the horizontal experimental model, usually used for invasion process studies, is to vary the effective acceleration due to gravity by using an inclined surface and varying

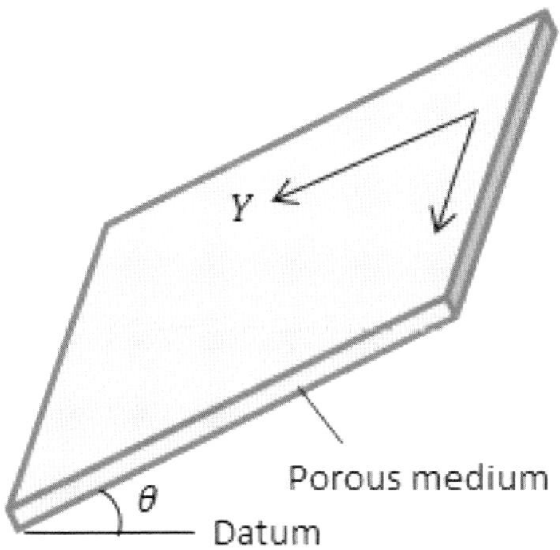

Figure 9.6. Sketch of the inclined porous medium used in Eq. (9.14) which shows the y coordinate and reference datum.

the angle of inclination (see Figure 9.6). For simulation purposes, we can account for gravity in the algorithm by replacing the threshold entry pressure (Eq. (9.5)) for invasion with

$$p(i) = \frac{A}{r_i} + B \cdot Y(i), \tag{9.14}$$

where $Y(i)$ is the y coordinate of the ith bond (or site) related to some datum (gravity is taken as acting in the y direction). The growth of the fluid–fluid displacement pattern can be described in terms of the propagation of the front into the porous medium. In this case, gravity can stabilise or destabilise the interface in the fluid displacement process, depending on the relative position of the denser fluid.

We first consider gravity-destabilised invasion percolation (for $B < 0$) which has been studied numerically and experimentally (Meakin *et al.*, 1992; Schmittbuhl *et al.*, 2000; Frette *et al.*, 1992; Auradou *et al.*, 1999; Masoum and Masihi, 2010; Bo *et al.*, 2006; Masson, 2016). Consider a low-density non-wetting fluid proceeding upwards through the medium filled initially with a denser wetting phase. The simulation results show that the interface progresses in a series of fingers of a finite width ξ (the root mean square of the front width in the direction of the flow) in the direction of gravity. These numerical results are consistent with experimental observations (Meakin *et al.*, 2000; Bo *et al.*, 2006). The pattern is described by a row of fractal blobs along the lower edge of the medium and a vertically oriented string of blobs that is a self-affine directed random walk (Figure 9.7). When the buoyancy forces increase, the probability for vertical displacement increases and the width of the fingers decreases.

In such a gradient-destabilised invasion-percolation process, the gradient prevents the sides of the finger from propagating too far and broadening the finger. So the active zone is confined to the highest blob in the string of blobs. The size of the blob ξ at the end of the finger (the characteristic length scale that is the width of finger and is perpendicular to the displacement) is related to the correlation length (Eq. (2.3)), which is also a measure of the size of finite clusters. The characteristic length ξ scales as (Wilkinson, 1984, 1986; Clement

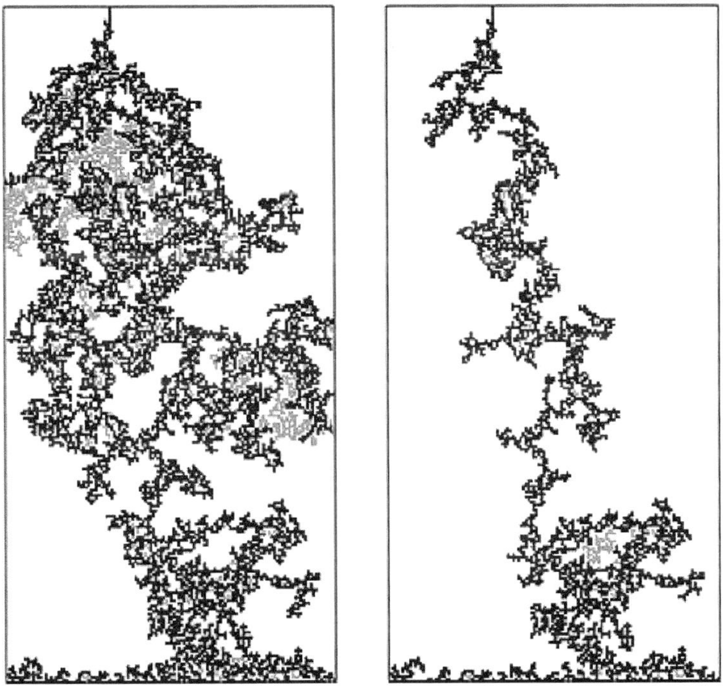

Figure 9.7. Two realisations of slow gravity-destabilised displacement of a wetting fluid by a non-wetting fluid in a site square lattice of size 128×256 with and without trapping and varying the influence of gravity: (a) $B = -0.0001$, (b) $B = -0.001$. The sites in grey show the trapped regions (Masson, 2016).

et al., 1987; Birovljev *et al.*, 1991; Meakin *et al.*, 1992, 2000; Frette *et al.*, 1992)

$$\frac{\xi}{l} \propto B^{-\nu/(1+\nu)}, \tag{9.15}$$

where ν is the correlation length exponent from Table 2.2 which gives the exponent -0.57 and -0.47 in 2D and 3D respectively. This power law emphasises that the relation of the width of the blobs and the gradient field is independent of the type of pore/throat size distribution (Meakin *et al.*, 1992). Figure 9.8 shows the finger width from simulations which indicates its power law dependency with the Bond number with exponent -0.58 in 2D.

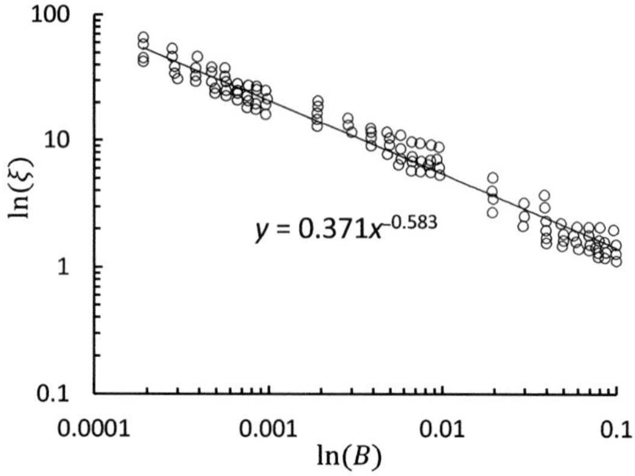

Figure 9.8. Variation of the width of fingers versus the Bond number on a log–log scale for 10 samples which shows a slope of −0.58.

This instability is further enhanced as the Bond number increases. This reduces the lateral width of the cluster ξ as shown in Figure 9.7. From Eq. (9.15), one order of magnitude increase in gravity reduces the finger width to $\xi_2 = 0.26\xi_1$ in 2D. Bo *et al.* (2006) showed that there is a competition between the field gradient and the width of pore/throat size distribution which may affect the correlation length scale and consequently results in either simple invasion percolation, gradient invasion percolation or pure vertical displacement.

As the displacement is like a chain of blobs with characteristic size (ξ), the number of filled bonds scales as $\xi^D(L/\xi)$; then at the scale of the whole medium, the saturation of the non-wetting phase (the volume fraction filled by the injected fluid) scales as (Frette *et al.*, 1992; Meakin *et al.*, 2000; Bo *et al.*, 2006)

$$S_{\mathrm{nw}} \propto \left(\frac{l}{L}\right) \cdot \left(\frac{1}{\lambda}B\right)^{-\frac{(D-1)\nu}{1+\nu}}. \tag{9.16}$$

Using the numerical values of fractal dimensions in Table 2.3 and critical exponents in Table 2.2, the exponent here becomes −0.47

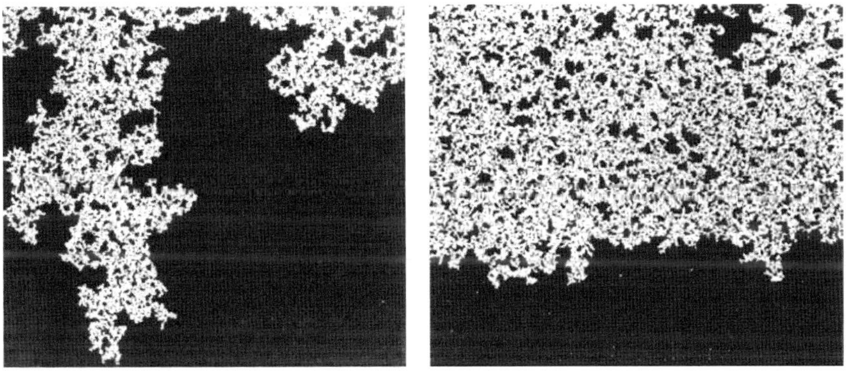

(a) (b)

Figure 9.9. Two patterns of slow gravity-stabilised displacement of a wetting fluid by a non-wetting one from the top end (a) without gravity in a horizontal cell and (b) with gravity (Meakin *et al.*, 2000).

and −0.71 in 2D and 3D respectively and $\lambda = P_c^{\max} - P_c^{\min}$ is the width of the capillary pressure distribution (or, alternatively, the pore throat distribution) (Bo *et al.*, 2006). This power law dependence of saturation on the Bond number has also been obtained from experiments (Birovljev *et al.*, 1991; Méheust *et al.*, 2002).

Alternatively we look at gravity-stabilised invasion percolation (for $B > 0$) (Wilkinson, 1984, 1986; Birovljev *et al.*, 1991; Méheust *et al.*, 2002). In this case, the denser phase rises and the active zone is stabilised by the effects of gravity. The displacement pattern can be described in terms of a compact packing of self-similar (fractal) blobs with characteristic length ξ which is the width of the advancing front. This characteristic length follows the same power law as before (Eq. (9.15)) (Wilkinson, 1984; Birovljev *et al.*, 1991; Meakin *et al.*, 1992; Méheust *et al.*, 2002). Note that in this regime, the active zone is the entire external perimeter and the effect of gravity is to prevent the trailing edge of the active zone from getting too far behind the leading edge; it decreases the front width (Figure 9.9). The trailing edge progresses by invading sites with thresholds that would be too large to occupy without the effect of gravity. There is also a shift in the residual saturation.

9.3.3 *Invasion percolation in the presence of viscous force*

The term viscous force loosely refers to the flow induced by the macroscopic pressure through Darcy's law as aopposed to flow driven by buoyancy of interfacial tension. To account for this, we use the dimensionless capillary number Ca (Eq. (9.1)). The typical values of the capillary number at the pore scale are in the range of 10^{-9} to 10^{-7} which shows that the viscous effects are small compared to the capillary effects. The mechanism is similar to that of gravity; the difference here is that the global pressure field rearranges as the large-scale interface moves.

We consider both stabilised and destabilised displacements (Saffman and Taylor, 1958). At low flow rates, the viscous pressure drop across the porous medium is negligible compared with the threshold capillary pressures inside the medium, so the regime is described by simple invasion percolation as before (Chandler *et al.*, 1982; Lenormand and Zarcone, 1985). At high flow rates, viscous forces dominate capillary forces which results in a fingering of the displacing fluid into the displaced fluid similar to diffusion-limited aggregation patterns (Paterson, 1984; Homsy, 1987; Chen and Wilkinson, 1985; Frette *et al.*, 1990).

Consider first the unstable displacement of a highly viscous wetting-phase fluid by a non-wetting fluid with low viscosity. The high viscosity contrast leads to a highly unstable displacement pattern with a rapid breakthrough of the displacing fluid into the displaced fluid called viscous fingering (Homsy, 1987; Frette *et al.*, 1990; Méheust *et al.*, 2002; Løvoll *et al.*, 2005). Experimentally, it is usually found that as the capillary number increases, the residual oil saturation decreases (Wilkinson, 1984).

The presence of a viscous pressure drop alters the local capillary pressure occupancy rules and causes finite width fingers conceptually similar to those discussed in the case of gravity. The capillary pressure is still governed by interfacial tension and the tube radii, but imposed on top of this is the pressure from the external boundary conditions which changes the local entry pressure required for a given pore. This is similar to the approach used with gravity. However, as pointed out by Meakin *et al.* (2000), the gravity is a body force that acts everywhere while the pressure field is modified by the fluid configuration and has the strongest effect at the tips of the fingers. To demonstrate this difference, Meakin *et al.* (2000) showed the experimental pattern of viscous displacements at various injection rates. At low capillary numbers, the patterns can be described in terms of a string of fractal blobs oriented along the flow direction similar to that in the gravity-destabilised invasion-percolation model (Figure 9.10(a)). At high capillary numbers, the pattern is more compact (Figure 9.10(b)).

From the definition of Darcy's law for multiphase flow (Dullien, 1992), and the capillary number (Eq. (9.1)), the gradient in capillary pressure caused by the non-wetting displacement in the flow direction (which we will label as x) is

$$\frac{dP_c}{dx} = -\frac{\sigma Ca_o}{K \cdot K_{r_o}}, \tag{9.17}$$

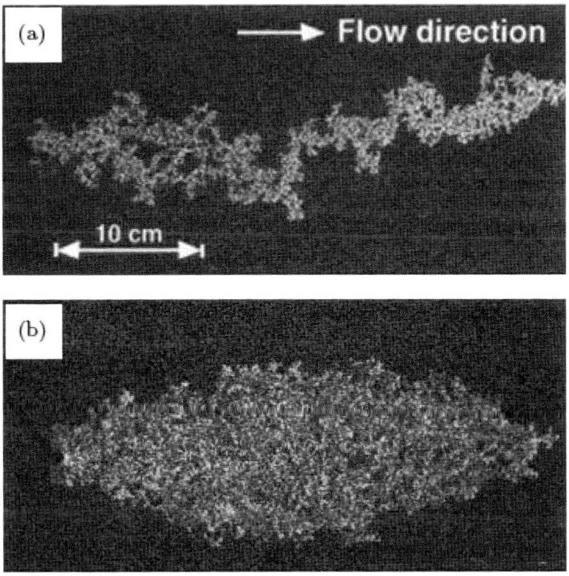

Figure 9.10. Two patterns of slow displacement of a dyed water/glycerin mixture by injecting air in a 2D horizontal porous medium where (a) $Ca = 0.0151$, (b) $Ca = 0.112$ (Meakin *et al.*, 2000).

where Ca_o is the capillary number of the non-wetting phase and the negative sign indicates that the capillary pressure decreases away from the inlet. We assume a model where the non-wetting flow rate (q_{nw}) is a fraction of the total flow rate (q_t) through $q_{\text{nw}} = S_{\text{nw}} q_t$ and the conductivity relation $K_{r_o} = S_{\text{nw}}^{\mu}$ (Eq. (2.7)). Then by using the J function ($J = \sigma\sqrt{K/\phi}$), Eq. (9.17) becomes

$$\frac{dP_c}{dx} = -\frac{Ca \cdot S_{\text{nw}}^{\mu}}{J \cdot \sqrt{K\phi}}, \tag{9.18}$$

where Ca is the capillary number based on total flow rate and ϕ is the rock porosity. For simulation purposes, we can account for this effect by replacing the threshold entry pressure (Eq. (9.5)) for invasion by

$$p(i) = \frac{A}{r_i} + Ca\,X(i), \tag{9.19}$$

where $X(i)$ is the x coordinate of the ith bond and the capillary number Ca is a gradient acting in the x direction. The first term in Eq. (9.19) is the capillary pressure term as it arises from the Young–Laplace law.

As in the case of gravity, the viscous forces result in a similar power law depending on the capillary number Ca. The correlation length ξ indicating the typical cluster size scales as (Wilkinson, 1986)

$$\frac{\xi}{l} \propto Ca^{-\nu/(1+\nu)}. \tag{9.20}$$

Moreover, the cluster mass of the displacement pattern or non-wetting saturation follows similar power law scaling (Meakin *et al.*, 2000):

$$S_{\text{nw}} \propto Ca^{-\frac{(D-1)\nu}{1+\nu}}. \tag{9.21}$$

As stated for the gravity case, there will also be a shift in the residual saturation at the end of the process due to the viscous forces which can again be described as a power law which will be demonstrated later.

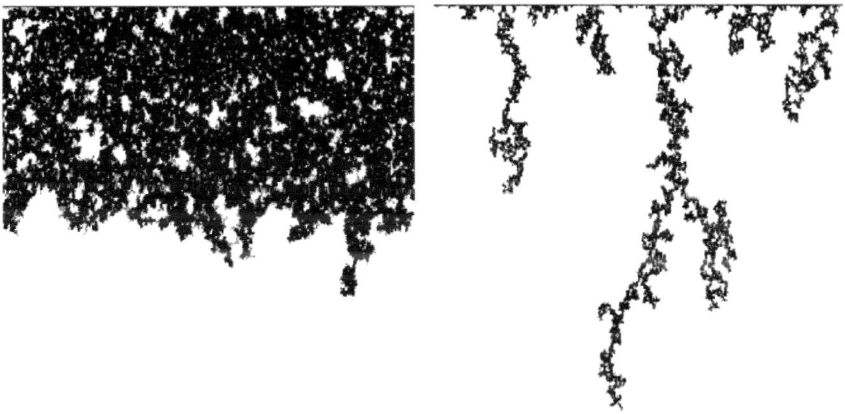

(a) (b)

Figure 9.11. Two experimentally observed displacement patterns obtained by injection of air at the top and pushing dyed mixture of glycerol and water: (a) slow displacement where the competition is between capillary and gravity forces and (b) viscous fingering under gravity (Løvoll *et al.*, 2005).

9.3.4 *Invasion percolation in the presence of both gravity and viscous forces*

Typically all three forces are present in real reservoirs, to a greater or lesser extent. Invasion percolation must be considered in the presence of all three driving forces. The combination of the viscous and gravitational effects can be described as follows. At high flow rates, where the capillary effects are negligible, the main competition is between the viscous forces and gravitational effects (Figure 9.11(a)). If the displaced fluid is more viscous, the displacement becomes unstable (viscous fingering) as shown in Figure 9.11(b); otherwise, when the displaced fluid has a lower viscosity, viscous effects stabilise the front (Méheust *et al.*, 2002; Løvoll *et al.*, 2005).

In the case where both viscous and gravitational forces contribute to the gradient in local capillary pressure, the correlation length scales as (Blunt and Scher, 1995)

$$\frac{\xi}{W} \propto \left[\frac{W}{J \cdot \sqrt{K\phi}} Ca + \frac{W}{l} B\right]^{-\nu/(1+\nu)}, \tag{9.22}$$

where W is the width of finite fingers. It is shown numerically and experimentally that the front width during drainage for both stable ($B > Ca$) and unstable ($B < Ca$) displacements scales as (Méheust *et al.*, 2002; Løvoll *et al.*, 2005)

$$\xi \propto |B - Ca|^{-\frac{\nu}{1+\nu}} , \tag{9.23}$$

with the exponent -0.57 in 2D.

9.4 Percolation and porous medium characterisation

Percolation concepts can be used to estimate macroscale flow properties from the pore-scale structure. Also, by using these properties, percolation can be used to determine pore-scale characteristics. In particular, it is used to infer the pore-size distribution and to determine the saturation-dependent functions such as relative permeabilities and capillary pressure.

It has long been known that capillary pressure and relative permeability are very important controls on reservoir recovery (Muskat, 1949). They enable us to coarse grain the pore-scale flow to develop continuum multiphase flow equations at larger scales (Muskat, 1949; Craft and Hawkins, 1959; Vafai, 2005; Sahimi, 2011). These saturation-dependent functions can be determined through displacement experiments or pore network modelling studies (Anderson, 1987; Blunt and King, 1991; Bryant and Blunt, 1992; Berkowitz and Ewing, 1998; Valvatne and Blunt, 2004; Piri and Blunt, 2005; Okasha *et al.*, 2007; Zhao *et al.*, 2010; Krevor *et al.*, 2012).

In this section, we first use percolation to determine the functional forms for the capillary pressure and relative permeability. We will also see how to analyse the drainage and imbibition processes to give good agreement with the experimental results (Wilkinson, 1984; Blunt *et al.*, 1992; Sahimi, 1993). We then show how network models and percolation can be combined in a statistical way to predict the saturation-dependent functions for these displacement processes. Finally we use the capillary pressure curves during drainage to give information about the pore-size distribution.

9.4.1 *Scaling functions for relative permeability and capillary pressure*

The scaling laws for invasion percolation provide a basis for developing appropriate scaling functions for macroscopic properties such as the capillary pressure and phase relative permeability functions.

We first consider drainage and assume that the displacement continues until breakthrough occurs. As the trapping is negligible in 3D, the spanning cluster is similar to that in ordinary percolation except there is just one cluster. In the case of slow displacement where the dominant force is capillary near the critical region, the dimensionless capillary pressure P_{cD} can be approximated by the increment in the non-wetting-phase fraction $P_{cD} \approx \Delta p = p - p_c$. Then around the percolation threshold, the non-wetting-phase saturation is the mass of the spanning cluster (Eq. (2.1)) and so scales as

$$S_{\text{nw}} \propto (p - p_c)^{\beta}. \tag{9.24}$$

From this, the scaling law for the capillary pressure at breakthrough becomes (Wilkinson, 1986; Sahimi, 1994)

$$P_{cD} \approx \Delta p \propto S_{\text{nw}}^{1/\beta}, \tag{9.25}$$

where the exponent $1/\beta$ is 2.44 in 3D using the critical connectivity exponent β given in Table 2.2. This capillary pressure function for drainage is shown in Figure 9.12(a). This indicates a sharp increase in the capillary pressure as the non-wetting saturation increases.

To develop a scaling relation for the phase relative permeability in drainage, we ignore the effect of the wetting phase in the corners of pores (as discussed by Blunt (2016). This allows us to use the power law relation for the conductance of the medium (Eq. (2.7)). The non-wetting-phase relative permeability $K_{r_{\text{nw}}}$ scales as,

$$K_{r_{\text{nw}}} \propto (\Delta p)^{\mu}. \tag{9.26}$$

This power law scaling can be combined with saturation equation (Eq. (9.24)) to obtain the saturation dependency of the

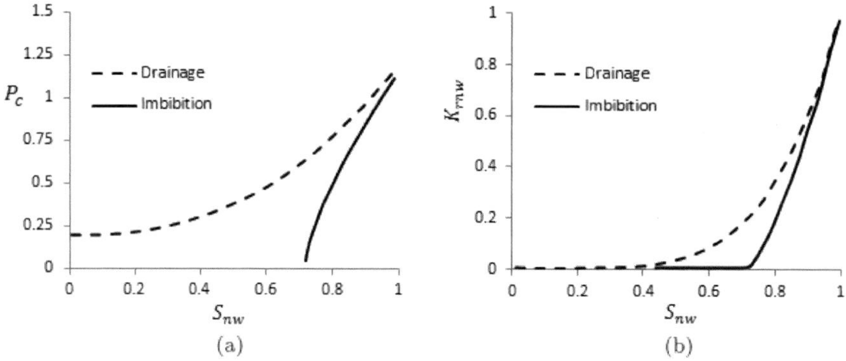

Figure 9.12. Illustration of (a) dimensionless capillary pressure function based on Eqs. (9.25) and (9.29) and (b) relative permeability of non-wetting phase from Eqs. (9.27) and (9.30).

non-wetting-phase relative permeability (Wilkinson, 1986; Blunt *et al.*, 1992),

$$K_{r_{\text{nw}}} \propto S_{\text{nw}}^{\mu/\beta}, \tag{9.27}$$

where the exponent μ/β is 4.87 in 3D using the exponent values from Table 2.2. This non-wetting-phase relative permeability for drainage is shown in Figure 9.12(b). Moreover, the relative permeability is zero at the beginning of the drainage process. There is a capillary entry pressure, P_c^{entry} for the non-wetting-phase to invade the largest throat at the inlet boundary with the throat radius $r_1 = 2\sigma\cos\theta/P_c^{\text{entry}}$ from Eq. (9.5).

We now consider imbibition where the wetting phase is injected and displaces the non-wetting phase. Assume that the displacement continues until the wetting phase forms a continuous pathway the the outlet. This is the second threshold (as described in Section 9.3) at which only small finite clusters of wetting phase exist. This is the residual non-wetting-phase saturation $S^*_{\text{nw}_r}$. Near this point, the change in the non-wetting-phase saturation by increasing the occupancy from a given p to the critical point p^* can be found from the integral $\int_{p^*}^{p} P(p)dp$. Then using the power law (Eq. (2.1)), it is easy to show that the change in the non-wetting-phase saturation scales as

$$\Delta S_{\mathrm{nw}} = S_{\mathrm{nw}} - S^*_{\mathrm{nw}_r} \propto (\Delta p)^{1+\beta}. \tag{9.28}$$

Hence, the scaling law for the capillary pressure at the second threshold becomes (Wilkinson, 1986; Sahimi, 1994)

$$\Delta p_c \approx \Delta p \propto \Delta S_{\mathrm{nw}}^{\frac{1}{1+\beta}}, \tag{9.29}$$

where the exponent $1/(1+\beta)$ is 0.71 in 3D. This is shown in Figure 9.12(a). This indicates again a sharp increase in the capillary pressure as the non-wetting saturation increases.

Similarly we can develop a scaling relation for the imbibition relative permeability. Using the power law relation for the conductance of the percolation medium (Eq. (2.7)) in conjunction with the saturation scaling equation (Eq. (9.29)) gives the non-wetting-phase relative permeability $K_{r_{\mathrm{nw}}}$ (Wilkinson, 1986; Blunt *et al.*, 1992),

$$K_{r_{\mathrm{nw}}} \propto (\Delta S_{\mathrm{nw}})^{\mu/(1+\beta)}, \tag{9.30}$$

where the exponent $\mu/(1+\beta)$ is 1.42 in 3D (Figure 9.12(b)).

In practice, it is necessary to incorporate other pore-scale details to get more physical predictions from percolation, especially for imbibition (Lenormand and Zarcone, 1984). For example, Blunt *et al.* (1992) suggested adding two pore-scale trapping mechanisms into the percolation model to account for the flow of the wetting phase along small crevices in the pore edges. They used finite-size scaling for the permeability of the non-wetting phase, $K_{r_{\mathrm{nw}}} \propto L^{-\mu/\nu}$, from percolation theory (Section 2.4), where the cluster size is of order of system size L. As shown by the power law (Eq. (9.30)), the relative permeability vanishes due to trapping after breakthrough. This is not realistic as there must be some wetting-phase flow in the small-scale irregularities (Blunt *et al.*, 1992).

9.4.2 *Percolation and residual-phase saturation*

Another important parameter in estimating recovery efficiency is the residual oil saturation at the end of imbibition. This is the amount of oil that has been trapped which cannot be recovered by water flooding alone.

Network models show that the amount of trapping at the end of displacement depends on the details of the pore-space connectivity (Ewing and Gupta, 1993; Sok *et al.*, 2002). Percolation has been used to estimate the distribution of clusters of the non-wetting phase trapped at the end of imbibition. These results are supported by experimental observations (Chatzis *et al.*, 1983; Mayer and Miller, 1993; Blunt and Scher, 1995; Iglauer *et al.*, 2010; Krummel *et al.*, 2013; Andrew *et al.*, 2014). For example, Ewing and Gupta (1993) examined the effects of lattice size and shape on the degree of residual non-wetting-phase saturation. Blunt and Scher (1995) investigated the wetting behaviour at the pore scale in a numerical model which describes competing flow in the centres of the pore spaces and along crevices. They compared the residual saturation calculated from their proposed numerical method with the prediction from scaling laws based on percolation theory. The distribution of trapped phases in water-wet and oil-wet (or mixed-wet) systems has also been investigated by using pore-scale imaging (Prodanovic *et al.*, 2007; Al-Raoush, 2009; Iglauer *et al.*, 2010; Herring *et al.*, 2013). A typical numerical value for the residual saturation of a sandstone rock sample is around one-third of the pore space ($S_{\mathrm{nw}_r} \cong 0.33$) as reported by Andrew *et al.* (2014). Their estimate was obtained by image analysis on the observed residual non-wetting-phase clusters of different sizes, from the scale of a single pore to multipore blobs.

Consider a cubic water-wet network of size L on which we study ordinary percolation by assuming that the non-wetting phase occupies a fraction p_{nw} of the bonds. We start the imbibition from a high value of p_{nw} as the initial state and decrease its value until the threshold p_{nw_c} occurs. This is when the non-wetting phase first becomes disconnected. Hence, as a first estimate, we use this threshold value as the non-wetting-phase residual saturation, i.e., $S_{\mathrm{nw}_r} \cong p_{\mathrm{nw}_c}$ (Larson *et al.*, 1981). However, the nature of two-phase displacement is more complex which results in different non-wetting-phase saturations depending on the pore–throat size distributions, topology of the medium, wettability preference and the flow pattern. For example, Blunt (2016) showed that the amount of the non-wetting-phase saturation is controlled by snap-off rather than piston-like advance and cooperative pore filling. Also the non-wetting-phase saturation increases with lower contact angles with wetting layer flow, poorly connected pore spaces and with a large difference between the pore and throat size distributions.

To present some quantitative analysis, we look at the configuration of the system just before the threshold where there is a distribution of finite-size clusters of trapped non-wetting phase. Then the non-wetting-phase clusters follow a power law size distribution $n(s) \propto s^{-\tau}$ with $n(s)$ as the number of clusters containing s pores (Chapter 2) and the Fisher exponent is $\tau = 2.19$ in 3D (Blunt *et al.*, 1992; Lorenz and Ziff, 1998; Tiggemann, 2001). This kind of power law distribution for the non-wetting-phase cluster size was also observed experimentally by using pore-scale image analysis (Georgiadis *et al.*, 2013).

The relevant length scales in the medium that control the cluster sizes are the characteristic pore size, l, the finger width of the wetting phase, W, and the correlation length, ξ, which represents the typical size of largest cluster ($\sim L$ at breakthrough). As the total number of pores in the network is $n = (L/l)^d$, the residual saturation can be estimated by integration over the non-wetting-phase clusters normalised by the total number of pores $S_{\mathrm{nw}} \cong (l/L)^d \int_0^\infty sn(s)ds$. However, to perform the integration, we should consider the lower and upper cut-offs for the cluster size distribution. Blunt (2016) described the lower (s_L) and upper (s_U) cut-offs for the cluster size by the respective power laws $s_L \sim (W/l)^d$ and $s_U \sim (W/l)^d(\xi/W)^D$ (with d the dimensionality of the network and D the fractal dimension of the spanning cluster, given in Table 2.3). Considering these cut-offs in the integration give the scaling law for residual non-wetting saturation for a large network as (Blunt and Scher, 1995; Blunt, 2016)

$$S_{\mathrm{nw}_r} \propto (W/l)^{-d(\tau-2)}, \tag{9.31}$$

with exponent $d(\tau - 2) \approx 0.57$ in 3D. This relation shows that the residual saturation depends not only on the pore-size distribution l

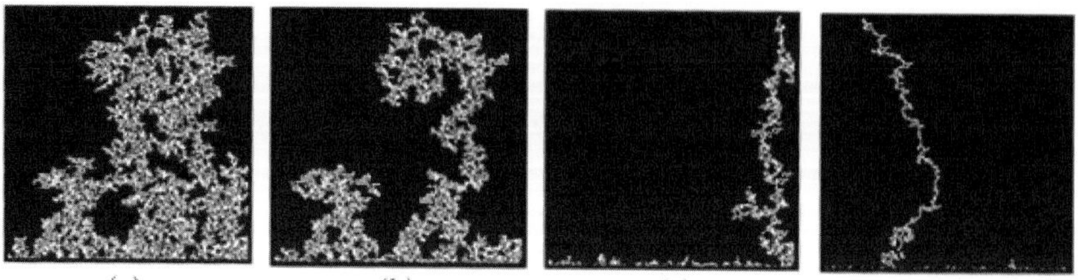

Figure 9.13. Four realisations of gradient-destabilised displacement of a wetting fluid by a non-wetting fluid on a square lattice of size 200 × 200 where Bond numbers are (a) $B = 0$, (b) $B = -0.0001$, (c) $B = -0.001$ and (d) $B = -0.01$.

but also on the finger width W. The finger width can be calculated numerically from the image by image filtering to separate the wetting from the non-wetting phase and then extracting the invasion front. Figure 9.13 shows four finger structures at the front with different widths; when the density gradient is increased, the structure is dominated by a single branch and the width of this branch decreases.

For example, a smaller residual saturation is predicted for the larger values of the finger width. It should be recalled that the finger size is controlled by the displacement pattern as shown in Figure 9.5. So the residual non-wetting saturation is also a function of the flow pattern. This scaling relation has been validated by Blunt (2016) who used the detailed image analysis of five rock samples and measured the distribution of trapped clusters of the non-wetting phase. For example, for a sandstone sample of typical pore size $l = 80\,\mu$m, finger width $W = 0.5\,$mm, Eq. (9.31) gives $S_{\mathrm{nw}} \approx 35\%$. This power law scaling is valid for water-wet porous media. However, as discussed by Blunt (2016), the distribution of non-wetting clusters in oil-wet networks is different. In this case, the oil resides in layers in the corners of pores which results in smaller and flatter clusters of trapped oil and less trapping. For example, $S_{\mathrm{nw}} \cong 19\%$ in oil-wet systems compared with 35% in water-wet systems.

The initial configuration of two phases in the porous medium can also affect the efficiency of the displacement. In particular, the initial wetting-phase saturation, S_{w_i}, at the beginning of the imbibition has an effect on the residual non-wetting-phase saturation S_{nw_r}. This effect can be investigated experimentally by stopping the primary drainage at some intermediate saturation and then reversing the displacement by performing a water flooding. It is observed that the amount of the residual oil saturations increases with the initial saturation of non-wetting phase S_{nw_i} and is less than the initial non-wetting-phase saturation, $S_{\mathrm{nw}_r} < S_{\mathrm{nw}_i}$ (Al-Mansoori *et al.*, 2010; Pentland *et al.*, 2011; Herring *et al.*, 2013). There are several empirical formulae to relate the amount of non-wetting-phase trapping to the initial saturation of non-wetting phase S_{nw_i}; among them is the Land empirical functional form $S_{\mathrm{nw}_r} = \frac{S_{\mathrm{nw}_i}}{1+cS_{\mathrm{nw}_i}/(1-S_{wc})}$ with fitting parameter c (Land, 1968). The dependency of the residual non-wetting phase, S_{nw_r}, on the initial saturation, S_{nw_i}, can also be inferred from percolation concepts. Blunt (2016) considered a coordination number z_{nw_i} for the non-wetting phase at the initial state of imbibition. This is lower than the coordination number of the whole system z as there is also the initial wetting-phase saturation in the medium. We can then estimate the percolation threshold p_{c_i} corresponding to this new coordination number by using $p_c \approx \frac{d}{z(d-1)}$ (Selyakov and Kadet, 1996), and again as the first estimate, we set this threshold as the residual saturation $S_{\mathrm{nw}_r} \cong p_{c_i}$.

To describe the effect of the initial wetting saturation qualitatively, we assume the non-wetting phase to be the spanning cluster at the beginning of imbibition. The spatial statistics of the pore space controls the configuration of the spanning cluster and the amount of pore space occupied by the non-wetting phase that is directly related to the initial wetting phase (i.e., $S_{w_i} = 1 - S_{\mathrm{nw}_i}$). Two extreme cases are either very low (below the percolation threshold) or very high (above the percolation threshold) non-wetting saturations at the beginning of imbibition process. First consider the low non-wetting saturation state in which the dangling ends are a significant fraction of the network. These exist along with some red links as described in Chapter 6. In this case, as the imbibition progresses the dangling ends are nibbled away by the filling process and there is a high possibility of filling a red link by the wetting phase which disconnects the non-wetting-phase cluster (Blunt, 2016). This results in trapping

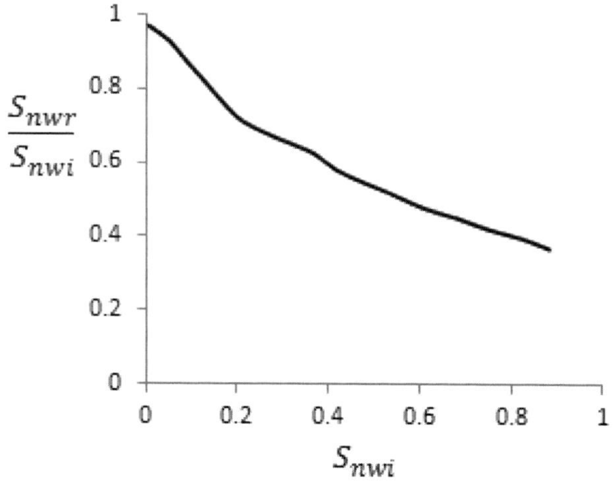

Figure 9.14. Illustration of the typical behaviour of the ratio $S_{\mathrm{nw}_r}/S_{\mathrm{nw}_i}$ against the initial non-wetting-phase saturation S_{nw_i}.

of most of the initial non-wetting-phase saturation during imbibition so that $S_{\mathrm{nw}_r}/S_{\mathrm{nw}_i} \approx 1$. On the other hand, at a high occupancy fraction, the dangling ends are a small part of the network with no red links. In this case, there are no such red links to be broken by the spanning cluster and so there will be a sequence of filling based on the size distribution during imbibition which results in displacing most of the initial non-wetting-phase saturation. Hence, the ratio $S_{\mathrm{nw}_r}/S_{\mathrm{nw}_i}$ decreases as the initial non-wetting-phase saturation S_{nw_i} increases. The trend depends on the heterogeneity of the medium (see Figure 9.14 for typical behaviour).

So far, the description of the residual non-wetting saturation has been for slow displacements on a very large network where the dominant force is capillary pressure. In practice, the effects of finite-size boundaries, gravity (characterised by the Bond number) and the viscous force (characterised by the capillary number) should be considered. The change in the residual non-wetting saturation of a finite-size system compared to that in an infinite porous medium depends on the increment in the occupied fraction Δp due to the change in cluster size around the threshold. The change in the residual non-wetting saturation may be calculated by multiplying the saturation in clusters spanning a region with a typical size of the correlation length ξ and the increment in the occupied fraction Δp (as given by Eq. (9.29) which can be rewritten in terms of the correlation length (with scaling law $\Delta p \propto \xi^{-1/\nu}$) (Blunt *et al.*, 1992),

$$\frac{\Delta S_{\mathrm{nw}_r}}{S^{\infty}_{\mathrm{nw}_r}} \propto \xi^{-(1+\beta)/\nu}, \tag{9.32}$$

where $S^{\infty}_{\mathrm{nw}_r}$ refers to the residual non-wetting saturation of an infinite system and the numerical value of the exponents $(1 + \beta)/\nu = 1.6$ in 3D. This scaling equation shows that a 10% increase in the correlation length (which is the characteristic size of the fingers) decreases the trapped saturation by about 30%. As discussed previously, the main effects of gravity and viscous forces in an invasion-type displacement is to increase the residual non-wetting-phase saturation which is considered as the first threshold in the drainage process and the second threshold in the imbibition process. The shift in the residual non-wetting-phase saturation depends on the magnitude of the Bond and capillary numbers. In the presence of both gravity and viscous forces, the relevant dimensionless number as suggested by Blunt (2016) is the effective capillary pressure $Ca_{\mathrm{eff}} = (B + \frac{l \cdot Ca}{\sqrt{k\phi}})$ with a pore-scale transition to viscous-dominated flow at $Ca_{\mathrm{eff}} \approx 1$. It is shown that the scaling for the shift in the residual non-wetting-phase saturation follows the power law (Wilkinson, 1986; Blunt *et al.*, 1992),

$$\frac{\Delta S_{\mathrm{nw}_r}}{S^{\infty}_{\mathrm{nw}_r}} \propto X^{\frac{1+\beta}{1+\nu}}, \tag{9.33}$$

where X stands for the Bond number, the capillary number or the effective capillary pressure Ca_{eff} and the numerical value of the exponent is $(1 + \beta)/(1 + \nu) = 0.75$. As an example for a typical Bond number 10^{-3} to 10^{-4}, this scaling equation predicts a decrease in critical water saturation of about 0.5–1%, compared with displacements with fluids of a similar density. The change in the residual non-wetting-phase saturation due to the presence of viscous and gravity effects should be considered in developing the relative permeability and capillary pressure functions (Figure 9.12), as in this case, the assumption of the slow displacements is no longer valid.

9.4.3 *Percolation and network models*

In this section, we use the combination of network models with percolation theory in a statistical framework to develop appropriate functions for the capillary pressure and relative permeabilities (Melrose and Brandner, 1974; Larson *et al.*, 1981; Diaz *et al.*, 1987; Heiba *et al.*, 1992; Sahimi, 1993; Selyakov and Kadet, 1996; Berkowitz and Ewing, 1998; Kadet and Galechyan, 2013). We describe the methodology by presenting few simple conceptual models. We start with the simplest capillary bundle model and then extend to a more general cubic network of sites (pores) and bonds (throats).

9.4.3.1 *Capillary bundle model*

We represent the pores by a bundle of parallel capillary tubes of the same length L and radii between $r_{\min}$ and $r_{\max}$ with probability denity function $f(r)$ (Burdine, 1953). This basic model neglects the fundamental topological property of porous media and is structurally anisotropic. The tubes in this model are only filled with one phase at a time with the largest filled first by the non-wetting phase and smallest filled first by the wetting phase. At each step, tubes with the minimum radius $r_{\min}$ to a radius r are filled with the wetting phase and tubes with radius r to the maximum $r_{\max}$ are filled with the non-wetting phase. Hence, the phase saturations can be calculated by

$$S_{\text{nw}}(r) = \frac{\int_r^{r_{\max}} r^2 f(r)dr}{\int_{r_{\min}}^{r_{\max}} r^2 f(r)dr}, \quad S_w = 1 - S_{\text{nw}}. \tag{9.34}$$

For slow displacement, Eq. (9.5) shows that the capillary dependence varies inversely with the radius r ($P_c \propto 1/r$) which can be used with Eq. (9.34) to obtain the capillary pressure curve.

To determine the relative permeabilities, we assume steady-state flow of an incompressible fluid under a fixed pressure gradient (∇P) along the tube. We use Poiseuille's law for each tube

$\left(Q = -\frac{\pi r^4}{8\mu}\nabla P\right)$; then the flow rate of wetting and non-wetting phases can be calculated by

$$Q_{\mathrm{nw}} = -\frac{\pi \nabla P}{8\mu_{\mathrm{nw}}} \int_{r}^{r_{\max}} r^4 f(r)dr, \tag{9.35a}$$

$$Q_w = -\frac{\pi \nabla P}{8\mu_w} \int_{r_{\min}}^{r} r^4 f(r)dr. \tag{9.35b}$$

Finally, we compare the flow rate of each phase (Eq. (9.35)) with the flow rate when all the tubes are filled with only one phase:

$$K_{r_{\mathrm{nw}}}(r) = \frac{\int_{r}^{r_{\max}} r^4 f(r)dr}{\int_{r_{\min}}^{r_{\max}} r^4 f(r)dr}, \quad K_{r_w} = 1 - K_{r_{\mathrm{nw}}}. \tag{9.36}$$

To illustrate this, we assume the throat radii follow the distribution, $f(r) \propto re^{-r/\lambda}$ where λ is a characteristic radius equivalent to the half average radius (from $r = 2\lambda$). Define a dimensionless radius $R = r/\lambda$. The fraction of tubes (p) that are allowed for the non-wetting phase with radius greater than a given R is

$$p = \frac{\int_{R}^{\infty} Re^{-R}dR}{\int_{0}^{\infty} Re^{-R}dR} = e^{-R}(1+R). \tag{9.37}$$

Inserting the throat radius distribution $f(r) \propto re^{-r/\lambda}$ into Eq. (9.34) and integrating gives the non-wetting-phase saturation,

$$S_{\mathrm{nw}}(R) = \frac{\int_{R}^{\infty} R^3 e^{-R}dR}{\int_{0}^{\infty} R^3 e^{-R}dR} = e^{-R}\left[1 + \sum_{n=1}^{3} \frac{R^n}{n!}\right]. \tag{9.38}$$

The capillary bundle model is essentially 1D, so there is no percolation threshold. To mimic the effect of a more complex topology that admits a percolation transition, we consider a simple cubic lattice in 3D. Then the critical threshold for the non-wetting phase is $p_c = 0.25$ and for the wetting phase $1 - p_c = 0.75$. These critical occupancies are equivalent to two critical dimensionless radii $R_c = 2.69$ and $R_c = 0.96$ (from Eq. (9.37)) which give the critical saturations of the wetting phase and non-wetting phases as

$$S_w(R_c = 0.96) = 0.017, \tag{9.39a}$$

$$S_{\rm nw}(R_c = 2.69) = 0.71. \tag{9.39b}$$

There is a considerable difference between the critical saturations estimated by this very simplified model and the numerical values determined from analysis of the pore-scale images or calculated from the detailed network models which consider the wettability conditions (Blunt, 2016). Alternative network models aim to relate the residual saturations to the interconnectivity in pore space or pore throat size and in agreement with the experimental results (Jerauld and Salter, 1990; Mayer and Miller, 1993).

We use a reference capillary pressure corresponding to an average tube radius $\langle r\rangle$ $\left(P_c^{\rm ref} = 2\sigma\cos\theta/\langle r\rangle\right)$ to define the dimensionless capillary pressure,

$$P_c' = P_c/P_c^{\rm ref} = 2/R. \tag{9.40}$$

Now we have determined both capillary pressure (Eq. (9.40)) and saturation (Eq. (9.38)). A typical plot of the dimensionless capillary pressure against saturation is shown in Figure 9.15(a).

Using the throat radii distribution $(f(r) \propto re^{-r/\lambda})$ in Eq. (9.36) and integrating gives the non-wetting-phase relative permeability,

$$K_{r_{\rm nw}}(R) = \frac{\int_R^\infty R^5 e^{-R} dR}{\int_0^\infty R^5 e^{-R} dR} = e^{-R}\left[1 + \sum_{n=1}^{5} \frac{R^n}{n!}\right]. \tag{9.41}$$

These are plotted in Figure 9.15(b).

Although this model gives interesting analytical expressions for the capillary pressure and relative permeability functions, it does not consider the wettability preference or the connectivity in the model which results in $K_{r_w} + K_{r_{\rm nw}} = 1$. Moreover, the results shown in Figure 9.15 indicate a convex curve for the non-wetting-phase relative permeability which is not supported by the capillary-controlled displacement experiments. This approach was useful in that it established a framework for later semi-analytical expressions

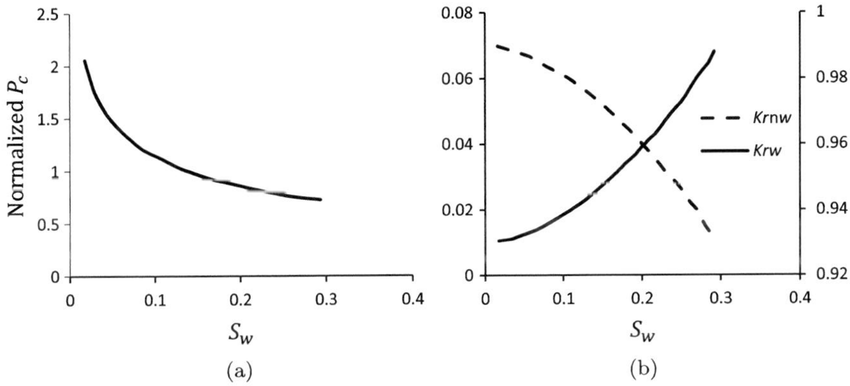

Figure 9.15. Schematic behaviour of (a) the dimensionless capillary pressure curve (by using Eqs. (9.38) and (9.40)) and (b) the phase relative permeability functions (by using Eqs. (9.38) and (9.41)) between the two critical saturation values $(0.017, 0.29)$.

for capillary pressure and relative permeability (Mualem, 1976; van Genuchten and Nielsen, 1985). For example, the above analysis uses cylindrical throats of radius r with the volume $V(r) \propto r^2$ and Poiseuille's law with the conductance $g(r) \propto r^4$. These could be generalised with power law relationships for the volume and conductance, with the powers (x and y) used as fitting parameters (Heiba *et al.*, 1992),

$$\begin{cases} V(r) \propto r^x & \text{for} \quad 0 \leq x \leq 2, \\ g(r) \propto r^y & \text{for} \quad 3 \leq y \leq 4. \end{cases} \tag{9.42}$$

The other issue with this model is the topology which can be handled with network models.

9.4.3.2 *Network models*

Consider a large 3D network model representing pore spaces of the medium with the throats connecting the pore bodies. The sizes of throats are often smaller than the pores themselves and both follow different distributions of sizes. On this network, we use a percolation approach to study the slow displacement (neglecting viscous and buoyancy forces) to determine the capillary pressure and relative permeability. In this section, we introduce the concepts with some example cases.

In the first example, assume a strongly water-wet medium and consider that the pore bodies do not play a role in flow except in the calculation of phase saturation (Heiba *et al.*, 1992; Sahimi, 1994). Two important percolation quantities of the network are the occupied fraction (denoted by X^o) and the accessible fraction to the flow (shown by X^A). This latter is the fraction of occupied sites belonging to the spanning cluster. These two elements play the same role as the occupancy probability p and the connected fraction P respectively in ordinary bond percolation. The key element of the network in this statistical approach is the distribution of throat radii $f(r)$. The other ingredients of the model such as the pore body radii, pore volume and conductance are assumed to be a function of the throat radii r. To implement this statistical method, it is necessary to specify the statistics of the occupied and accessible pore fractions at each stage. This requires a simplification of the physics of the displacement.

First consider drainage where the non-wetting phase invades those throats with radius $r > r_{\text{min}}$. The fraction of throats available to the injected fluid is (Sahimi, 1994)

$$p_{\text{nw}}(r_{\text{min}}) = \int_{r_{\text{min}}}^{\infty} f(r)dr. \tag{9.43}$$

Then the occupied fraction is assumed to be $X^o_{\text{nw}} = X^A(p_{\text{nw}})$. The distribution of throat radii of the occupied $f^o_{\text{nw}}(r)$ and accessible $f^A_{\text{nw}}(r)$ fractions becomes (Heiba *et al.*, 1992)

$$f^o_{\text{nw}}(r) = f^A_{\text{nw}}(r) = \begin{cases} \dfrac{f(r)}{p_{\text{nw}}} & \text{for} \quad r \geq r_{\text{min}}, \\ 0 & \text{for} \quad r < r_{\text{min}}. \end{cases} \tag{9.44}$$

Drainage continues until the allowed fraction of the non-wetting phase at the end of primary drainage becomes $p_{\text{nw},t}$.

In imbibition the non-wetting phase is displaced from the pores with $r < r_o$ by the wetting phase. The pore fraction available to the non-wetting phase becomes (Heiba *et al.*, 1992)

$$p_{\mathrm{nw}}(r_o) = \int_{r_{\min,t}}^{r_o} f(r)dr - \frac{X^A(p_{w,t})}{p_w}\int_{r_{\min,t}}^{r_o} f(r)dr, \tag{9.45}$$

where the subscript t refers to the end of injection. In this process, the distribution of throat radii of the occupied $f_w^o(r)$ and accessible $f_w^A(r)$ parts becomes (Heiba *et al.*, 1992)

$$f_{\mathrm{nw}}^o(r) = f_{\mathrm{nw}}^A(r) = \begin{cases} \dfrac{f(r)}{p_{\mathrm{nw}}} & \text{for} \quad r > r_o, \\ \dfrac{f(r)}{p_{\mathrm{nw}}}\left[1 - \dfrac{X^A(p_{w,t})}{p_w}\right] & \text{for} \quad r_{\min,t} < r < r_0. \end{cases} \tag{9.46}$$

Other expressions can be obtained for the wetting phase during drainage and imbibition (Heiba *et al.*, 1992). The saturation of the non-wetting phase (S_{nw}) is calculated by

$$S_{\mathrm{nw}} = X_{\mathrm{nw}}^o \frac{\int_0^\infty f_{\mathrm{nw}}^o(r)V(r)dr}{\int_0^\infty f(r)V(r)dr}. \tag{9.47}$$

Then the difference between wetting and non-wetting occupation (from Eq. (9.47) is the capillary pressure which can be calculated by using Eq. (9.43). Similarly, the relative permeability can be determined by comparing the phase conductance of the non-wetting phase $G_{\mathrm{nw}}(g)$ and single-phase conductance $G(g)$ defined as follows (Heiba *et al.*, 1992):

$$G_{\mathrm{nw}}(g) = (1 - p_{\mathrm{nw}})\delta(g) + p_{\mathrm{nw}} f_{\mathrm{nw}}^A(r)|dr/dg|, \tag{9.48a}$$

$$G(g) = f(r)|dr/dg|. \tag{9.48b}$$

The second example for developing an analytical expression for capillary pressure and relative permeability functions in both drainage and imbibition was suggested by Selyakov and Kadet (1996). Their derivation is based on the assumption that the backbone of the percolation cluster which carries the flow is an irregular network with correlation length R made by a number of parallel capillary chains with no interflow between them (Figure 9.16).

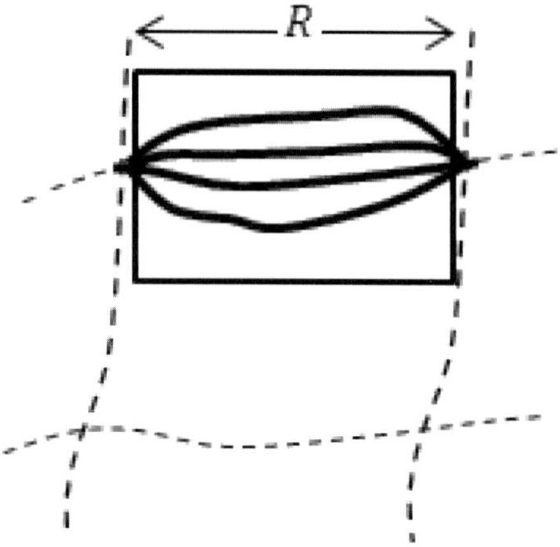

Figure 9.16. Schematic of the backbone structure of an infinite percolation cluster with correlation length R made by parallel capillary chains shown in the rectangle.

Then we can use the flow equation through capillaries (Eq. (9.35)) along with the scaling law of the correlation length (Eq. (2.3)). This approach gives the two-phase relative permeabilities as (Selyakov and Kadet, 1996; Kadet and Galechyan, 2013)

$$K_{ro}(r_k) = \frac{1}{K}\int_{r_k}^{r_{c1}} \frac{\left(\int_{r_1}^{r_{c1}} f(r)dr\right)^{\nu} f(r_1)}{\frac{8\int_{r_1}^{\infty}\frac{f(r)}{r^4}dr}{\pi\int_{r_1}^{\infty} f(r)dr}} dr_1, \tag{9.49a}$$

$$K_{rw}(r_k) = \frac{1}{K}\int_{0}^{r_{c2}} \frac{\left(\int_{r_1}^{r_{c2}} f(r)dr\right)^{\nu} f(r_1)}{\frac{8\int_{r_1}^{\infty}\frac{f(r)}{r^4}dr}{\pi\int_{r_1}^{\infty} f(r)dr}} dr_1, \tag{9.49b}$$

where ν is the correlation length exponent, r_k is the minimum radius of a capillary from which the wetting phase can be displaced under capillary pressure, r_{c1} and r_{c2} are obtained from $\int_{r_{c1}}^{\infty} f(r)dr = p_c$ for the non-wetting phase and $\int_{r_{c2}}^{r_k} f(r)dr = p_c$ for the wetting phase. The absolute permeability K is obtained by assuming that $r_k = 0$ in Eq. (9.49) (Selyakov and Kadet, 1996; Kadet and

Galechyan, 2013):

$$K = \int_0^{r_{c1}} \frac{\left(\int_{r_1}^{r_{c1}} f(r)dr\right)^{\nu} f(r_1)}{\frac{8\int_{r_1}^{\infty} \frac{f(r)}{r^4}dr}{\pi \int_{r_1}^{\infty} f(r)dr}} dr_1. \tag{9.50}$$

In this model, the radius distribution for the capillaries filled by the non-wetting phase is $f(r)$ with $r \geq r_k$ and by the wetting phase is $f(r)$ with $\leq r_k$. Also, the capillary pressure and wetting saturation are

$$\Delta p(r_k) = 2\sigma \cos\theta / r_k, \tag{9.51}$$

$$S_w(r_k) = \int_0^{r_k} f(r)dr. \tag{9.52}$$

Given a radius distribution for the capillaries, Eqs. (9.49)–(9.52) give the capillary pressure and phase relative permeabilities. Similar analytical expressions can be found for imbibition and can be extended to modelling hysteresis in relative permeability (Kadet and Galechyan, 2013). Take a simple illustrative example in which the capillaries have the radius distribution $f(r) = \frac{r_{\min} \cdot r_{\max}}{r_{\max} - r_{\min}} r^{-2}$ with $r_{\min}$ and $r_{\max}$ as the minimum and maximum capillary radii (Selyakov and Kadet, 1996). Then from Eq. (9.52), the linear water saturation can be obtained $S_w(r_k) = \frac{r_k - r_{\min}}{r_{\max} - r_{\min}}$. Then from Eq. (9.51), the capillary pressure scales as $P_c \propto \frac{1}{r_{\min} + S_w(r_{\max} - r_{\min})}$. Also, the phase relative permeabilities can be obtained from Eq. (9.49).

9.4.4 *Percolation and pore-structure characterisation*

The characterisation of the pore spaces is of prime importance as it strongly affects the flow and transport within the system. The morphology of a pore space is described by its pore-size distribution, its connectivity and its surface roughness. The connectivity of the pores, which affects the flow behaviour, and pore-size distribution, which controls the conductivity of the porous medium, may be analysed from pore-scale images obtained from micro-CT scanning of rock samples or through the interpretation of two-phase flow or transport experiments (Zhou and Stenby, 1993; Spanne *et al.*, 1994; Dong and Blunt, 2009).

We have already described pore-scale imaging in Section 9.1. This can be used to determine the pore-size distribution. In this section, we concentrate on one specific two-phase displacement process to infer the pore-size distribution. There are various techniques for evaluating the pore-size distribution of a porous medium such as mercury porosimetry or flow permporometry which uses displacement of a wetting fluid in the pore space by a non-wetting phase. Also adsorption–desorption isotherm experiments (Cao *et al.*, 1993; Lowell *et al.*, 2004; Mourhatch *et al.*, 2011) can be used. In mercury porosimetry, the injected fluid is mercury, while in flow permporometry, a non-condensable gas is blown through a liquid-filled porous medium. In the adsorption–desorption test, the injected fluid (such as nitrogen) is adsorbed as a film (its thickness depending on the injection pressure) on the pore walls. Desorption occurs by decreasing the pressure. As will be discussed later, the pore throat sizes control the drainage process in the mercury injection and desorption process in the adsorption–desorption test. The percolation concepts discussed in previous section can be used to analyse the results of these experiments in order to find the pore-size distribution (Chatzis and Dullien, 1977; Wall and Brown, 1981; Seaton, 1991; Sahimi, 1993). In this section, we concentrate on the interpretation of the results of mercury porosimetry tests.

The mercury porosimetry test is a two-phase slow displacement process. It is conceptually a simple drainage experiment by which a rock sample is first cleaned, dried and then placed in a vacuum at zero pressure. Mercury is then injected. Mercury has a higher interfacial tension with the solid than the vacuum, so it has a contact angle greater than 90° and acts as a non-wetting phase. If we assume a slow displacement, then we can neglect any pressure changes in mercury. The external pressure imposed on mercury is similar to the capillary pressure across a meniscus. Hence, this process is controlled by the threshold entry pressures for invasion of the throat with the smallest radius and a local maximum capillary pressure.

The test is continued by increasing the external pressure on mercury gradually. Consequently, mercury invades the accessible pore space in order from the largest to the narrowest throats. This is identical to invasion percolation. Since the displaced fluid is vacuum, there is no trapping. The volume of mercury that enters the rock sample is recorded as a function of the imposed external pressure which is then analysed to get a reasonable pore-size distribution (Washburn, 1921; Chatzis and Dullien, 1985). The details of how to determine the pore-size distribution from an example mercury injection test will be described later. Typical behaviour for the volume of injected mercury against the external pressure is shown in Figure 9.17. The two points P_1 and P_2 shown in Figure 9.17 correspond to respectively the point when mercury first enters the porous medium and the point when the intruded mercury breaks through (first forms a connected spanning cluster) (Chatzis and Dullien, 1977; Thompson *et al.*, 1987b). The detailed information on the pore structure can be extracted from the network simulations (Toledo *et al.*, 1989).

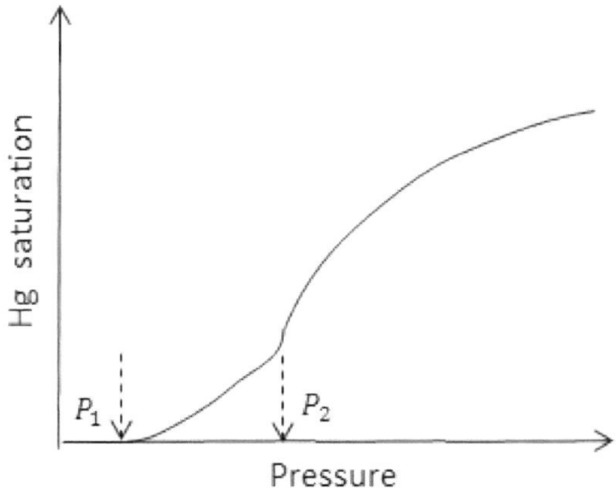

Figure 9.17. Illustration of typical behaviour for the volume of injected mercury against the external pressure in mercury injection test.

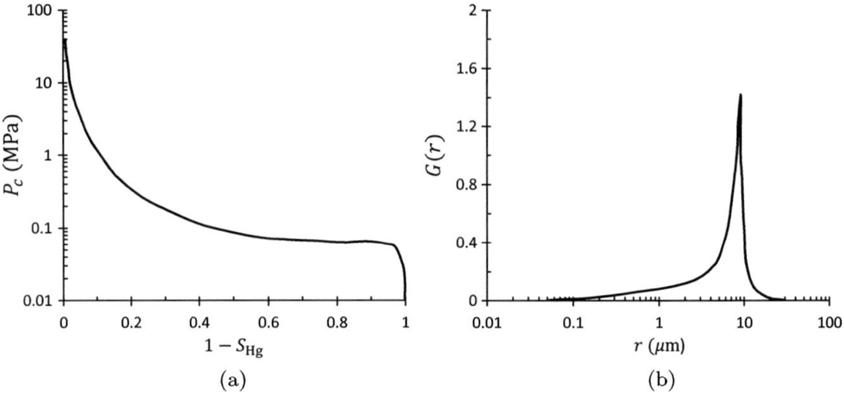

Figure 9.18. A schematic of (a) the capillary pressure versus saturation curve and (b) throat-size distribution from mercury injection test.

Moreover, the pressure can be lowered back to the initial state which causes the mercury to retract from the pores. In this process, the pore body radius controls the capillary pressure. A schematic of the capillary pressure curve from mercury injection test during drainage is shown in Figure 9.18.

We can use the capillary pressure measurements to interpret the throat radius. Assuming an interfacial tension of $\sigma = 485\,\mathrm{mN/m}$ for mercury and contact angle of $\theta = 45°$ as measured through the wetting phase, from Eq. (9.5), we get $P_c = 686/r$ which can be used with the capillary pressure measurements to interpret the throat radius. In Figure 9.18, the entry capillary pressure is around 0.06 MPa which corresponds to a radius of $12\,\mu\mathrm{m}$, and the capillary pressure at which 50% (and 90%) of the wetting phase is displaced is around 0.1 MPa (and 1 MPa), corresponding to the throat radii of $7\,\mu\mathrm{m}$ (and $0.7\,\mu\mathrm{m}$).

These capillary pressure measurements are routinely used for estimating the throat-size distribution. We assume the pore space to be represented by a bundle of capillary tubes of the same length L and radii between $r_{\min}$ and $r_{\max}$ with probability density function $f(r)$. As the pressure increases gradually, the tubes only get filled with mercury one at a time with the largest filled first according to Eq. (9.5). Hence, the saturation of wetting phase becomes

$$S_w = 1 - S_{\text{Hg}} = \frac{\int_{r_{\min}}^{r} r^2 f(r)dr}{\int_{r_{\min}}^{r_{\max}} r^2 f(r)dr}. \tag{9.53}$$

Then the inferred throat-size distribution is estimated from the derivative of saturation with respect to the logarithm of capillary pressure (Blunt, 2016):

$$G(r) = \frac{r^3 f(r)}{\int_{r_{\min}}^{r_{\max}} r^2 f(r)dr} = \frac{-dS_w}{d\ln P_c}. \tag{9.54}$$

A typical throat-size distribution based on the capillary pressure curve in Figure 9.18(a) is shown in Figure 9.18(b).

However, the real pore space is not as simple as a bundle of capillary tubes. This analysis neglected the shapes of the pores and most importantly the connectivity of the throats (Larson and Morrow, 1981; Thompson *et al.*, 1987a; Tsakiroglou and Payatakes, 2000). Invasion percolation is used to account for this (Rigby *et al.*, 2002; Sahimi, 1994; Mourhatch *et al.*, 2011). First a pore-size distribution $f(r)$ is assumed. Then the saturation function from Eq. (9.47) is iterated with the experimental data until it is satisfied (Sahimi, 1994). To illustrate the dependency of the inferred curve on the sample type, the throat-size distribution inferred in a commercial laboratory for four previously shown carbonate rock samples in Figure 9.1 is compared in Figure 9.19.

As discussed before, just above the percolation threshold (assumed to be $p_c = 0.4$), a small change in capillary pressure results in a very large shift in saturation (directly related to the number density of the filled throats) as estimated from Eq. (9.24). For example, for $\Delta p = 0.03$, this gives $S_{\text{nw}} = 0.03^{0.41} = 23\%$. This is 13% more than the typical non-wetting saturation at the threshold (a typical saturation at the threshold is around 10% from Blunt (2016)). This rapid filling of the mercury near the percolation threshold helps us to infer the throat-size distribution as it indicates the critical throat radius necessary to span the medium. For example,

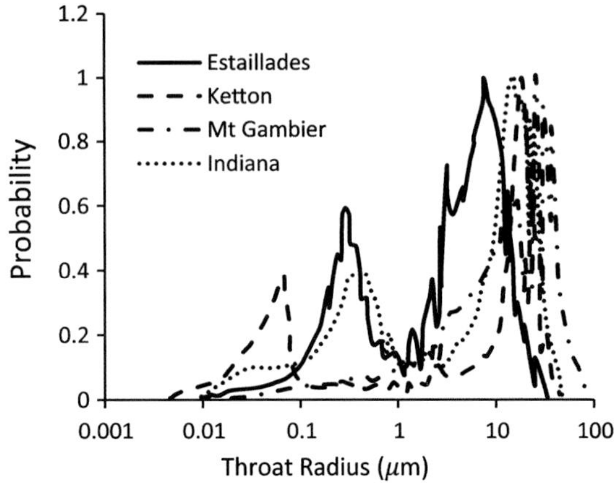

Figure 9.19. Comparison of the throat-radius distributions inferred from the MICP measurements for four carbonate rock samples: (1) Ketton limestone, (2) Mount Gambier limestone, (3) Estaillades limestone and (4) Indiana limestone (data from Bijeljic *et al.*, 2013).

from Figure 9.18, we can find the critical throat radius necessary to span the medium is $r_c = 10\,\mu\text{m}$.

9.5 Diffusion and dispersion

Diffusion and dispersion in porous media are both important for many practical problems. Examples include contaminant transport in aquifers, mixing of fresh and salty water in ground water and also in miscible recovery in hydrocarbon reservoirs (such as miscible solvents, enriched gas, high-pressure lean gas process), CO_2 storage and sequestration and so on (Bear, 1972; Lake, 1989; Gelhar *et al.*, 1992; Sahimi, 1994; Berkowitz and Ewing, 1998; Green and Willhite, 1998; Boustani and Maini, 2001; Jahangiri and Zhang, 2011; Bijeljic *et al.*, 2013). The analysis of diffusion and dispersion is more complicated in heterogeneous carbonate formations which hold about half of the world's conventional oil reserves (Chilingar *et al.*, 1972; Ahlbrandt *et al.*, 2005).

In this section, we consider diffusion and dispersion as two simple transport processes in porous media and see how percolation concepts can be used to describe them. The conventional description for the pure diffusion is Fick's law (Bird *et al.*, 1960; Perkins and Johnston, 1963). This relates the flux of diffusing particles to the gradient of their concentration ($v = -D\nabla C$). Along with the mass conservation law $\left(\frac{\partial C}{\partial t} = \nabla \cdot v\right)$, it leads to the standard diffusion equation,

$$\frac{\partial C}{\partial \mathrm{t}} = D \cdot \nabla^2 C, \tag{9.55}$$

where D is the molecular diffusion coefficient and C is the solute concentration.

If, in addition, there is flow in the medium, there is additional mixing of two fluids due to variations in the velocity field; this is called dispersion. At small scales in porous media, the dispersion is largely controlled by the variation in flow velocities and paths as the fluid moves through the pore space. At larger scales, the velocity fluctuations are due to the permeability variations. This can have a significant role.

We define a longitudinal dispersion coefficient D_L for the spread in solute concentration in the direction of the macroscopic flow and a transverse dispersion coefficient D_T in the direction perpendicular to the mean flow (Taylor, 1953; Saffman, 1959; Bear, 1972). Typically longitudinal dispersion is larger leading to an anisotropy in the dispersion characterised by D_L/D_T. This leads to the conventional advection diffusion equation (Ogata and Banks, 1961; Bear, 1972),

$$\frac{\partial C}{\partial t} + u \cdot \nabla C = D_L \cdot \nabla_L^2 C + D_T \cdot \nabla_T^2 C, \tag{9.56}$$

where u is the macroscopic mean velocity and ∇_L^2 and ∇_T^2 are Laplacians in the longitudinal and transverse directions respectively. Dispersion has been studied by various methods including volume averaging methods (Perkins and Johnston, 1963; Bear, 1972; Sahimi, 1994; Vafai, 2005), analytical methods and simulations of the governing equations on, e.g., network models (Saffman, 1960; Sahimi *et al.*,

1986; Green and Willhite, 1998; Bijeljic *et al.*, 2004; Rhodes *et al.*, 2008; Koplik *et al.*, 1988; Berkowitz and Braester, 1991), method of moments (Theodoropoulou, 2007), stochastic analysis with particle tracking and continuous random walk (Taylor, 1953; Sahimi and Imdakm, 1988; Sahimi, 2012; Stalgorova and Babadagli, 2012), percolation and critical path models (Sahimi, 1994; Selyakov and Kadet, 1996; Hunt and Ewing, 2005). Dispersion can be measured experimentally from the breakthrough curve (Corapcioglu *et al.*, 1997; Green and Willhite, 1998).

9.5.1 *Diffusion*

First we concentrate on diffusion in a network. The medium is modelled by a lattice on which a random walk (to represent the diffusing particles) takes place. The random walkers can jump at each step randomly between nearest neighbour sites. We then measure the mean squared displacement of the random walker, $r^2(t)$, after t time steps, which scales as

$$\langle r^2(t)\rangle \propto t^{2/d_w}, \tag{9.57}$$

where d_w is called the random walk diffusion exponent. We can also define the effective diffusion coefficient D of a random walker as

$$D = \frac{1}{2}\frac{d\langle r^2(t)\rangle}{dt}. \tag{9.58}$$

Homogeneous systems. In a homogeneous medium, where the diffusion is described by Eq. (9.55) with a constant diffusion coefficient, the mean squared displacement linearly varies with time in both 2D and 3D (Haile, 1992),

$$\langle r^2(t)\rangle \propto D\,t. \tag{9.59}$$

Inhomogeneous systems. We use simple percolation as a model of diffusion on inhomogeneous systems. We assume that the walkers are confined to the clusters of occupied sites and cannot jump to unoccupied ones. The diffusion depends on whether the system is below or above the percolation threshold and the length scale of displacement, i.e.,

$$L \sim \langle r^2(t) \rangle^{1/2}. \tag{9.60}$$

(A) For occupancy above the threshold $p > p_c$ and length scale $L > \xi$, the medium looks homogeneous and Eq. (9.59) holds. This means that $d_w = 2$. Particles travelling on the spanning cluster are limited in their speed of drift by the percolation conductivity K (Eq. (2.7)). It must also be inversely proportional to the flow carrying cluster density P (Eq. (2.1)), so the drift speed is $u \propto K/P$. Hence, the effective diffusion coefficient becomes (Gefen *et al.*, 1983; Nakayama and Yakubo, 1994)

$$D \propto u \propto (p - p_c)^{\mu-\beta} \quad \text{or} \quad D \propto \xi^{\frac{\mu-\beta}{-\nu}}. \tag{9.61}$$

So from Eq. (9.59), the mean squared displacement becomes

$$\langle r^2(t) \rangle \propto t \cdot (p - p_c)^{\mu-\beta}. \tag{9.62}$$

(B) For occupancy below the threshold $p < p_c$, the random walker moves on finite-size clusters. This means that the long-time motion of particles is limited to this finite size (the correlation length) $\langle r^2(t) \rangle^{1/2} \propto \xi$. So the long-time behaviour becomes

$$\langle r^2(t) \rangle \propto |p - p_c|^{-2\nu}. \tag{9.63}$$

(C) At the threshold, due to the presence of dangling ends and the fractal nature of the percolation cluster, the motion of random walker is slower than that in homogeneous systems (Gefen *et al.*, 1983; Bacri *et al.*, 1990). The asymptotic behaviour is obtained by comparing Eqs. (9.62) and (9.63) which gives the crossover time $\tau \propto (p - p_c)^{-2\nu-\mu+\beta}$. Again, the mean squared displacement in the medium which is controlled by the correlation length, ξ, scales as

$$\langle r^2(t) \rangle \propto t^{\frac{2\nu}{2\nu+\mu-\beta}}, \tag{9.64}$$

which means that $d_w = 2 + (\mu - \beta)/\nu$ with d_w greater than 2 emphasising slower motion of a random walker in percolating systems. The numerical values of d_w in 2D and 3D are $d_w = 2.87$ and 3.75 respectively (Hong *et al.*, 1984; Havlin and Bunde, 1991; Movshovitz and Havlin, 1988).

9.5.2 *Dispersion*

Now consider the effects of a flow field. In this case, dispersion is described as the macroscopic outcome of actual movement of individual tracer particles through pore spaces that is controlled by the morphology of the medium (Bear, 1972). The conventional advection dispersion equation (Eq. (9.56)) does not consider the dependency of D_L and D_T on the morphology of the pore space. It is sometimes found in experiments and field observations that the dispersion is not correctly described by the simple model (Silliman and Simpson, 1987; Adams and Gelhar, 1992; Hatano and Hatano, 1998; Levy and Berkowitz, 2003; Gouze *et al.*, 2009). This is due to the presence of dangling ends and the fractal nature of percolation cluster (Aronofsky and Heller, 1957; Scher *et al.*, 2002; Bijeljic *et al.*, 2013).

Consider a network model and release of a large number of random walkers in the flow field using a Monte Carlo approach. Each particle follows a different path and consequently has a different travel time to the exit. One definition of the dispersion coefficient D (like the diffusion coefficient) is from the time derivative of the variance of the distance travelled by particles,

$$D = \frac{1}{2}\frac{d\sigma^2(t)}{dt}, \tag{9.65}$$

where $\sigma^2 = \langle (x_i - \bar{x}_l)^2 \rangle$ is the variance of the particle displacement $x_i(t)$.

Dispersion can be characterised by a dimensionless Peclet number which may be interpreted as the ratio of the diffusion time to the convection time (Perkins and Johnston, 1963; Sahimi, 1987),

$$Pe = \frac{u \cdot l}{D}, \tag{9.66}$$

where l is the characteristic length scale (like the pore or grain size d_g), u is the mean velocity and D is the molecular diffusivity. For percolation, the characteristic length scale is the correlation length ξ, so the macroscopic Peclet number is defined as $Pe = \frac{u \cdot \xi}{D}$. The characteristic length scales estimated by Bijeljic *et al.* (2013) for four samples shown in Figure 9.1 are 320 μm (Ketton limestone), 72 μm (Mount Gambier limestone), 158 μm (Estaillades limestone) and 299 μm (Indiana limestone).

The type of dispersion depends on both the morphology of the porous medium and the velocity field. Experimental studies show that there is pure diffusion at small Peclet numbers $Pe < 5$ (with $D_L \sim D$), both convection and diffusion at intermediate Peclet numbers (with $D_L/D \propto Pe^{1.2}$) and there will be pure convective mixing with no diffusion at high Peclet numbers $Pe > 300$ (also called mechanical dispersion with $D_L/D \propto Pe$). However, it should be noted that the long arrival time tails must be influenced by molecular diffusion. So-called holdup dispersion (Charlaix *et al.*, 1988; Gist *et al.*, 1990; Sahimi, 1994) occurs when solute is trapped in the dead ends or stagnant regions and can only escape by molecular diffusion (with $D_L/D \propto Pe^2$).

Consider dispersion on a system with a broad permeability distribution, on which the transport is largely controlled by high permeability regions, effectively the percolating cluster. We define the mean squared displacement as used in Eq. (9.57) in the direction of the flow,

$$\langle r_L^2(t) \rangle \propto t^{2/d_w^L}, \tag{9.67}$$

with d_w^L as the longitudinal random walk power law exponent. The scaling for this exponent, and thus the dispersion coefficient in the percolation critical region, depends on the Peclet number.

(A) Very small Peclet number. This is the regime of negligible flow and so molecular diffusion dominates. In this regime, $D_L = D$ and so $d_w^L = d_w = 2$ and Eq. (9.61) holds.

(B) Very high Peclet number. This is the regime where the solute particles are assumed to be travelling only on the backbone, and we can neglect the solute movement within the dead end regions (Bijeljic *et al.*, 2011; Sahimi, 2012). Then the average velocity of the solute along the backbone is proportional to K/B which is

$u_B \propto (p - p_c)^{\mu-\beta_B}$. As mechanical dispersion is valid in this regime, we have that $D_L \propto u_B \cdot \xi$. So the longitudinal dispersion coefficient D_L becomes (Sahimi *et al.*, 1986)

$$D_L \propto \xi^{\frac{\mu-\beta_B-\nu}{-\nu}} \propto (p - p_c)^{\mu-\beta_B-\nu}. \tag{9.68}$$

This shows that the longitudinal dispersion coefficient depends on the morphology of the porous medium. The dispersion coefficients contain the information about the heterogeneity. This is not represented by the differential equation (Eq. (9.56)) but is observed in field investigations (Gelhar *et al.*, 1992). The effective diffusivity given by D_L is also given by the definition in Eq. (9.58). Integration of this by replacing D_L from Eq. (9.68) and the fact that the relevant length scale around the threshold is the correlation length $\xi^2 \sim r^2(t)$ gives

$$\langle r^2(t)\rangle \propto t^{\frac{2\nu}{\nu+\mu-\beta_B}}. \tag{9.69}$$

This shows that the mean squared displacement of a solute particle is a power law given by Eq. (9.67) with the longitudinal random walk power law exponent $d_w^L = 1 + \frac{\mu-\beta_B}{\nu}$. Using the numerical percolation exponents in Table 2.2, we find $d_w^L = 1.58$ and 2.2 in 2D and 3D respectively.

(C) Dispersion and diffusion are comparable in magnitude. This is the regime where the solute particles are assumed to travel over the whole spanning cluster, the possible movement of solute within the dead ends (Sahimi, 2012). Then the average velocity of the solute along the spanning cluster is proportional to K/P which is $u \propto (p - p_c)^{\mu-\beta}$ (Sahimi *et al.*, 1986). In this regime, there is holdup dispersion (when $D_L/D \propto (u \cdot \xi)^2$), so the longitudinal dispersion coefficient D_L becomes (de Gennes, 1983; Sahimi, 1987)

$$D_L \propto \xi^{\frac{\mu-\beta-2\nu}{-\nu}} \propto (p - p_c)^{\mu-\beta-2\nu}. \tag{9.70}$$

This shows again that the longitudinal dispersion coefficient depends on the morphology of the porous medium. Equation (9.58) and integration using D_L from Eq. (9.70) gives

$$\langle r^2(t)\rangle \propto t^{\frac{2\nu}{\mu-\beta}}. \tag{9.71}$$

This shows that the mean squared displacement of a solute particle is a power law (Eq. (9.67)) with exponent $d_w^L = \frac{\mu-\beta}{\nu}$. Using the numerical vaues from Table 2.2 gives $d_w^L = 0.87$ and 1.78 in 2D and 3D respectively. As emphasised by Sahimi (2012), the small discrepancy of the percolation power law predictions (Eqs. (9.68) and (9.70)) from experimental observations (Cortis and Berkowitz, 2004) is due to finite-size effects which were not considered in deriving these scaling relations.

9.6 Problems

9.1. Use the published sets of experimental relative permeability and capillary pressure curves in the literature and investigate the percolation power law scalings for the residual saturation and the relative permeability and capillary pressure functions.

9.2. Start with the integration $S_{\text{nw}} \cong (l/L)^d \int_0^\infty sn(s)ds$ and consider appropriate cut-offs for the non-wetting-phase cluster-size distribution, $n(s)$, to derive the scaling law for residual non-wetting saturation of a large-size network as $S_{\text{nw}_r} \propto (W/l)^{-d(\tau-2)}$.

9.3. Use various percolation models introduced in this book to find a reasonable method of relating the coordination number z of a pore space to the percolation quantities.

9.4. Use the relevant scaling laws to show the change in the residual-phase saturation in absence of viscous force, in presence of weak, intermediate and large viscous forces.

9.5. Compare the diffusion of a solute particle in a homogeneous and an inhomogeneous percolation medium by plotting the typical scaling of the mean squared displacements of motion of a random walker against time.

9.6. Compare the dispersion process in a homogeneous and an inhomogeneous percolation medium by plotting the typical scaling of the mean squared displacements of motion of a random walker against time at low, intermediate, high and very high Peclet numbers.

9.7. Using scaling laws developed in this book for transport properties, propose the possible methods of inferring the effective permeability of the inhomogeneous percolation medium from the test results of dispersion of solute particles.

References

Adams, E.E. and Gelhar, L.W., Field study of dispersion in a heterogeneous aquifer: 2. Spatial moments analysis. *Water Resource Research*, 28: 3293, 1992.

Adler, P.M., Jacquin, C.G. and Quiblier, J.A., Flow in simulated porous media. *International Journal of Multiphase Flow*, 16(4): 691–712, 1990.

Ahlbrandt, T.S., Charpentier, R.R., Klett, T.R., Schmoker, J.W., Schenk, C.J. and Ulmishek, G.F., *AAPG Memoir 86: Global Resource Estimates from Total Petroleum Systems*. American Association of Petroleum Geologists, Tulsa, OK, 2005.

Al-Mansoori, S.K., Itsekiri, E., Iglauer, S., Pentland, C.H., Bijeljic, B. and Blunt, M.J., Measurements of non-wetting phase trapping applied to carbon dioxide storage. *International Journal of Greenhouse Gas Control*, 4(2): 283–288, 2010.

Al-Raoush, R.I., Impact of wettability on pore-scale characteristics of residual nonaqueous phase liquids. *Environmental Science and Technology*, 43(13): 4796–4801, 2009.

Alava, M., Dube, M. and Rost, M., Imbibition in disordered media. *Advances in Physics*, 53(2): 83–175, 2004.

Anderson, W.G., Wettability literature survey — Part 4: effects of wettability on capillary pressure. *Journal of Petroleum Technology*, 39: 1283–1300, 1987.

Andrew, M.G., Bijeljic, B. and Blunt, M.J., Pore-scale imaging of trapped supercritical carbon dioxide in sandstones and carbonates. *International Journal of Greenhouse Gas Control*, 22: 1–14, 2014.

Angel, O., Goodman, J., den Hollander, F. and Slade, G., Invasion percolation on regular trees. *The Annals of Probability*, 36(2): 420–466, 2008.

Aronofsky, J.S. and Heller, J.P., A diffusion model to explain mixing of flowing miscible fluids in porous media. *Transactions of the AIME*, 210: 345–349, 1957.

Auradou, H., Måløy, K.J., Schmittbuhl, J., Hansen, A. and Bideau, D. Competition between correlated buoyancy and uncorrelated capillary effects during drainage. *Physical Review E*, 60(6): 7224–7234, 1999.

Babalievski, F., Cluster counting: the Hoshen–Kopelman algorithm versus spanning tree approaches. *International Journal of Modern Physics*, C9(01): 43–60, 1998.

Bacri, J.-C., Rakotomalala, N. and Salin, D., Anomalous dispersion and finite-size effects in hydrodynamic dispersion. *Physics of Fluids A: Fluid Dynamics*, 2(5): 674–680, 1990.

Bakke, S. and Øren, P.E. 3-D pore-scale modelling of sandstones and flow simulations in the pore networks. *SPE Journal*, 2(2): 136–149, 1997.

Baldwin, C.A., Sederman, A.J., Mantle, M.D., Alexander, P. and Gladden, L.F., Determination and characterization of the structure of a pore space from 3D volume images. *Journal of Colloid and Interface Science*, 181(1): 79–92, 1996.

Bear, J., *Dynamics of Fluids in Porous Media*. Elsevier, New York, 1972.

ben-Avraham, D. and Havlin, S., *Diffusion and Reactions in Fractals and Disordered Systems*. Cambridge University Press, Cambridge, UK, 2000.

Berkowitz, B. and Braester, C., Dispersion in sub-representative elementary volume fracture networks: percolation theory and random walk approaches. *Water Resources Research*, 27: 3159–3164, 1991.

Berkowitz, B. and Ewing, R.P., Percolation theory and network modelling applications in soil physics. *Surveys in Geophysics*, 19: 23–72, 1998.

Bijeljic, B., Muggeridge, A.H. and Blunt, M.J., Pore-scale modelling of longitudinal dispersion, *Water Resources Research*, 40: W11501, doi: 10.1029/2004WR003567, 2004.

Bijeljic, B., Mostaghimi, P. and Blunt, M.J., Signature of non-Fickian solute transport in complex heterogeneous porous media. *Physical Review Letters*, 107(20): 204502, 2011.

Bijeljic, B., Mostaghimi, P. and Blunt, M.J., Insights into non-Fickian solute transport in carbonates. *Water Resources Research*, 49(5): 2714–2728, 2013.

Bird, R.B., Stewart, W.E. and Lightfoot, E.N., *Transport Phenomena*. Wiley, New York, 1960.

Birovljev, A., Furuberg, L., Feder, J., Jøssang, T., Måløy, K.J. and Aharony, A. Gravity invasion percolation in two dimensions: experiment and simulation. *Physical Review Letters*, 67(5): 584–587, 1991.

Biswal, B., Øren, P.E., Held, R.J., Bakke, S. and Hilfer, R., Stochastic multiscale model for carbonate rocks. *Physical Review E*, 75(6): 061303, 2007.

Blunt, M. and King, P., Relative permeabilities from two- and three-dimensional pore-scale network modelling. *Transport in Porous Media*, 6: 407–433, 1991.

Blunt, M., King, M.J. and Scher, H., Simulation and theory of two-phase flow in porous media. *Physical Review A*, 46(12): 7680–7699, 1992.

Blunt, M.J. and Scher, H., Pore-level modeling of wetting. *Physical Review E*, 52(6): 6387–6403, 1995.

Blunt, M.J., Bijeljic, B., Dong, H., Gharbia, O., Iglauer, S., Mostaghimi, P., Paluszny, A. and Pentland, C., Pore-scale imaging and modeling. *Advances in Water Resources*, 51: 197–216, 2013.

Blunt, M.J., *Multiphase Flow in Permeable Media: A Pore-Scale Perspective*. Cambridge University Press, Cambridge, UK, pp. 544, 2016.

Bo, Z., Loggia, D., Xiaorong, L., Vassueur, G. and Ping, H., Numerical studies of gravity destabilized percolation in 2D porous media. *European Physics Journal B*, 50: 631–637, 2006.

Boustani, A. and Maini, B.B., The role of diffusion and convective dispersion in vapour extraction process. *Journal of Canadian Petroleum Technology*, 40: 68–77, 2001.

Bryant, S. and Blunt, M.J., Prediction of relative permeability in simple porous media. *Physical Review A*, 46(4): 2004–2011, 1992.

Bryant, S.L., Mellor, D.W. and Cade, C.A., Physically representative network models of transport in porous media. *AIChE Journal*, 39(3): 387–396, 1993.

Burdine, N.T., Relative permeability calculations from pore size distribution data. *Journal of Petroleum Technology*, 5(3): 71–78, 1953.

Caers, J., *Petroleum Geostatistics*. Society of Petroleum Engineers, Richardson, TX, 2005.

Cao, G.Z., Meijerink, J., Brinkman, H.W. and Burggraaf, A.J., Permporometry study on the size distribution of active pores in porous ceramic membranes. *Journal of Membrane Science*, 83: 221, 1993.

Chandler, R., Koplik, J., Lerman, K. and Willemsen, J.F., Capillary displacement and percolation in porous media. *Journal of Fluid Mechanics*, 119: 249–267, 1982.

Charlaix, E., Hulin, J.-P., Leroy, C. and Zarcone, C., Experimental study of tracer dispersion in flow through two-dimensional networks of etched capillaries. *Journal of Physics D: Applied Physics*, 21(12): 1727–1732, 1988.

Chatzis, I. and Dullien, F.A., Modelling pore structure by 2D and 3D networks with application to sandstones. *Journal of Canadian Petroleum Technology*, 16(1): 97–108, 1977.

Chatzis, I. and Dullien, F.A.L., The modeling of mercury porosimetry and the relative permeability of mercury in sandstones using percolation theory. *International Chemical Engineering*, 25: 47–66, 1985.

Chatzis, I. and Morrow, N.R., Correlation of capillary number relationships for sandstones. SPE paper 10114, 56th Annual Fall Technical Conference and Exhibition, San Antonio, TX, October 5–7, 1981.

Chatzis, I., Morrow, N.R. and Lim, H.T., Magnitude and detailed structure of residual oil saturation. *SPE Journal*, 23(2): 311–326, 1983.

Chen, J.-D. and Wilkinson, D., Pore-scale viscous fingering in porous media. *Physical Review Letters*, 55: 1892–1895, 1985.

Chilingar, G.V., Mannon, R.W. and Rieke, H.H. III, *Oil and Gas Production from Carbonate Rocks*. Elsevier, New York, 1972.

Cieplak, M. and Robbins, M.O., Dynamical transition in quasistatic fluid invasion in porous media. *Physical Review Letters*, 60(20): 2042–2045, 1988.

Clement, E., Baudet, C., Guyon, E. and Hulin, J.P., Invasion front structure in a 3-D model porous medium under a hydrostatic pressure gradient. *Journal of Physics D: Applied Physics*, 20: 608, 1987.

Coker, D.A., Torquato, S. and Dunsmuir, J.H., Morphology and physical properties of Fontainebleau sandstone via a tomographic analysis. *Journal of Geophysical Research: Solid Earth*, 101(B8): 17497–17506, 1996.

Corapcioglu, M.Y., Chowdhury, S. and Roosevelt, S.E., Micromodel visualization and quantification of solute transport in porous media. *Water Resources Research*, 33(11): 2547–2558, 1997.

Cortis, A. and Berkowitz, B., Anomalous transport in classical soil and sand columns. *Soil Science Society of America Journal*, 68(5): 1539–1548, 2004.

Craft, B.C. and Hawkins, M., *Applied Petroleum Reservoir Engineering*. Prentice Hall, Englewood Cliffs, NJ, 1959.

de Gennes, P.G., Hydrodynamic dispersion in unsaturated porous media. *Journal of Fluid Mechanics*, 136: 189–200, 1983.

Deutsch, C.V., *Geostatistical Reservoir Modelling*. Oxford University Press, New York, pp. 384, 2002.

Dias, M.M. and Wilkinson, D., Percolation with trapping. *Journal of Physics A: Mathematical and General*, 19(15): 3131–3146, 1986.

Diaz, C.E., Chatzis, I. and Dullien, F.A.L., Simulation of capillary pressure curves using bond correlated site percolation on a simple cubic network. *Transport in Porous Media*, 2: 215–240, 1987.

Dong, H. and Blunt, M.J., Pore-network extraction from micro-computerized tomography images. *Physical Review E*, 80(3): 036307, 2009.

Dullien, F.A.L., *Porous Media: Fluid Transport and Pore Structure, Second Edition*. Academic Press, London, 1992.

Dunsmuir, J.H., Ferguson, S.R., D'Amico, K.L. and Stokes, J.P., X-ray microtomography: a new tool for the characterization of porous media. Paper SPE 22860, Proceedings of the 66th SPE Annual Technical Conference and Exhibition, Dallas, TX, 6–9 October, 1991.

Ewing, R.P. and Gupta, S.C., Modeling percolation properties of random media using a domain network. *Water Resources Research*, 29: 3169–3178, 1993.

Fetter, C.W., *Contaminant Hydrogeology*. Macmillan, New York, 1992.

Flannery, B.P., Deckman, H.W., Roberge, W.G. and D'Amico, K.L., Three-dimensional X-ray microtomography. *Science*, 237(4821): 1439–1444, 1987.

Fredrich, J.T., 3D imaging of porous media using laser scanning confocal microscopy with application to microscale transport processes. *Physics and Chemistry of the Earth, Part A: Solid Earth Geodesy*, 24(7): 551–561, 1999.

Frette, V., Måløy, K.J., Bogaer, F., Feder, J., Jøssang, T. and Meakin, P., Diffusion limited aggregation like displacement structures in a three dimensional porous medium. *Physical Review A*, 42(6): 3164, 1990.

Frette, V., Feder, J., Jøssang, T. and Meakin, P., Buoyancy-driven fluid migration in porous media. *Physical Review Letters*, 68(21): 3164, 1992.

Gefen, Y., Aharony, A. and Alexander, S., Anomalous diffusion on percolating clusters. *Physical Review Letters*, 50: 77, 1983.

Gelhar, L.W., Welty, C. and Rehfeldt, K.R., A critical review of data on field-scale dispersion in aquifers. *Water Resources Research*, 28: 1955–1974, 1992.

Georgiadis, A., Berg, S., Makurat, A., Maitland, G. and Ott, H., Pore-scale micro-computed-tomography imaging: nonwetting-phase cluster-size distribution during drainage and imbibition. *Physical Review E*, 88(3): 033002, 2013.

Gielen, T., Hassanizadeh, S.M., Leijnse, A. and Nordhaug, H.F., Dynamic effects in multiphase flow: a pore scale network approach. In: Das, D.B. and Hassanizadeh, S.M. (eds.), *Upscaling Multiphase Flow in Porous Media: From Pore to Core and Beyond*. Springer, Dordrecht, 2005.

Gist, G.A., Thompson, A.H., Katz, A.J. and Higgins, R.L., Hydrodynamic dispersion and pore geometry in consolidated rock. *Physics of Fluids A: Fluid Dynamics*, 2: 1533, 1990.

Gouze, P., Leprovost, R., Poidras, T., Le Borgne, T., Lods, G. and Pezard, P.A., CoFIS and TELog: new downhole tools for characterizing dispersion processes in aquifers by single-well injection-withdrawal tracer tests. *Comptes Rendus Geoscience*, 341(10): 965–975, 2009.

Grassberger, P., Spreading and backbone dimensions of 2D percolation. *Journal of Physics A: Mathematical and General*, 25: 5475, 1992.

Grassberger, P., Directed percolation: results and open problems. In: Puri, S. and Dattagupta, S. (eds.), *Nonlinearities in Complex Systems: Proceedings of the 1995 Shimla Conference on Complex Systems.* Narosa Publishing, New Delhi, 1997.

Green, D.W. and Willhite, G.P., *Enhanced Oil Recovery.* Society of Petroleum Engineers, Richardson, TX, pp. 545, 1998.

Guises, R., Xiang, J., Latham, J.P. and Munjiza, A., Granular packing: numerical simulation and the characterisation of the effect of particle shape. *Granular Matter*, 11(5): 281–292, 2009.

Haile, J.M., *Molecular Dynamics Simulation.* John Wiley & Sons, New York, 1992.

Hassanizadeh, M. and Gray, W.G., General conservation equations for multi-phase systems: 1. Averaging procedure. *Advances in Water Resources*, 2: 131–144, 1979.

Hatano, Y. and Hatano, N., Dispersive transport of ions in column experiments: an explanation of long-tailed profiles. *Water Resources Research*, 34(5): 1027–1033, 1998.

Havlin, S. and ben-Avraham, D., Diffusion in disordered media. *Advances in Physics*, 36: 695–798, 1987.

Havlin, S. and Bunde, S., Percolation II. In: Bunde, A. and Havlin, S. (eds.), *Fractals and Disordered Systems.* Springer, Berlin, 1991.

Hazlett, R.D., Simulation of capillary-dominated displacements in microtomographic images of reservoir rocks. *Transport in Porous Media*, 20(1–2): 21–35, 1995.

Hazlett, R.D., Statistical characterization and stochastic modeling of pore networks in relation to fluid flow. *Mathematical Geology*, 29(6): 801–822, 1997.

Heiba, A.A., Sahimi, M., Scriven, L.E. and Davis, H.T., Percolation theory of two-phase relative permeability. *SPE Reservoir Engineering*, 7: 123–132, 1992.

Herring, A.L., Harper, E.J., Andersson, L., Sheppard, A., Bay, B.K. and Wildenschild, D., Effect of fluid topology on residual nonwetting phase trapping: implications for geologic CO_2 sequestration. *Advances in Water Resources*, 62: 47–58, 2013.

Herrmann, H.J., Hong, D.C. and Stanley, H.E., Backbone and elastic backbone of percolation clusters obtained by the new method of 'burning'. *Journal of Physics A: Mathematical and General*, 17: L261–L266, 1984.

Hilfer, R. Review on scale dependent characterization of the microstructure of porous media. *Transport in Porous Media*, 46(2–3): 373–390, 2002.

Hilfer, R. and Zauner, T., High-precision synthetic computed tomography of reconstructed porous media. *Physical Review E*, 84(6): 062301, 2011.

Homsy, G.M., Viscous fingering in porous media. *Annual Review of Fluid Mechanics*, 19(1): 271–311, 1987.

Hong, D.C., Havlin, S., Herrmann, H.J. and Stanley, H.E., Breakdown of Alexander-Orbach conjecture for percolation: exact enumeration of random walks on percolation backbones. *Physical Review B*, 30: 4083, 1984.

Hoshen, J. and Kopelman, R., Percolation and cluster distribution: I. Cluster multiple labelling technique and critical concentration algorithm. *Physical Review B*, 14(8): 3438–3445, 1976.

Hunt, A.G. and Ewing, R., *Percolation Theory for Flow in Porous Media*. Lecture Notes in Physics, vol. 771. Springer, Berlin, 2005.

Iglauer, S., Favretto, S., Spinelli, G., Schena, G. and Blunt, M.J., X-ray tomography measurements of power-law cluster size distributions for the nonwetting phase in sandstones. *Physical Review E*, 82(5): 056315, 2010.

Jahangiri, H.R. and Zhang, D., Effect of spatial heterogeneity on plume distribution and dilution during CO_2 sequestration. *International Journal of Greenhouse Gas Control*, 5: 281–293, 2011.

Jerauld, G.R. and Salter, S.J., The effect of pore-structure on hysteresis in relative permeability and capillary pressure: pore-level modeling. *Transport in Porous Media*, 5(2): 103–151, 1990.

Journel, A.G., *Fundamentals of Geostatistics in Five Lessons*. American Geophysical Union, Washington, DC, 1989.

Kadet, V.V. and Galechyan, A.M., Percolation model of relative permeability hysteresis. *Journal of Applied Mechanics and Technical Physics*, 54(3): 423–432, 2013.

Ketcham, R.A. and Carlson, W.D., Acquisition, optimization and interpretation of X-ray computed tomographic imagery: applications to the geosciences. *Computational Geosciences*, 27: 381–400, 2001.

Koplik, J., Redner, S. and Wilkinson, D., Transport and dispersion in random networks with percolation disorder. *Physical Review A*, 37: 2619–2636, 1988.

Krevor, S., Pini, R., Zuo, L. and Benson, S., Relative permeability and trapping of CO_2 and water in sandstone rocks at reservoir conditions. *Water Resources Research*, 48(2): W02532, 2012.

Krummel, A.T., Datta, S.S., Munster, S. and Weitz, D.A., Visualizing multiphase flow and trapped fluid configurations in a model three-dimensional porous medium. *AIChE Journal*, 59(3): 1022–1029, 2013.

Lake, L.W., *Enhanced Oil Recovery*. Prentice Hall, Upper Saddle River, NJ, 1989.

Land, C.S., Calculation of imbibition relative permeability for two- and three-phase flow from rock properties. *SPE Journal*, 8(2): 149–156, 1968.

Larsen, J.K., Bech, N. and Winter, A., Three-phase immiscible WAG injection: micromodel experiments and network models. Paper SPE 59324, Proceedings of the SPE/DOE Improved Oil Recovery Symposium, Tulsa, Oklahoma, 3–5 April, 2000.

Larson, R.G. and Morrow, N.R., Effects of sample size on capillary pressure in porous media. *Powder Technology*, 30: 123, 1981.

Larson, R.G., Scriven, L.E. and Davis, H.T., Percolation theory of two phase flow in porous media. *Chemical Engineering Science*, 36(1): 57–73, 1981.

Latief, F.D.E., Biswal, B., Fauzi, U. and Hilfer, R. Continuum reconstruction of the pore scale microstructure for Fontainebleau sandstone. *Physica A: Statistical Mechanics and its Applications*, 389(8): 1607–1618, 2010.

Lenormand, R. and Bories, S., Description d'un mécanisme de connexion de liaison destiné à l'étude du drainage avec piégeage en milieu poreux. *Comptes Rendus des Séances de l'Académie des Sciences B*, 291: 279–282, 1980.

Lenormand, R. and Zarcone, C., Role of roughness and edges during imbibition in square capillaries. Paper SPE 13264, Proceedings of the 59th SPE Annual Technical Conference and Exhibition, Houston, Texas, 16–19 September, 1984.

Lenormand, R. and Zarcone, C., Invasion percolation in an etched network: measurement of a fractal dimension. *Physical Review Letters*, 54(20): 2226–2229, 1985.

Lenormand, R., Touboul, E. and Zarcone, C. Numerical models and experiments on immiscible displacements in porous media. *Journal of Fluid Mechanics*, 189: 165–187, 1988.

Levy, M. and Berkowitz, B., Measurement and analysis of non-Fickian dispersion in heterogeneous porous media. *Journal of Contaminant Hydrology*, 64(3–4): 203–226, 2003.

Lindquist, W.B., Lee, S.M., Coker, D.A., Jones, K.W. and Spanne, P., Medial axis analysis of void structure in three-dimensional tomographic images of porous media. *Journal of Geophysical Research: Solid Earth*, 101(B4): 8297–8310, 1996.

Lorenz, C.D. and Ziff, R.M., Precise determination of the bond percolation thresholds and finite-size scaling corrections for the sc, fcc, and bcc lattices. *Physical Review E*, 57(1): 230–236, 1998.

Løvoll, G., Méheust, Y., Måløy, K.J., Aker, E. and Schmittbuhl, J., Competition of gravity, capillary and viscous forces during drainage in a two-dimensional porous medium: a pore scale study. *Energy*, 30(6): 861–872, 2005.

Lowell, S., Shields, J.E., Thomas, M.A. and Thommes, M., *Characterization of Porous Solids and Powders: Surface Area, Pore Size and Density*. Kluwer Academic Publishers, Dordrecht, 2004.

Mani, V. and Mohanty, K.K., Pore-level network modeling of three-phase capillary pressure and relative permeability curves. *SPE Journal*, 3: 238–248, 1998.

Masoum, S. and Masihi, M., Invasion percolation in the presence of gravity. *Iranian Journal of Chemistry and Chemical Engineering*, 29(1): 71–83, 2010.

Masson, Y. and Pride, S.R., A fast algorithm for invasion percolation. *Transport in Porous Media*, 102: 301–312, 2014.

Masson, Y., A fast two-step algorithm for invasion percolation with trapping. *Computers and Geosciences*, 90: 41–48, 2016.

Mayer, A.S. and Miller, C.T., An experimental investigation of pore-scale distributions of nonaqueous phase liquids at residual saturation. *Transport in Porous Media*, 10(1): 57–80, 1993.

Meakin, P., Feder, J., Frette, V. and Jossang, T., Invasion percolation in a destabilizing gradient. *Physical Review A*, 46(6): 3357–3368, 1992.

Meakin, P., Wagner, G., Vedvik, A., Amundsen, H., Feder, J. and Jøssang, T., Invasion percolation and secondary migration: experiments and simulations. *Marine and Petroleum Geology*, 17(7): 777–795, 2000.

Méheust, Y., Løvoll, G., Måløy, K.J. and Schmittbuhl, J. Interface scaling in a 2D porous medium under combined viscous, gravity and capillary effects. *Physical Review E*, 66: 51603–51615, 2002.

Melrose, J.C. and Brandner, C.F., Role of capillary forces in determining microscopical displacement efficiency for oil recovery by water flooding. *The Journal of Canadian Petroleum Technology*, 13: 54, 1974.

Mourhatch, R., Tsotsis, T.T. and Sahimi, M., Determination of the true pore size distribution by flow permporometry experiments: an invasion percolation model. *Journal of Membrane Science*, 367: 55–62, 2011.

Movshovitz, D. and Havlin, S., Structural and dynamical properties of random walk clusters. *Journal of Physics A: Mathematical and General*, 21(12): 2761–2774, 1988.

Mualem, Y., A new model for predicting the hydraulic conductivity of unsaturated porous media. *Water Resources Research*, 12(3): 513–522, 1976.

Muskat, M., *Physical Principles of Oil Production*. McGraw Hill, New York, 1949.

Nakayama, T. and Yakubo, Y., Dynamical properties of fractal networks: scaling, numerical simulations and physical realizations. *Review of Modern Physics*, 66(2): 381–442, 1994.

Ogata, A. and Banks, R.B., A solution of the differential equation of longitudinal dispersion in porous media. Professional Paper No. 411-A, U.S. Geological Survey, Washington, DC, 1961.

Okabe, H. and Blunt, M.J., Prediction of permeability for porous media reconstructed using multiple-point statistics. *Physical Review E*, 70(6): 066135, 2004.

Okabe, H. and Blunt, M.J., Pore space reconstruction using multiple-point statistics. *Journal of Petroleum Science and Engineering*, 46(1–2): 121–137, 2005.

Okasha, T.M., Funk, J.J. and Rashidi, H.N., Fifty years of wettability measurements in the Arab-D carbonate reservoir. Paper SPE 105114, Proceedings of the SPE Middle East Oil and Gas Show and Conference, 11–14 March 2007, Manama, Kingdom of Bahrain, 2007.

Paterson, L., Diffusion limited aggregation and two fluid displacements in porous media. *Physical Review Letters*, 52: 1621, 1984.

Peaceman, D.W., *Fundamentals of Numerical Reservoir Simulation*. Elsevier, Amsterdam, 1977.

Pentland, C.H., El-Maghraby, R., Iglauer, S. and Blunt, M.J., Measurements of the capillary trapping of super-critical carbon dioxide in Berea sandstone. *Geophysical Research Letters*, 38(6): L06401, 2011.

Perkins, T.K. and Johnston, O.C., A review of diffusion and dispersion in porous media. *SPE Journal*, 3: 70–84, 1963.

Piri, M. and Blunt, M.J., Three-dimensional mixed-wet random pore-scale network modeling of two- and three-phase flow in porous media. I. Model description. *Physical Review E*, 71(2): 026301, 2005.

Prodanovic, M., Lindquist, W.B. and Seright, R.S., 3D image-based characterization of fluid displacement in a Berea core. *Advances in Water Resources*, 30(2): 214–226, 2007.

Rhodes, M.E., Bijeljic, B. and Blunt, M.J., Pore-to-field simulation of single-phase transport using continuous time random walks. *Advances in Water Resources*, 31: 1527–1539, 2008.

Rigby, S.P., Fletcher, R.S. and Riley, S.N., Determination of the multiscale percolation properties of porous media using mercury porosimetry. *Industrial & Engineering Chemistry Research*, 41: 1205, 2002.

Saffman, P.G., A theory of dispersion in a porous medium. *Journal of Fluid Mechanics*, 6: 321–349, 1959.

Saffman, P.G., Dispersion due to molecular diffusion and macroscopic mixing in flow through a network of capillaries. *Journal of Fluid Mechanics*, 7(2): 194–208, 1960.

Sahimi, M., Hydrodynamic dispersion near the percolation threshold: scaling and probability densities. *Journal of Physics A: Mathematical and General*, 20: L1293–L1298, 1987.

Sahimi, M., Flow phenomena in rocks: from continuum models to fractals, percolation, cellular automata, and simulated annealing. *Reviews of Modern Physics*, 65(4): 1393–1534, 1993.

Sahimi, M., *Applications of Percolation Theory.* Taylor and Francis, London, pp. 258, 1994.

Sahimi, M., *Flow and Transport in Porous Media and Fractured Rock, Second Edition.* Wiley-VCH, Weinheim, 2011.

Sahimi, M., Dispersion in porous media, continuous-time random walks, and percolation. *Physical Review E*, 85: 016316, 2012.

Sahimi, M., Hughes, B.D., Scriven L.E. and Davis, H.T., Dispersion in flow through porous media: 1. One-phase flow. *Chemical Engineering Science*, 41(8): 2103–2122, 1986.

Sahimi, M. and Imdakm, A.O., The effect of morphological disorder on hydrodynamic dispersion in flow through porous media. *Journal of Physics A: Mathematical and General*, 21: 3833–3870, 1988.

Scher, H., Margolin, G. and Berkowitz, B., Towards a unified framework for anomalous transport in heterogeneous media. *Chemical Physics*, 284: 349–359, 2002.

Schluter, S., Sheppard, A., Brown, K. and Wildenschild, D., Image processing of multiphase images obtained via X-ray microtomography: a review. *Water Resources Research*, 50(4): 3615–3639, 2014.

Schmittbuhl, J., Hansen, A., Auradou, H. and Måløy, K.J., Geometry and dynamics of invasion percolation with correlated buoyancy. *Physical Review E*, 61(4): 3985–3995, 2000.

Seaton, N.A., Determination of the connectivity of porous solids from nitrogen sorption measurements. *Chemical Engineering Science*, 46(6): 1895–1909, 1991.

Selyakov, V.I. and Kadet, V.V., *Percolation Models for Transport in Porous Media with Applications to Reservoir Engineering*. Kluwer Academic Publishers, London, pp. 241, 1996.

Sheppard, A.P., Knackstedt, M.A., Pinczewski, W.V. and Sahimi, M. Invasion percolation: new algorithms and universality classes. *Journal of Physics A: Mathematical and General*, 32(49): L521–L529, 1999.

Silin, D. and Patzek, T., Pore space morphology analysis using maximal inscribed spheres. *Physica A: Statistical Mechanics and its Applications*, 371(2): 336–360, 2006.

Silliman, S.E. and Simpson, E.S., Laboratory evidence of the scale effect in dispersion of solutes in porous media. *Water Resource Research*, 23(8): 1667–1673, 1987.

Sok, R.M., Knackstedt, M.A., Sheppard, A.P., Pinczewski, W.V., Lindquist, W.B., Venkatarangan, A. and Paterson, L., Direct and stochastic generation of network models from tomographic images: effect of topology on residual saturations. *Transport in Porous Media*, 46(2–3): 345–371, 2002.

Spanne, P., Thovert, J.F., Jacquin, C.J., Lindquist, W.B., Jones, K.W. and Adler, P.M., Synchrotron computed microtomography of porous media: topology and transports. *Physical Review Letters*, 73(14): 2001–2004, 1994.

Stalgorova, E. and Babadagli, T., Modeling miscible injection in fractured porous media using random walk simulation. *Chemical Engineering Science*, 74: 93–104, 2012.

Stauffer, D. and Aharony, A. *Introduction to Percolation Theory, Second Edition*. Taylor & Francis, London, 1991.

Taylor, G.I., Dispersion of soluble matter in solvent flowing slowly through a tube. *Proceedings of the Royal Society of London A: Mathematical, Physical and Engineering Sciences*, 219: 186–203, 1953.

Theodoropoulou, M.A., Dispersion of dissolved contaminants in groundwater: from visualization experiments to macroscopic simulation. *Water, Air, and Soil Pollution*, 181: 235–245, 2007.

Thompson, A.H., Katz, A.J. and Krohn, C.E., The micro geometry and transport properties of sedimentary rock. *Advances in Physics*, 36(5): 625–694, 1987a.

Thompson, A.H., Katz, A.J. and Rashke, R.A., Mercury injection in porous media: a resistance devil's staircase with percolation geometry. *Physical Review Letters*, 58: 29, 1987b.

Thovert, J.F., Salles, J. and Adler, P.M., Computerized characterization of the geometry of real porous media: their discretization, analysis and interpretation. *Journal of Microscopy*, 170(1): 65–79, 1993.

Tiggemann, D., Simulation of percolation on massively parallel computers. *International Journal of Modern Physics C*, 12(6): 871–878, 2001.

Toledo, P.G., Scriven, L.E. and Davis, H.T., Pore space statistics and capillary pressure curves from volume controlled porosimetry. Paper SPE 19618, 64th Annual Technical Conference and Exhibition of the SPE, October 8–11, San Antonio, Texas, 1989.

Tomutsa, L. and Silin, D., Analysis of chalk petrophysical properties by means of submicron-scale pore imaging and modeling. *SPE Reservoir Evaluation Engineering*, 10(3): 285–293, 2007.

Toussaint, R., Løvoll, G., Méheust,Y., Måløy, K.J. and Schmittbuhl, J., Influence of pore-scale disorder on viscous fingering during drainage. *Europhysics Letters*, 71(4): 583, 2005.

Toussaint, R., Måløy, K.J., Méheust, Y., Løvoll, G., Jankov, M., Schäfer, G. and Schmittbuhl, J., Two-phase flow: structure, upscaling, and consequences for macroscopic transport properties. *Vadose Zone Journal*, 11: 3, 2012.

Tsakiroglou, C.D. and Payatakes, A.C., Characterization of the pore structure of reservoir rocks with the aid of serial sectioning analysis, mercury porosimetry and network simulation. *Advances in Water Resources*, 23: 773, 2000.

Vafai, K., *Handbook of Porous Media, Second Edition.* Taylor and Francis, London, pp. 747, 2005.

Valvatne, P.H. and Blunt, M.J., Predictive pore-scale modeling of two-phase flow in mixed wet media. *Water Resources Research*, 40(7): W07406, 2004.

van Genuchten, M.T. and Nielsen, D.R., On describing and predicting the hydraulic properties of unsaturated soils. *Annales Geophysicae*, 3(5): 615–628, 1985.

Wall, G.C. and Brown, R.J.C., The determination of pore size distribution from sorption isotherm and mercury penetration in interconnected pores. The application of percolation theory. *Journal of Colloid and Interface Science*, 82: 141–149, 1981.

Washburn, E.W., Note on a method of determining the distribution of pore sizes in a porous material. *Proceedings of the National Academy of Sciences*, 7: 115–116, 1921.

Wildenschild, D., Vaz, C., Rivers, M., Rikard, D. and Christensen, B., Using X-ray computed tomography in hydrology: systems, resolutions, and limitations. *Journal of Hydrology*, 267: 285–297, 2002.

Wildenschild, D. and Sheppard, A.P., X-ray imaging and analysis techniques for quantifying pore-scale structure and processes in subsurface porous medium systems. *Advances in Water Resources*, 51: 217–246, 2013.

Wilkinson, D., Percolation model of immiscible displacement in the presence of buoyancy forces. *Physical Review A*, 30(1): 520–531, 1984.

Wilkinson, D., Percolation effects in immiscible displacement. *Physical Review A*, 34(2): 1380–1391, 1986.

Wilkinson, D. and Willemsen, J.F., Invasion percolation: a new form of percolation theory. *Journal of Physics A: Mathematical and General*, 16(16): 3365–3376, 1983.

Yeong, C.L.Y. and Torquato, S., Reconstructing random media II. Three-dimensional media from two-dimensional cuts. *Physical Review E*, 58(1): 224–233, 1998.

Yortsos, Y.C., Xu, B. and Salin, D., Phase diagram of fully developed drainage in porous media. *Physical Review Letters*, 79(23): 4581–4584, 1997.
Zhao, X., Blunt, M.J. and Yao, J., Pore-scale modeling: effects of wettability on waterood oil recovery. *Journal of Petroleum Science and Engineering*, 71(3–4): 169–178, 2010.
Zhou, D. and Stenby, E.H., Interpretation of capillary pressure curves using invasion percolation theory. *Transport in Porous Media*, 11: 17–33, 1993.

Index

D

G

H

I

J

K

L

M

Y